Dry Etching for VLSI

UPDATES IN APPLIED PHYSICS AND ELECTRICAL TECHNOLOGY

Series Editor: P. J. Dobson
Oxford University
Oxford, England

CRYSTAL GROWTH: Principles and Progress
A. W. Vere

DRY ETCHING FOR VLSI
A. J. van Roosmalen, J. A. G. Baggerman, and S. J. H. Brader

OPTICAL FIBRES AND SOURCES FOR COMMUNICATIONS
M. J. Adams and I. D. Henning

PHOTODETECTORS: An Introduction to Current Technology
P. N. J. Dennis

Dry Etching for VLSI

A. J. van Roosmalen
Philips Components
Eindhoven, The Netherlands

J. A. G. Baggerman
Philips Research Laboratories
Eindhoven, The Netherlands

S. J. H. Brader
Philips Research Laboratories
Eindhoven, The Netherlands

Plenum Press • New York and London

ISBN 0-306-43835-6

A Division of Plenum Publishing Corporation
233 Spring Street, New York, N.Y. 10013

Printed in the United States of America

This book is dedicated to

Margo

Truus and Anton

Mieke

for their support and understanding

Preface

This book has been written as part of a series of scientific books being published by Plenum Press. The scope of the series is to review a chosen topic in each volume. To supplement this information, the abstracts to the most important references cited in the text are reprinted, thus allowing the reader to find in-depth material without having to refer to many additional publications.

This volume is dedicated to the field of dry (plasma) etching, as applied in silicon semiconductor processing. Although a number of books have appeared dealing with this area of physics and chemistry, these all deal with parts of the field. This book is unique in that it gives a compact, yet complete, in-depth overview of fundamentals, systems, processes, tools, and applications of etching with gas plasmas for VLSI. Examples are given throughout the fundamental sections, in order to give the reader a better insight in the meaning and magnitude of the many parameters relevant to dry etching. Electrical engineering concepts are emphasized to explain the pros and cons of reactor concepts and excitation frequency ranges. In the description of practical applications, extensive use is made of cross-referencing between processes and materials, as well as theory and practice. It is thus intended to provide a total model for understanding dry etching.

The book has been written such that no previous knowledge of the subject is required. It is intended as a review of all aspects of dry etching for silicon semiconductor processing. The level of the book is suitable to final year undergraduates and postgraduates. The book is well suited as an overview for new-comers to plasma etching, but may also serve as a source of review material for experienced workers in this extensive field. The text can be used as course material, but is also suitable for individual reading.

The material as it appears now, is to a large extent based on the manual from the international Philips course on "Plasma Etch Science and Technology" held every Spring at the Philips Research Laboratories in Eindhoven, The Netherlands.

Acknowledgements

The idea for writing this book originated back in 1984 at Racal Research in Reading, Great Britain. For their support and encouragements in the early days, Andy Blunt, Sion Quinlan and Barry Darby are hereby gratefully acknowledged.

The material as it appears now, is to a large extent based on the manual from the International Philips Course on Plasma Etch Science and Technology held every Spring in Eindhoven, The Netherlands. This instruction in etching was started in 1982 on an initiative from Fred van Roosmalen and Wilbert van den Hoek. At first, it was only meant for scientists and engineers from the thin film divisions at the Philips Research Laboratories. Gradually, however, the course expanded to a general introduction for graduate newcomers, active in the field of dry etching and related subjects. Through the years many people have presented parts of the course, and by doing so implicitly contributed to the preparation of the present book. Lecturers until 1989 were – in alphabetical order – Maritza Heijman, Wilbert van den Hoek, Harrie van Houtum, Henk Kalter, Gerard van Veen, Robert-Jan Visser, Quint van Voorst Vader, Carol de Vries, and Peer Zalm. We thank them here for their contributions.

Contents

List of symbols

Symbol	Units	Description
a_o	m	First Bohr orbital for hydrogen (= 5.292×10^{-11})
c	m s^{-1}	Speed of light (= 2.998×10^{8})
e	C	Elementary charge (= 1.602×10^{-19})
ϵ	F m^{-1}	Permittivity of vacuum (= 8.854×10^{-12})
h	J s	Planck constant (= 6.626×10^{-34})
k	J K^{-1}	Boltzmann constant (= 1.381×10^{-23})
m_e	kg	Electron rest mass (= 9.110×10^{-31})
a	m	Particle diameter
A	m^2	Area
α		Degree of ionization
B	G	Magnetic field
C	F	Capacitance
d	m	Sheath thickness
D	m^2 s^{-1}	Diffusion coefficient
δ	m	Penetration depth
Δ	m^{-3}	Small density fluctuation
E	V m^{-1}	Electric field
$\mathcal{E}$	J, eV	Energy
$\mathcal{E}_{th}$	eV	Threshold energy
η	eV	Effective energy loss per ionizing collision
f_e		Electron energy distribution function
F	N	Force
$\vec{F}$	N	Directional force (vector)
Φ	m^{-2}	Flux density
γ		Surface recombination coefficient
I	A	Current
J	A m^{-2}	Current density

Symbol	Units	Description
J_e	A m^{-2}	Electron current density
J_i	A m^{-2}	Ion current density
k_r	s^{-1}, $m^3\ s^{-1}$, $m^6\ s^{-1}$	Reaction rate coefficient
κ	$m^{3/2}\ V^{-1/2}\ s^{-1}$	Ion mobility proportionality factor
l	m	Length
L	H	Inductance
λ	m	Mean free path
λ_e	m	Electron mean free path
λ_D	m	Debye length
m	kg	Mass
m_i	kg	Ion mass
μ	$m^2\ V^{-1}\ s^{-1}$	Mobility
μ_e	$m^2\ V^{-1}\ s^{-1}$	Electron mobility
n_R		Refractive index
n	m^{-3}	Density
n_e	m^{-3}	Electron density
n_i	m^{-3}	Ion density
n_o	m^{-3}	Density at infinity, the plasma density
n_g	m^{-3}	Neutral gas density
N_D	m^{-3}	Number of particles in a Debye sphere
ν	s^{-1}	Collision frequency
ν_e	s^{-1}	Electron collision frequency
ω	s^{-1}	Angular frequency
ω_e	s^{-1}	Electron plasma frequency
ω_i	s^{-1}	Ion plasma frequency
ω_{ec}	s^{-1}	Electron cyclotron frequency
p	Pa	Pressure
P	W	Power
Ψ		Vibrational wave function
r	m	Radius, internuclear distance
R	Ω	Resistance
s	$m^{-3}\ s^{-1}$	Reaction rate
S_n		Nuclear stopping power
σ	m^2	Collision cross section
t	s	Time
T	K, °C	Temperature

Symbol	Units	Description
T_e	K, eV	Electron temperature
T_g	K, eV	Neutral gas temperature
τ	s	Life time, decay time
θ	deg	Angle
U	J	Kinetic energy
U_o	eV	Surface barrier energy
v	m s^{-1}	Velocity
$\overline{v}$	m s^{-1}	Mean velocity
$\vec{v}$	m s^{-1}	Directional velocity (vector)
v_e	m s^{-1}	Electron velocity
v_i	m s^{-1}	Ion velocity
V	V	Voltage, potential difference
V_{dc}	V	DC potential
V_f	V	Floating potential
V_F	V	Forward voltage
V_p	V	Plasma potential
V_{rf}	V	RF potential
V_R	V	Reverse voltage
V_s	V	Sheath potential
X		Imaginary part of impedance
Y		Sputter yield
Z		Impedance
$\mathcal{Z}$		Atomic number

Chapter 1

INTRODUCTION

In the early days of silicon Integrated Circuit technology circuit dimensions were large (10 μm and up) and wafers were small (2 inch and down). Mask patterns were transferred mostly by means of hard contact exposure of a photoresist. After developing the resist pattern, the wafers were immersed in some aggressive liquid to remove the material not covered by resist. Since wet chemical etching is in principle isotropic, the pattern etched is wider than the original hole. This so-called undercutting is shown in figure 1.1.

As packing density and wafer sizes increased, it became apparent that the conventional wet etchants were not reliable enough for all applications. This was true especially in the critical processing needed for the definition of poly-crystalline silicon MOS gates and the silicon nitride oxidation barriers in LOCOS isolation. For wet etching of these layers highly aggressive reagents are needed, such as mixtures of concentrated HF/HNO_3 and hot H_3PO_4, respectively. Both media often exhibit irreproducible delays in the initiation of the etching

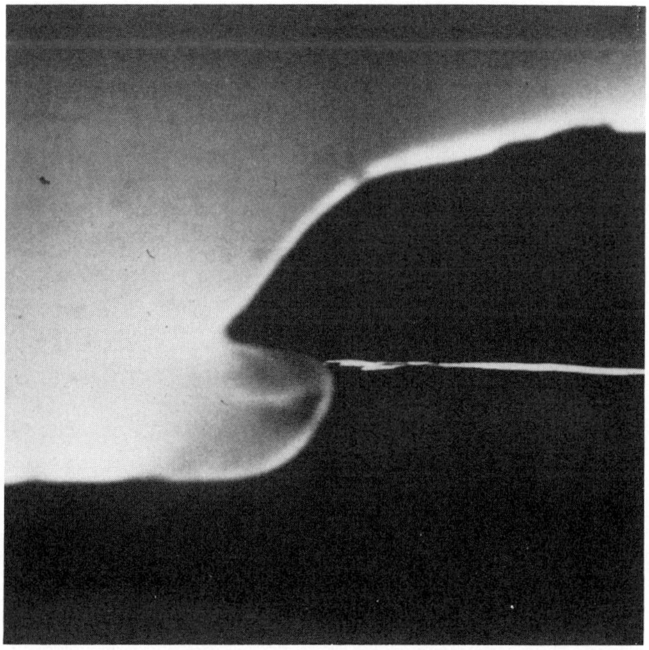

Figure 1.1. Photograph of isotropic etching resulting in undercutting.

process. An even worse phenomenon is the severe attack of conventional photoresists by these etchants.

A major achievement for IC processing, was the discovery that silicon-containing materials can be etched conveniently by exposure to an rf glow discharge containing halocarbon (Freon) compounds [1,2,3,4]. Systems came on the market consisting of what basically was a large glass jar evacuated by a rotary vane pump. Freon gas mixed with oxygen was fed into the system and high-frequency power was applied by means of external coils or electrodes strapped to the jar. This action not only leads to light emission, but also to the formation of large amounts of fluorine atoms. The fluorine etches all materials that can form volatile fluorides, such as silicon. These "barrel" etchers are a category of dry etch equipment which is still extensively used in virtually all IC production lines. This approach led to significant improvements in process control, but pattern transfer was still isotropic; the liquid agent had simply been replaced by a gaseous one.

Towards the end of the 70's, feature sizes gradually shrunk from 6 to 3 μm, wafer sizes increased from 3 to 4 inch, and people started to replace the original name LSI (Large Scale Integration) by the now more familiar VLSI (Very Large Scale Integration). At these dimensions the absolute precision of pattern control becomes a dominant item. In lithography, contact aligners were replaced by projection equipment. At the same time, the urge was growing for a 1:1 transfer process from resist pattern to the underlying material. This meant that etching would need to be anisotropic and preferably in one direction only; perpendicular to the wafer surface.

By now, the time had come for a next step in dry processing. This involved getting rid of the idea that electrical discharges should be kept at a safe distance from sensitive IC devices. By exposing the wafers directly to the electric field, it was found that etching could be made completely non-undercutting [5,6]. Also, layers previously considered as not being fit for discharge etching, such as aluminum, could now be patterned [7]. One of the earliest commercial systems used for these purposes was the "pancake" reactor, see figure 1.2 after the original patent [8]. Interestingly, this reactor type had originally been developed for the purpose of plasma-assisted chemical vapor deposition of passivation and scratch protection layers; it just happened to etch just as well [9].

Both the barrel and the pancake reactor are examples of what will be referred to in this book as "dry etching" systems. By definition, it involves etching through interaction with weakly ionized reactive gases produced by striking a discharge at a reduced pressure. Other names frequently used are plasma etching, reactive ion etching, and reactive sputtering. There was a time when these terms were used for what were considered to be different processes; by now, it is known that they are all the same, see chapter 5. Having an electrical discharge and a reactive gas at the same time is, by the way, not a prerequisite for etching in a gaseous medium. At elevated temperatures it is, for example, possible to etch SiO_2 in anhydrous HF [10] and to etch Si in Cl_2 [11]. Instead of heating the whole sample, it is also possible to apply the thermal energy locally, for example by exposing the surface to a focussed laser beam in the presence of reactive gases [12]. There is evidence that some of these processes are based on light-induced rather than thermal etching [13,14]. At this moment, etching with light or hot gases is not nearly as wide spread as etching with reactive discharges, and certainly not present in wafer production environments. Furthermore, the processing examples in this book will cover the area of Si VLSI only. Applications in solar cell and display technology [15], III-V compound semiconductors [16,17,18], mask-making [19], and

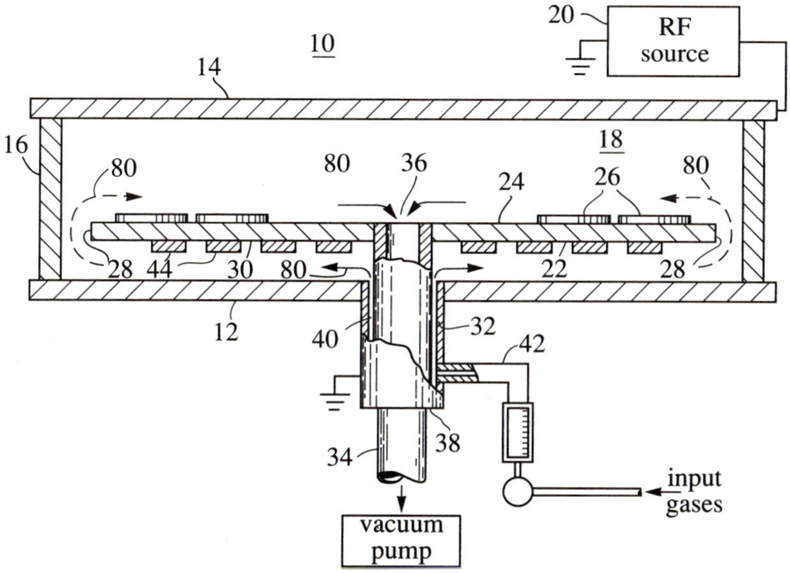

Figure 1.2. The "radial flow reactor" – the first pancake etcher.

other areas where materials are patterned with dry etching [20,21], are not dealt with in this book. The references cited are only meant as an introduction to those fields.

After the introduction of the parallel-plate reactor the dry etch system family expanded explosively. A large variety of designs evolved, featuring multiple as well as concentric electrodes, small single-wafer and large multi-wafer systems, square and circular reactors. In order to keep dry etching interesting from a cost point of view, etch rates, which originally were seldom higher than 1 nm s^{-1}, were accelerated to 10 nm s^{-1}, and more. Excitation frequencies ranging from a few kHz to several GHz have been used. Presently most reactors run at 13.56 MHz, which happens to be an allowed industrial frequency. Although this choice may have been arbitrary, there are some interesting implications, see chapter 3. Gas mixtures also underwent a revolution. From the simple Freon-14 (CF_4) of the barrels and the carbon tetrachloride (CCl_4) of the first pancakes, a huge family has been developed of sometimes highly complicated gas mixtures. The large majority contains a halogen, but almost any commercially available gas bottle appears to have been connected to some dry etcher somewhere. In the meanwhile, applications shifted from MOS gate and LOCOS isolation to interconnects, contact windows, trench isolation, planarization, and many others. Where at first 0.5 μm undercutting was acceptable, 0.05 μm has now become an achievable necessity; the etch-rate ratio of the material to be etched to the masking and underlying materials, less than 5 at first, has been raised almost to infinity under certain conditions.

In the process of developing applications, it became increasingly clear that etch discharges are extremely complicated systems. A large variety of technical disciplines is involved, the system as a whole is far from chemical equilibrium, and all processes influence each other. Even the interpretation of a simple diagnostic experiment can lead to confusion as almost all measurement set-ups interact with the discharge to some extent. A useful by-product of the many successful and not so successful attempts, is the current availability of a large number of etch end-point detection systems. Early theoretical models set up to predict dry etch plasma behavior had only limited success, based as they were on "brute force" simulation from regular plasma physics. The latter treats the plasma as an infinite bulk medium, whereas the most important part of any real system is the reactor wall. Only recently models emphasizing the plasma boundary have been developed. A popular experimental method for fundamental studies is to bombard selected surfaces with controlled particle beams. Under carefully chosen conditions, this can indeed simulate dry etching. Still, actual etch rates can be orders of magnitude different for "real" systems sharing the same chemistry. Even the etch reactor itself has been found to be an important parameter in the determination of the process characteristics. The background of the latter lies in the high frequencies used for excitation.

Looking back over the past two decades, it is amazing to see the pace at which dry etching has been introduced in the world of VLSI. Although the first commercial parallel plate reactor for anisotropic etching did not come on the market until 1976, and the first cassette-to-cassette system followed not earlier than 1979, dry etching is dominating all wafer patterning steps in VLSI production lines since 1985. Modern wafer fabs now use wet chemical processing in clean and strip steps only. More often than not, processes have been forced into production before backgrounds and mechanisms were known in any detail. Evidently this has resulted in many unexplainable results and irreproducible data, as well as regular nervous breakdowns in the engineering crew. Such events, combined with the unearthly light and foul smell of etch discharges, led people in those days to argue that dry etching relied on witch craft rather than science. One late Friday afternoon at the IC laboratory of Philips Research in Eindhoven, it was even suggested that dry process engineers should change their regular cleanroom suits for black ones with golden stars, and wear a conical hat instead of the usual cleanroom cap.

Much trial and error has gone into the development of dry etching processes, but by now the technique can be considered to be mature. At last, dry etching may be called a science rather than "just" magic.

Chapter 2

THE PLASMA STATE

2.1 Introduction

The term "plasma", in relation to the low-pressure ionized gas in dry etching reactors, has already been introduced casually in chapter 1. Plasmas can be considered as the fourth state of existence, together with solid, liquid and gas. It is a considerably less exclusive medium than one might think. On earth, it can be seen in lightning bolts and the Polar Lights; on a galactic scale, it is found in stars and much of the interstellar hydrogen. Some authors state that 99% of the matter in the universe is actually a plasma, which may indeed be close to reality. Still, it would be too simple to call just any ionized gas a plasma. Ordinary air under normal conditions is also ionized, although the degree of ionization can be calculated to be extremely low, namely about 10^{-122} at room temperature. Not much, but it is there, and room-temperature air is certainly not a plasma. A useful definition for the plasma state has been postulated by Chen [23],

> *A plasma is a quasi-neutral gas of charged and neutral particles which exhibits collective behavior.*

This definition contains two important terms, namely "collective behavior" and "quasi-neutral". These aspects will first be treated in a phenomenological way; a more quantitative description follows later in this chapter.

Consider the forces between the particles in ordinary air. Since molecules are electrically neutral, and the force of gravity is negligible on a molecular scale, they will move undisturbed until they collide with other molecules. These – almost – elastic mutual collisions determine the motion of the particles. In a plasma the situation is entirely different, as it is an ionized gas and contains charged particles too. While moving around, these particles may generate local concentrations of positive and negative charge, thus giving rise to electric fields. Motion of charge means electrical current, and currents induce magnetic fields. Both the electric and the magnetic fields affect the motion of other charged particles far away by long-range Coulomb interaction. It is the Coulomb force that causes "collective behavior" in a real plasma; that is, the phenomenon that motion depends not only on local conditions, but also on the status of remote regions.

Coulomb interactions make it improbable that deviations from charge neutrality within a plasma persist for more than a very short period. Although a plasma contains many charged particles, as a whole it need not be charged relative to the outside world. In itself it will even almost be completely field free; hence, the term "quasi-neutral" in Chen's definition. There is similarity between this effect and the observation that spontaneous pressure fluctuations in an ordinary gas are also highly unlikely.

2.2 Gas kinetics

A good understanding of plasma behavior requires some fundamental knowledge of gas kinetics, especially with respect to the temperature concept. Consider an ideal gas where the individual particles are solid spheres interacting through elastic collisions only. The distribution function $f(v)$, describing the absolute velocity of these particles, can then be written in terms of a Maxwell-Boltzmann velocity distribution function,

$$v.f(v) = \frac{4}{\sqrt{\pi}} \left(\frac{m\,v^2}{2\,k\,T} \right)^{\frac{3}{2}} \exp\left(-\frac{m\,v^2}{2\,k\,T} \right) \tag{2.1}$$

where v is the particle velocity, m the particle mass, k Boltzmann's constant, and T the absolute temperature. Note that v is the length of a vector. The mean velocity can be found by direct integration of equation 2.1, and yields

$$\overline{v} = \int_0^\infty v\,f(v)\,dv = \sqrt{\frac{8\,k\,T}{\pi\,m}} \tag{2.2}$$

The parameter thus obtained is the speed of sound in the medium under consideration. For Ar at 400 K it is found that $\overline{v} = 460$ m s^{-1}, or about 1000 miles per hour.

Two other important parameters describing collision processes are the mean free path between collisions λ and the collision frequency ν. They are related with the mean velocity through

$$\lambda\,\nu = \overline{v} \tag{2.3}$$

This equation will be needed in the definition of quantitative criteria for the plasma state. In chapter 4 these and other collision parameters are discussed in detail.

Since the Maxwellian velocity distribution is not symmetrical, the mean velocity is not the same as the most probable or the root mean square velocity. Figure 2.1 gives an illustration of this effect. The root mean square velocity is important since it is related to the kinetic energy U of the gas particles by

$$U = \tfrac{1}{2}\,m\,\overline{v^2} = \tfrac{1}{2}\,m \int_0^\infty v^2\,f(v)\,dv = \tfrac{3}{2}\,k\,T \tag{2.4}$$

In plasma physics – and dry etching – it is customary to give temperatures in units of energy rather than in degrees Centigrade or Kelvin. To avoid confusion on the number of dimensions involved, it is not U (as derived in equation 2.4), but the energy corresponding to kT that is used to denote the temperature. The energy gained by a particle with unit charge e moving across a potential difference V,

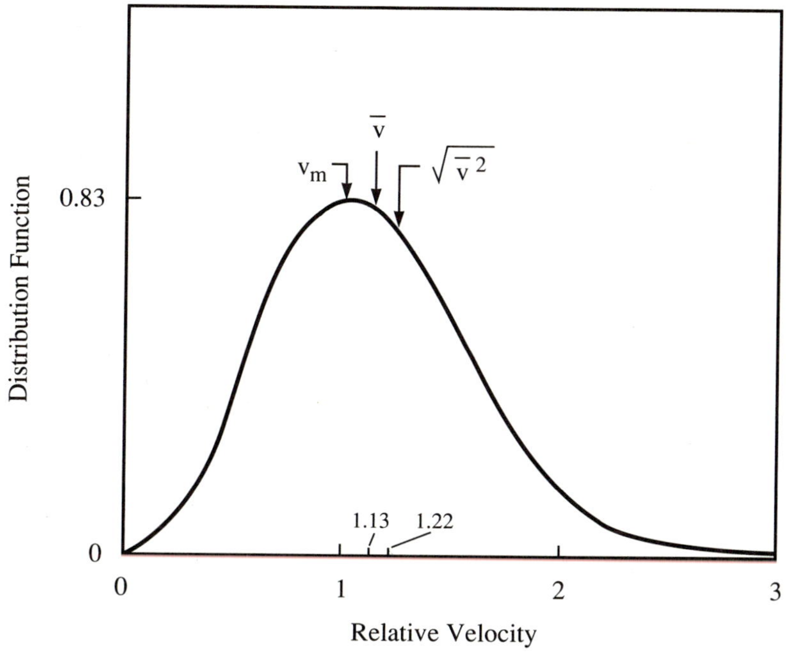

Figure 2.1. Maxwellian velocity distribution function. Velocities normalized relative to the most probable velocity, v_m.

$$U = e\,V \tag{2.5}$$

which is conveniently written in electronvolts (eV). Combining this potential energy with the kinetic energy kT leads to the following equivalence

$$1\ (\mathrm{eV}) = 1.6 \times 10^{-19}\ (\mathrm{J}) = 11600\ (\mathrm{K})$$

For example, room temperature corresponds to 0.026 eV. This is not the average kinetic energy, see equation 2.4!

What has been derived here for an ideal gas can in principle also be applied to plasmas. Since a plasma contains populations of quite a variety of particles, it can have several "temperatures" at the same time. Ions and electrons, for example, usually have more or less independent velocity distributions with different average temperatures. This is caused by the fact that elastic particle collisions result in appreciable impulse transfer only if the colliding species are of comparable mass. For an impression of the importance of this effect consider a collision between two point masses m_1 and m_2. Assume that m_2 is initially at rest, and that m_1 hits it with a velocity v_1 and an angle θ relative to the line joining the mass centers at the moment of collision. After the collision m_2 has acquired a velocity v_2 and is moving along the mass center line; at the same time m_1 speeds away at an angle θ' with a velocity v_3, see figure 2.2.

From conservation of linear momentum in the case of a perfectly elastic collision it follows that

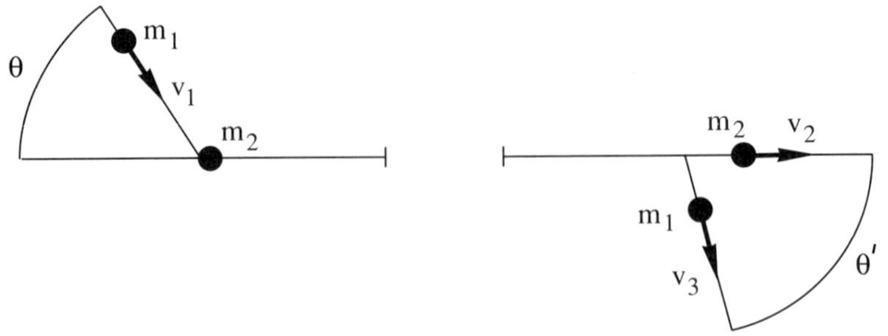

Figure 2.2. Schematic of a two-particle collision.

$$m_1 \; v_1 \; \sin\theta \; = \; m_1 \; v_3 \; \sin\theta' \tag{2.6}$$

$$m_1 \; v_1 \; \cos\theta \; = \; m_1 \; v_3 \; \cos\theta' \; + \; m_2 \; v_2 \tag{2.7}$$

and from conservation of kinetic energy

$$\tfrac{1}{2} \, m_1 \, v_1^2 \; = \; \tfrac{1}{2} \, m_1 \, v_3^2 \; + \; \tfrac{1}{2} \, m_2 \, v_2^2 \tag{2.8}$$

From these equations the fractional energy transfer from m_1 to m_2 can be found by elimination of v_2 and θ'. The value of this parameter can be expressed as

$$\frac{\frac{1}{2} \, m_2 \, v_2^2}{\frac{1}{2} \, m_1 \, v_1^2} \; = \; \frac{4 \, m_1 \, m_2 \, \cos^2\theta}{(m_1 \; + \; m_2)^2} \tag{2.9}$$

It is clear that the maximum efficiency is equal to $\cos^2\theta$, and that this is obtained for $m_1 = m_2$, i.e., particles of equal mass. Electrons and ions, however, differ in mass by at least four orders of magnitude. With equation 2.9 it can be calculated that in that case very little energy is transferred; even in the case of a head-on collision ($\theta = 0$) only about 0.01%.

A further aspect one should be aware of is the large difference in response to electric field for the various charged particles in a plasma. The acceleration of single-charged particles in an electric field is given by

$$F \; = \; q \; E \; = \; m \, \frac{dv}{dt} \tag{2.10}$$

where F is the force exerted by the electric field E, t is the time, and q the charge. Thus, because of the difference in mass, the negative charges (electrons) will move much faster than the positive charges (ions) when exposed to an electric field. This will be dealt with when discussing the formation of sheaths.

Although the temperature of the charged particles in a plasma is usually quite high, especially for the electrons, this does not mean that all plasmas are physically hot. It is the heat capacity that counts, and for that parameter the density of the "hot" particles must be considered as well. At a charged particle density of 10^{18} m^{-3}, which is an appreciable value for many plasmas, even one million K does not hurt. For comparison, ordinary air has a density of approximately 3×10^{25} m^{-3}.

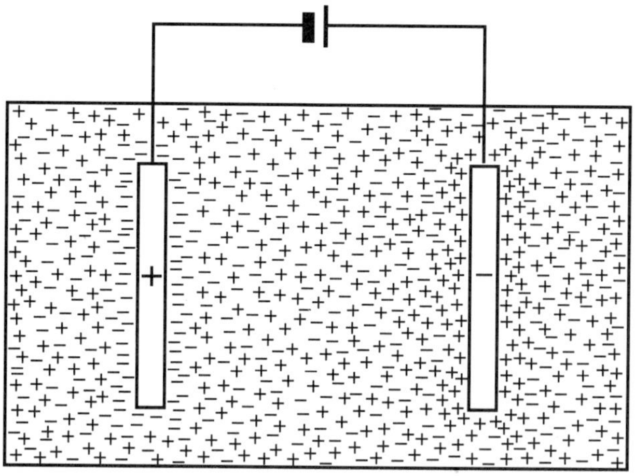

Figure 2.3. Schematic of potential shielding in a plasma.

2.3 Debye shielding and the plasma frequency

Imagine a hypothetic infinite free-floating plasma. We will now try to establish an electric field inside the plasma by inserting two charged plates connected to a battery, see figure 2.3. Immediately after insertion, the plates will start to attract particles of the opposite charge. Very soon a sheath of positive ions will cover the negative plate, while a sheath of electrons covers the positive one. If there were no thermal motion, there would be just as many charges in the sheath as in the plate on either side. The plasma would have shielded itself out completely. However, for any finite temperature, at least some particles at the outer edge of the sheaths have enough kinetic energy to escape from the potential which keeps them attracted to the plates. This "edge" occurs at a distance from the plate where the local potential energy relative to infinity is of the order of the thermal energy kT. In other words, potentials of the order of kT/e can leak into the plasma and cause finite electric fields to exist there. Hence, the term quasi-neutrality in the definition of the plasma state.

The approximate thickness of the sheaths can be derived as follows. Consider a plane at $x = 0$ in a plasma with a potential of $V(0)$ relative to infinity. For simplicity, assume all ions to be positive, singly-charged, and having an infinite mass. The only moving particles are then the electrons; the ions form an immobile uniform background of positive charge. The potential as a function of x is given by Poisson's equation. This equation relates the net charge density to the potential gradient and can be written in one dimension as

$$\frac{d^2V}{dx^2} = -\frac{e\,(n_i - n_e)}{\epsilon} \tag{2.11}$$

where ϵ is the dielectric constant and n_i and n_e are the densities of ions and electrons, respectively. Let the electron density at infinity be n_o. The ion density in the system is then

$$n_i = n_o \tag{2.12}$$

since the undisturbed plasma is electrically neutral. The parameter n_o is usually called "the" plasma density. In the presence of a potential energy $-eV$ the Maxwell-Boltzmann velocity distribution function for the electrons becomes

$$v\ f(v)\ \propto\ \exp\left(\frac{-\frac{1}{2}\ m_e\ v_e^2\ +\ e\ V}{k\ T_e}\right) \tag{2.13}$$

The effect of V on the pre-exponential factor from equation 2.1 has been neglected here. Equation 2.13 simply states that there are less particles at places where the potential energy is high, since not all particles have enough kinetic energy to get there. Integrating $f(v)$ over all values of v, noting that n_e becomes equal to n_o at infinity, it is found that

$$n_e\ \simeq\ n_o\ \exp\left(\frac{e\ V}{k\ T_e}\right) \tag{2.14}$$

Substituting for n_i and n_e in equation 2.11 with the aid of equations 2.12 and 2.14, leads to a differential equation in V,

$$\frac{d^2V}{dx^2}\ =\ -\ \frac{e\ n_o}{\epsilon_0}\left[1\ -\ \exp\left(\frac{e\ V}{k\ T_e}\right)\right] \tag{2.15}$$

For those values of x where $eV/kT_e \ll 1$ the exponential part of this expression can be linearized to give

$$\frac{d^2V}{dx^2}\ =\ \frac{e^2\ n_o\ V}{\epsilon\ k\ T_e} \tag{2.16}$$

No simplification is possible near $x\ =\ 0$, but this region does not contribute much to the thickness of the cloud because the potential falls off very rapidly there ($\frac{d}{dx}V$ is high). In other words, that region does not contribute much to the shielding length. For positive x the solution of equation 2.16 can be written as

$$V\ =\ V_0\ \exp\left(-\ \frac{x}{\lambda_D}\right) \tag{2.17}$$

where

$$\lambda_D\ =\ \sqrt{\frac{\epsilon\ k\ T_e}{e^2\ n_o}} \tag{2.18}$$

The term λ_D is called the Debye length. It is a measure of the distance it takes for the plasma to screen out external potentials.

Another way to quantify the ability of a plasma to protect its neutrality is to evaluate the way it reacts on movement of charged particles. Consider an external force displacing the plasma electrons over some length and disappearing immediately after that. The electron displacement induces an electric field pulling the electrons back to their original position. On reaching their origin the inertia of the electrons will cause them to overshoot and oscillate around the equilibrium situation. An expression for the frequency of this oscillation can be derived as follows. Assume a plasma where neither the ions nor the electrons undergo thermal motion. As in the previous thought experiment, the ions are forming an immobile positive background. Now imagine a small electron density fluctuation in the x direction,

$$n_e = n_o + \Delta \tag{2.19}$$

where the absolute value of Δ is much smaller than n_o. The electric field E and the acceleration $\frac{dv}{dt}$ of the electrons by that field are related through

$$m_e \frac{dv}{dt} = -e E \tag{2.20}$$

The total number of electrons remains constant. Therefore, the continuity equation for the electrons can be written as

$$\frac{dn_e}{dt} + \frac{d(n_e v)}{dx} = 0 \tag{2.21}$$

The amplitude of the density oscillation was assumed to be small. In the second term of the above equation, n_e can therefore be equated to n_o. Substituting equation 2.19 for n_e in the left-hand term gives

$$\frac{d\Delta}{dt} + n_o \frac{dv}{dx} = 0 \tag{2.22}$$

With a deviation from charge neutrality Δ, Poisson's equation in the x-dimension can be written as

$$\frac{dE}{dx} = -\frac{e \Delta}{\epsilon} \tag{2.23}$$

Differentiating equation 2.20 relative to x, followed by the elimination of $\frac{d}{dx}E$ in the result with the aid of equation 2.23, gives

$$m_e \frac{d^2 v}{dx dt} = \frac{e^2 \Delta}{\epsilon} \tag{2.24}$$

Equation 2.22 can be differentiated relative to t, and inserted in equation 2.24 to eliminate $\frac{d^2}{dx dt} v$. This results in a differential equation where Δ is the only variable,

$$\frac{d^2 \Delta}{dt^2} + \frac{e^2 n_o \Delta}{m_e \epsilon} = 0 \tag{2.25}$$

Equation 2.25 describes a harmonic oscillation in electron density with an angular frequency ω_e given by

$$\omega_e = \sqrt{\frac{e^2 n_o}{m_e \epsilon}} \tag{2.26}$$

This characteristic parameter is known as the electron plasma frequency. It is an indication of how much time is needed for charge equilibrium to be restored. The plasma frequency and the Debye length derived earlier are related through the mean electron velocity. Multiplication of equation 2.18 with equation 2.26, and comparing the result with equation 2.2 for electrons, shows that

$$\lambda_D \, \omega_e \simeq \overline{v_e} \tag{2.27}$$

This expression is correct but for a factor of $\sqrt{8/\pi}$, which is consistent with the description of λ_D and ω_e as the extent of perturbation of charge neutrality in space and time, respectively. Note the similarity between equation 2.27 and equation 2.3.

2.4 Criteria and domains

It is now possible to set up quantitative criteria for the identification of those ionized gases that can truly be termed plasmas. The first criterion takes into account the fact that any practical plasma is finite. In order to exhibit the effect of quasi-neutrality, potentials and local charge concentrations should be shielded out on a scale which is small relative to the dimensions of the system under consideration. This means that

$$\lambda_D \ll l \tag{2.28}$$

where l is the smallest dimension of the system.

For the second criterion, the thought experiment used for the derivation of the Debye length has to be reviewed. There, a sheath of particles screens out the applied potential. A certain fraction of the particles manages to escape from the potential well as a result of their thermal energy. Because a velocity distribution function was used, enough charged particles should be present in that sheath or it would not have been a statistically sound concept. The crucial parameter here is the number of particles in a Debye sphere N_D. In other words, the number of particles within a globe of plasma having a radius of λ_D. This number must be much larger than unity,

$$N_D = \tfrac{4}{3} \pi \, n_o \, \lambda_D^3 \simeq 1000 \sqrt{\frac{T_e^3}{n_o}} \gg 1 \tag{2.29}$$

The third criterion is a statement expressing the fact that particle motion in a plasma is controlled by electromagnetic forces rather than by ordinary fluid mechanics. The electrons should, in other words, interact with each other more frequently than with the neutral gas. In quantitative term

$$\omega_e > \nu_e \tag{2.30}$$

where ν_e is the electron-neutral collision frequency. It is also possible to express this criterion in terms of characteristic lengths. Because of equations 2.27 and 2.3, equation 2.30 can be written as

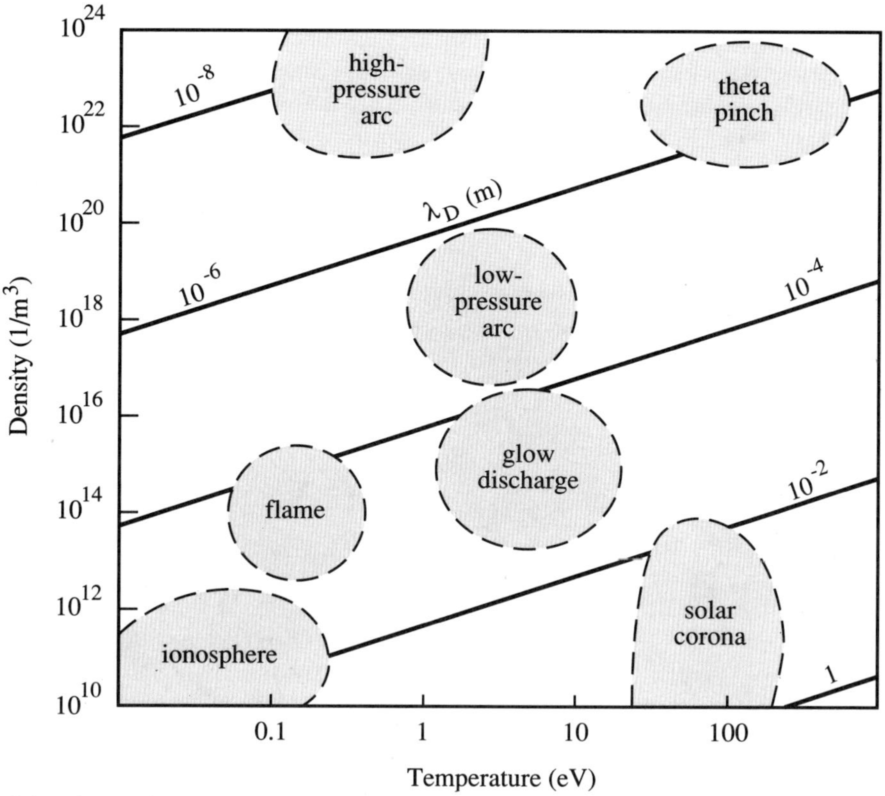

Figure 2.4. Approximate plasma domains on a n_e versus kT_e plot. Lines of constant λ_D are also shown.

$$\lambda_D < \lambda_e \tag{2.31}$$

where λ_e is the electron mean free path in the gas under consideration. Because the Debye length is a function of the electron density, and the electron mean free path is a function of the neutral particle density, equation 2.31 can be used to determine the minimum degree of ionization for a system to remain in the plasma state.

With the characteristic parameters and criteria for the plasma state now defined, a classification can be set up. The most important plasma parameters were found to be the Debye length and the plasma frequency. The expressions for these parameters contain two variables only, namely the plasma density and the electron temperature. Hence, it appears that plasmas can be characterized in principle by the latter two variables only. Figure 2.4 is a log-log plot of n_e versus kT_e, illustrating some of the more important plasma domains, after Hollahan and Bell [22]. Note that the graph covers an extremely wide range, 14 orders of magnitude in density and 5 orders in temperature! The tremendous range in density can be appreciated even better if it is realized that air and water differ in density by only a factor of thousand. Even neutron stars, probably the most dense material apart from black holes, are only 10^{15} times denser than water. Yet, all plasmas follow the same principal behavior and the same set of equations. The upper left hand corner of the figure is the boundary for the second criterion of the plasma state, that is, the line where $N_D = 1$ is crossed.

So far the discussion has been focussed primarily on the charged particles in plasmas. Nevertheless, the neutrals should not be neglected. In many practical applications, including dry etching, the large majority of the particles in the plasma is not charged. The parameter relating charged and neutral particles in a gas is the fractional degree of ionization α, defined as

$$\alpha = \frac{n_o}{n_g} \tag{2.32}$$

In this equation n_o is the plasma density – here, equal to the concentration of ionized molecules – and n_g the neutral gas density. For the domains in figure 2.4, the ionization factor ranges from 10^{-6} to 10^{6}, while the neutral gas temperature varies from near the absolute zero to several million K. It should be realized that ionization and dissociation are not necessarily related phenomena; a dry etch system with an ionization of 10^{-5} can easily hold up to a few percent of dissociated neutrals.

A major division in classifying families of plasmas lies at a fractional ionization of 10^{-2}. Below this value, the charged species are spread so thin that collisions of ions and electrons with neutral species are more frequent than collisions with each other. This takes into account the fact that the electrical – Coulomb – interaction ranges much further than the physical diameter of the respective particles. At the same time, the neutral species collide mostly with other neutrals. Since impulse transfer is very efficient between particles of comparable mass only, the ion and neutral populations will have almost identical temperatures. By the same argument the electron temperature can be quite different, and usually considerably higher. Since the heat capacity of the system is determined by the most abundant species – here, the neutral gas – plasmas with an α of less then 1% are called "cold". Typical examples are the glow discharges in fluorescent lamps, the earth ionosphere, many gas lasers, and almost all dry etch systems. In the diagram of figure 2.4 the division between cold and hot plasmas can be found roughly by drawing a line from the top-left to the bottom-right hand corner; the cold plasmas then occupy the bottom left half.

It will be clear that the species in cold plasmas are not in thermodynamic equilibrium, as a result of their rather limited interactions. Consequently, the assumption of a Maxwellian temperature distribution for the charged particles, see figure 2.1, becomes questionable. This is of importance especially for the electrons, as many important collision processes – such as ionization – are caused by the high-energy tail of the electron distribution function. A frequently used solution to this problem is to construct a velocity distribution out of several independent Maxwellian distributions, each with its own independent temperature. Another effect of the non-equilibrium situation in cold plasmas is that the lifetime of the charged particles is often quite high – of the order of milliseconds – especially at lower pressures. This is caused by the fact that the charged particles can be removed only by mutual recombination, and for that reaction either a 3-body collision or a wall is required to get rid of the excess momentum.

If α becomes more than 1%, Coulomb collisions between electrons and ions start to outweigh the interactions of charged particles with neutrals. The electron and ion energy distribution approach each other rapidly as the degree of ionization increases. Since the ions continue to transfer their kinetic energy to the neutrals, the overall gas temperature will also rise. These systems are called "hot" or thermal plasmas. The high interaction between ions, electrons, and neutrals in hot plasmas allows the assumption of thermodynamic equilibrium, at least locally (LTE). Nuclear fusion reactors, the central part of the sun, and the arcs used

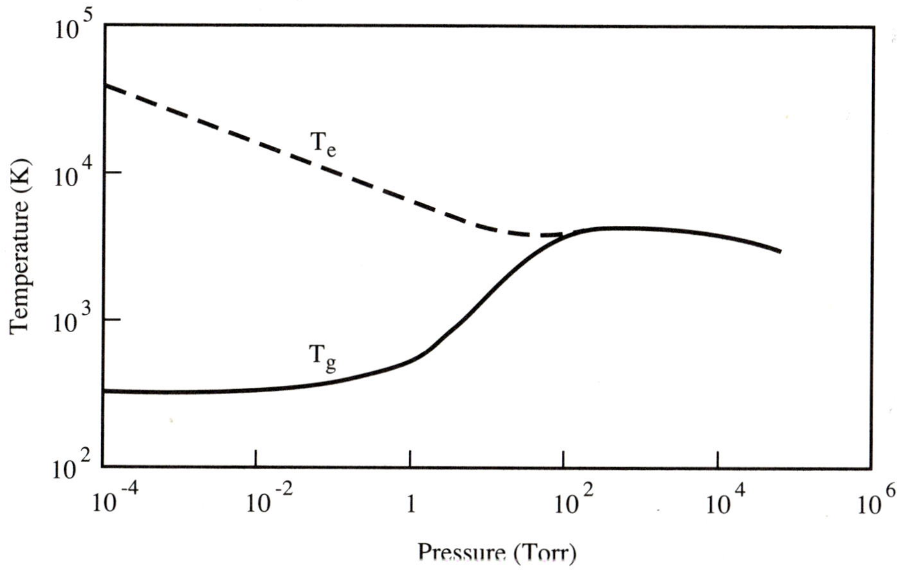

Figure 2.5. Variation of electron and neutral gas temperature with pressure in a Hg arc at constant current.

in welding and spray coating are all hot plasmas. The transition from a cold to a hot plasma can be seen clearly in figure 2.5, after reference [24]. In this example, the transition was brought about by increasing the gas pressure in a mercury vapor discharge. The pressure rise improves the efficiency of energy transfer, and thereby increases the degree of ionization.

Chapter 3

THE AC DISCHARGE

3.1 Introduction

Current dry etching systems use glow discharges with gas pressures ranging typically from 0.5 to 500 Pa. Electrical power supply is by ac only, with frequencies of 50 kHz and up. High-frequency excitation is applied as it takes out the necessity to draw direct current through the system. This allows for the use of non-conducting substrates and electrodes. In many cases, interactions at or near the wafer surface, such as ion bombardment, play an important role. Hence, a large part of the following text is dedicated to the space charge regions ("sheaths") formed between the plasma and the surrounding surfaces.

In this chapter, a standard system is introduced which allows calculation of parameter values representative for the situation in the average dry etching reactor. The standard system is a uniform Ar plasma operating at a pressure of 20 Pa, enclosed between two circular plates with radius r = 30 cm and spacing l = 5 cm, see figure 3.1. The selection of realistic values for the crucial parameters electron density and temperature in ac-operated systems is not trivial, mainly due to the fact that almost all techniques for measuring these quantities are disturbed by the presence of high-frequency fields, see chapter 7. For the standard system, it is assumed that the electron temperature T_e = 23000 K (2 eV), the neutral gas temperature T_g = 400 K (0.03 eV), and that the ion temperature is equal to that of the neutral gas, T_i = 400 K. The plasma density is assumed to have a magnitude $n_o = 10^{16}$ m^{-3}. Single-charged positive ions are considered only; hence, $n_o = n_i = n_e$, where n_i and n_e are the ion and electron densities

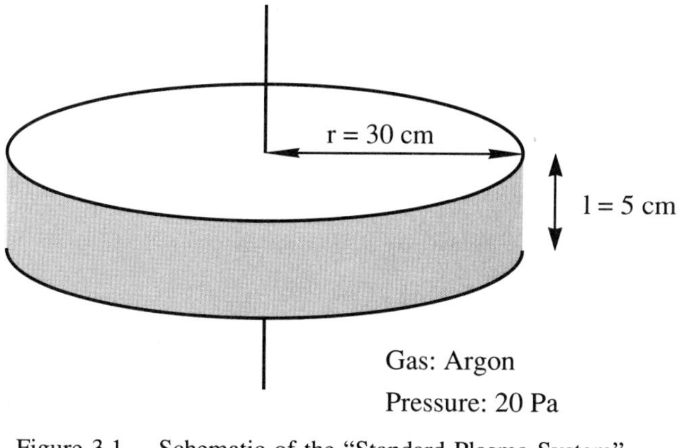

Figure 3.1. Schematic of the "Standard Plasma System".

in the plasma, respectively. All densities and temperatures are assumed to be independent of time and place within the plasma.

The mean velocity in a Maxwellian distribution and the Debye length are given by

$$\overline{v} = \sqrt{\frac{8\,k\,T}{\pi\,m}} \tag{3.1}$$

$$\lambda_D = \sqrt{\frac{\epsilon\,k\,T_e}{e^2\,n_o}} \tag{3.2}$$

where k is the Boltzmann constant, e the electron charge, ϵ the dielectric constant, and T and m the particle temperature and mass, respectively, see chapter 2. For the standard system the ions are Ar^+ with $m_i = 6.6 \times 10^{-26}$ kg, therefore $\lambda_D = 0.1$ mm, $\overline{v_e} = 9 \times 10^5$ m s^{-1} and $\overline{v_i} = 4 \times 10^2$ m s^{-1}.

The mean free path of the electrons under these conditions follows from the electron collision frequency ν_e,

$$\lambda_e = \frac{\overline{v_e}}{\nu_e} \tag{3.3}$$

which in its turn can be estimated from the electron mobility μ_e [25],

$$\mu_e = \frac{e}{m_e\,\nu_e} \tag{3.4}$$

In Ar at 20 Pa the mobility can be calculated to be about 300 m^2 V^{-1} s^{-1} [26]; thus $\nu_e = 6 \times 10^8$ s^{-1} and $\lambda_e = 2$ mm.

The parameters derived above can be used to verify – just in case – whether the standard system is really a plasma. Criteria 1 and 3 from chapter 2 require the Debye length to be small relative to both the smallest dimension of the system and the electron mean free path. Since $l = 5$ cm and $\lambda_e = 2$ mm, while $\lambda_D = 0.1$ mm, these conditions are evidently satisfied. For a statement in terms of the second criterion, the number of electrons in a globe with radius λ_D must be calculated. With equation 2.29 from chapter 2 this is readily found to be 35; much larger than one, hence, sufficient. A pressure of 20 Pa at a temperature of 400 K implies a neutral gas density of 4×10^{21} m^{-3}. The degree of ionization is then 3×10^{-6}, which classifies the standard system as a true, cold, gas plasma. A comparison with figure 2.4 shows that this plasma does indeed belong to the family of glow discharges.

3.2 Sheath formation and the floating potential

The thermal current density in one direction for any single charged species is

$$J = \frac{e\,n\,\overline{v}}{4} \tag{3.5}$$

Because of the large difference in temperature (hence, velocity) between electrons and ions, much more electrons than ions will pass through a given surface area per unit time. For the standard system introduced in the previous section, the current can be calculated to be 320 A m^{-2} and 0.16 A m^{-2} for the electrons and ions, respectively. Apparently an electrically floating object suspended in a cold discharge charges very rapidly negative with respect to its surroundings. The steady-state potential difference between the plasma and the floating object is called the floating potential, V_f. The magnitude of V_f can be deduced from the following argument.

In chapter 2 it was demonstrated that in a Maxwell-Boltzmann velocity distribution, the number of electrons that can pass a potential barrier of height V_f is predicted by

$$n_e = n_o \exp\left(\frac{e\,V_f}{k\,T_e}\right) \tag{3.6}$$

with n_o being the bulk plasma density. Note that V_f has a negative value, as expected. Because of the exponential decrease in electron density when approaching a negative potential – the floating object – the plasma ceases to exist and a positive space charge region called the ion sheath is built up. The ion sheath can, under suitable conditions, be observed by eye as a dark space, since most gas phase excitation processes involve electrons, and the optical emission of the discharge is largely caused by radiative decay of excited species. The net electron current density to the wall is found by inserting equations 3.1 and 3.6 in equation 3.5,

$$J_e = 0.4\,e\,n_o \sqrt{\frac{k\,T_e}{m_e}}\,\exp\left(\frac{e\,V_f}{k\,T_e}\right) \tag{3.7}$$

To obtain the ion current density at the wall, it must be realized that electrons do not screen out potential deviations small relative to their thermal potential, i.e. below kT_e/e. Therefore, the electric field from the space charge region can extend somewhat into the plasma. For ions such a potential is very large because $T_i \ll T_e$. This means that while the electrons are repelled, the (positive) ions are accelerated out of the plasma towards the floating object. It can be derived from mathematic continuity considerations that the ions must enter the sheath region with an energy of at least $\frac{1}{2}kT_e$, which is equivalent to a potential of at least $\frac{1}{2}kT_e/e$, or half the electron temperature in volts. This is known as the Bohm criterion [23,27]. Under assumption of collisionless motion, and neglecting the thermal motion of the ions, the directed ion velocity v_i follows from energy conservation:

$$\tfrac{1}{2}\,m_i\,v_i^2 = \tfrac{1}{2}\,k\,T_e \tag{3.8}$$

$$v_i = \sqrt{\frac{k\,T_e}{m_i}} \tag{3.9}$$

This velocity is considerably higher than the ion velocity in the bulk of the plasma, namely by a factor of approximately $\sqrt{T_e/T_i}$, see equation 3.1.

It has already been mentioned that the potential at the plasma to ion sheath boundary is $-\frac{1}{2}kT_e/e$. From equation 3.6 the electron density is then $n_o\exp(-\frac{1}{2}) = 0.6\,n_o$. The ion density J_i at the boundary is numerically equal to this value, such that the ion current density in the sheath is

$$J_i = e\,n_i\,v_i = 0.6\,e\,n_o\,\sqrt{\frac{k\,T_e}{m_i}} \tag{3.10}$$

For an object at floating potential $J_e = J_i$; there is no net charge accumulation at equilibrium. This means that the right-hand sides of equations 3.7 and 3.10 are numerically equal. By combining the two equations, the floating potential can be written as

$$V_f = -\frac{k\,T_e}{2\,e}\,\ln\left(\frac{m_i}{2.3\,m_e}\right) \tag{3.11}$$

Since the ion mass is 3-4 orders of magnitude higher than that of an electron, the floating potential will have a (negative) value of several times the electron temperature in volts. For the standard system used in this chapter it then follows that $J_i = 2$ A m^{-2} and $V_f = -10$ V.

3.3 The cathode fall

The interaction of a plasma with a perfectly floating object was analyzed in the previous section. For reasons which will become clear later on, such a situation will never happen for any practical dry etching system; there is always some external potential.

Therefore, as a preamble to the discussion on ac-excited systems, the sheath at an unheated cathode of a dc-operated low-pressure lamp discharge will be described, see also reference [28]. In the operating range usual for cold-cathode lamps, the electric current flows through only part of the electrode surface. Evidence for this effect comes directly from the observation that the visible part of the discharge has a smaller cross section than the cathodic electrode. This light-emitting part is the actual plasma. Raising the current only increases the electrode coverage, while leaving the current density and, hence, the voltage across the system, virtually unaffected. This situation is known as the "normal" glow discharge. As soon as the plasma covers the cathode completely, an abrupt change occurs in the discharge I-V characteristics. Because the current density must now increase with increasing current, the operating voltage also rises. At the same time the light emission of the plasma close to the cathode becomes more intense and the dark space – called cathode fall – becomes darker. The system has now become an "abnormal" discharge.

Figure 3.2 presents a schematic view at the relevant regions in an "abnormal" dc-operated tube discharge. A reasonable cathode-to-glow potential in this situation is -500 V. With voltages of this magnitude, no plasma electrons can reach the cathode surface; a net positive ion current is extracted from the glow. At the anode on the opposite side of the discharge, the electrode voltage will be much lower. Because a net negative current is required equal to the current at the cathode, the anode potential is less negative than the floating potential, sometimes even slightly positive. This allows more electrons than ions to leave the discharge at that point.

In the abnormal discharge two quite distinct glow regions can be observed, called the "negative glow" close to the cathode and the "positive column" in the remaining part. The two can, but need not be, separated by a region of reduced brightness, the "Faraday dark space" [28]. The positive column in a dc discharge is very close to the ideal plasma state. By altering the length of the tube, the glow can be given any required length, affecting the

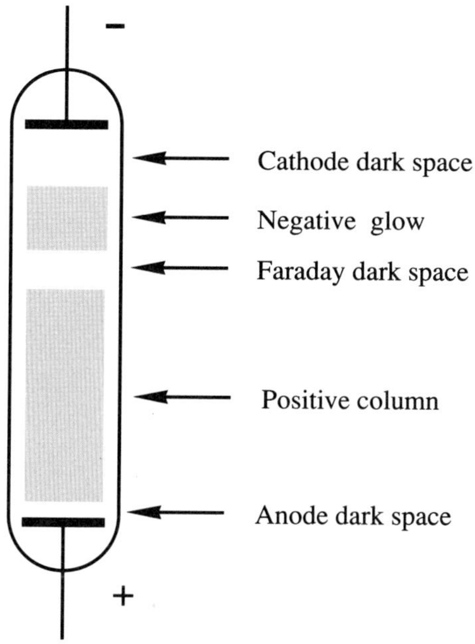

Figure 3.2. Bright and dark zones in a dc tube discharge.

length of the positive column only. In fact, the anode can be moved in as far as the edge of the negative glow without leading to significant changes in I-V characteristics. As soon as the negative glow is compressed, the voltage across the discharge rises sharply and the discharge eventually extinguishes. This indicates that the negative glow plays an important role in sustaining the plasma.

A qualitative description of the processes behind the glow formation is as follows. The bombardment of the cathode with energetic ions initiates secondary electron emission. These secondary electrons are accelerated back across the cathode and into the negative glow. The transition from cathode fall to negative glow is determined by the point where the secondary electrons lose most of their energy, i.e, at a distance from the cathode surface equal to the mean free path for electron-impact ionization. Transition from negative glow to positive column occurs when all secondary electrons are thermalized by collisions. While the positive column is thus an almost perfect plasma, the negative glow is strongly disturbed by the presence of at least two electron populations of which the hottest one certainly does not have a Maxwellian distribution. Increasing the pressure lowers the mean free path and, hence, brings the onset of cascade ionization closer to the cathode surface. It is, in fact, observed that the length of the cathode dark space is inversely proportional to the gas pressure, at least in a dc discharge [29]. For pressures of 50 Pa and up, this is also the case for ac-excited plasmas [30,31]. At lower pressures, other mechanisms have to be taken into account to explain the behavior of the ac sheath. This will be discussed in the next section.

With some simplification it is possible to derive expressions which allow calculation of the potential and density gradients in the cathode fall. Consider a positive x-axis perpendicular to the cathode surface, with the origin at the plasma-sheath boundary. Assume that single-charged ions are the only charge carriers, and that these can move without collisions from plasma to wall. Furthermore, let the ion temperature be negligible. Anywhere along the positive x-axis the ion current density can then be written as

$$J_i = e\, n_i\, v_i \tag{3.12}$$

where density n_i and ion velocity v_i are both a function of x. Since collisions that could lead to ionization or recombination are neglected, J_i is a constant given by the Bohm criterion, see equation 3.10. By energy conservation it follows that

$$\tfrac{1}{2}\, m_i\, v_i^2 = e\, V \tag{3.13}$$

such that

$$v_i = \sqrt{\frac{2\, e\, V}{m_i}} \tag{3.14}$$

where V is now the potential relative to the plasma-sheath boundary. This potential is, of course, also a function of x. Note the similarity between these expressions and equations 3.8 and 3.9.

Eliminating v_i in equation 3.12 with the aid of equation 3.14 gives

$$n_i = \frac{J_i}{e} \sqrt{\frac{m_i}{2\, e\, V}} \tag{3.15}$$

Since ions are considered to be the only charged particles, Poisson's equation for the space-charge induced potential gradient in one dimension is

$$\frac{d^2 V}{dx^2} = -\frac{e\, n_i}{\epsilon} \tag{3.16}$$

By substituting equation 3.15 for n_i in the above expression, a differential equation in V and x is obtained. With the boundary conditions $\frac{d}{dx} V = 0$ and $V = 0$ at $x = 0$ this equation can be solved by direct integration. After rearrangement, the result can be shown to be

$$J_i = 0.6\, \epsilon \sqrt{\frac{e}{m_i}}\, \frac{V^{\frac{3}{2}}}{x^2} \tag{3.17}$$

This is the Child-Langmuir equation for collisionless space-charge limited current [32].

As the derivation holds for all positive x values, V may be substituted by the wall potential V_s and the sheath thickness d for x. In the standard system $J_i = 2$ A m^{-2} and $V_s = -500$ V for 20 Pa Ar, resulting in $d = 7$ mm. This agrees well with optical observations of dark spaces for dc systems [33].

For the sheath at an electrically floating wall, $V_s = V_f = -10$ V. The sheath thickness is then $d = 0.2$ mm, which indicates that a floating sheath cannot normally be seen by the unaided eye. Similarly, visible sheaths cannot be assumed to be covering a floating wall. The floating wall sheath thickness can be converted to a dimensionless form by inserting equations 3.2, 3.10 and 3.11 in equation 3.17. Rearrangement gives

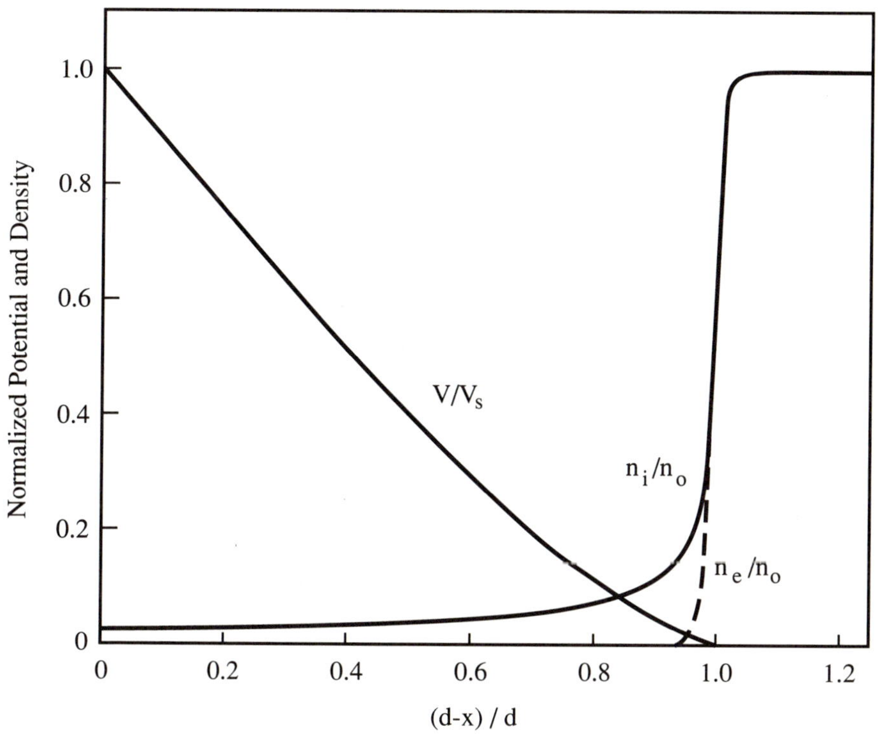

Figure 3.3. Dimensionless potential (V/V_s) and electron / ion density (n_e/n_o, n_i/n_o) gradients in a collisionless sheath. See text for x and d.

$$\frac{d}{\lambda_D} = 0.6 \left(\ln \frac{m_i}{2.3\ m_e} \right) \tag{3.18}$$

With the appropriate values for the various constants it can be derived that the ion sheath thickness at floating surfaces is approximately twice the Debye length.

Density and potential gradients calculated with equation 3.12, 3.15, and 3.17 are given in figure 3.3. The length axis is expressed here as $(d - x)/d$ in order to have the origin at the electrode surface. Since the ion density varies as $1/x$, it should rise to infinity at the plasma-sheath interface (distance from wall d; $x = 0$). This is caused by the assumed boundary condition $V = 0$ at $x = 0$ in the integration of Poisson's equation, implying that the ions are considered to have no thermal energy at all while entering the sheath region. Although this is not very realistic, the differential equation cannot be solved analytically for a finite ion temperature. The magnitude of the ion energy at the origin – $kT_e/2e$ – is negligible relative to the sheath potential; therefore the overall effect of this simplification is small. In the figure the density curve is drawn as an interpolation between the gradients calculated from the Child-Langmuir equation and the Bohm criterion, i.e., a decrease from the bulk density to $n_i(0) = n_e(0) = 0.6\ n_o$ over a length of λ_D.

If the ions have to cross a 7-mm thick sheath, as derived above, the assumed concept of collisionless motion may be doubted. The mean free path of Ar^+ in Ar is only 0.4 mm [33]. This will have consequences for the ion impact energy. Figure 3.4 shows the energy distribu-

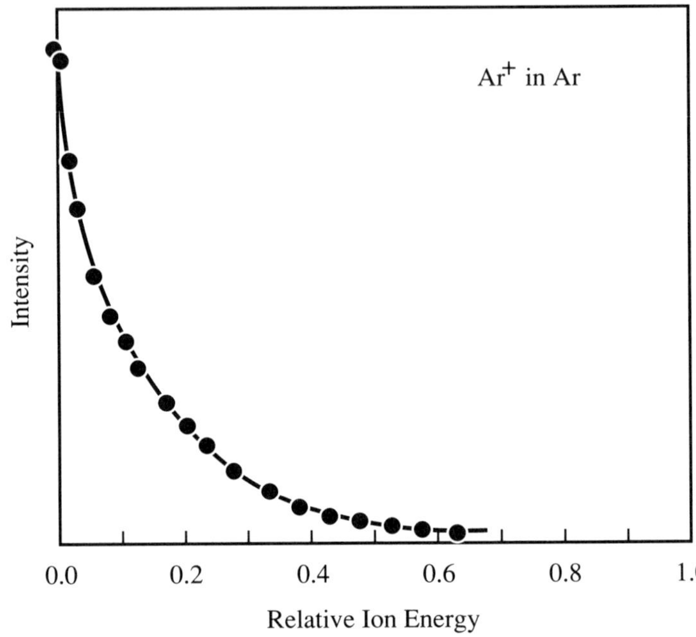

Figure 3.4. Energy distribution for Ar^+ from an Ar dc-discharge cathode fall.

tion for 600-eV Ar^+ ions after passing a 13-mm cathode fall at 8 Pa [33]; instead of one single energy for all ions, a wide range of energies is observed here. It would appear, therefore, that a mobility dominated model for the ion movement is more appropriate in the derivation of sheath gradients. Under collision-dominated conditions the drift of positive ions in strong electric fields is proportional to the square root of the reduced field (i.e., field strength divided by gas pressure) [34]. The directed velocity to the wall can then be described by

$$v_i = \kappa \sqrt{\frac{dV}{dx}} \tag{3.19}$$

where κ is a proportionality factor. Following the same procedure as with the Child-Langmuir theory, but using equation 3.19 rather than equation 3.14 it can be derived that

$$J_i = 1.4\,\epsilon\,\kappa\,\frac{V^{\frac{3}{2}}}{x^{\frac{5}{2}}} \tag{3.20}$$

This expression is surprisingly similar to that found for collisionless current in equation 3.17. For constant ion current the voltage varies now as $x^{5/3}$ instead of $x^{4/3}$, but that difference is small. Even the numerical agreement is considerable. The value of κ for Ar^+ ions in the $\sqrt{E/p}$ drift range is 20 $m^{3/2}\ V^{-1/2}\ s^{-1}$ at 20 Pa [34]. With 500 V and 2 A m^{-2} for V_s and J_i respectively, the calculated sheath thickness is now 5 mm, which is only 30% lower than what is obtained assuming no collisions at all. It appears that 20 Pa is already in a transition region; one has to go to either much higher or much lower pressures to differentiate between the two models [30]. At sufficiently high pressure (>150 Pa), it has been shown from direct application of fluid dynamics [35] as well as from Monte-Carlo simulation [36] that the energy of ions arriving at the electrode surface is proportional to $\sqrt{E/p}$ rather than to the total sheath potential; it amounts to several tens of eV only for typical high-pressure dry etch conditions. For lower pressures the corrections are found to be small, with the regular Child-Langmuir equation giving satisfactory results [37,38].

So far, expressions have been obtained for J_i both in terms of internal (plasma density and temperature, ion mass; Bohm) as well as external parameters (dark-space thickness and potential; Child-Langmuir). It is possible to eliminate J_i from equations 3.10 and 3.17 to get an expression for the plasma density. Inserting the boundary conditions for the electrode surface, i.e., V_s and d, it is found that

$$n_o = \frac{\epsilon}{\sqrt{e\,k\,T_e}} \frac{V_s^{\frac{3}{2}}}{d^2} \tag{3.21}$$

This equation contains only three variables of which one can be realistically estimated – the electron temperature – while the other two can be measured directly by relatively simple optical and electrical methods. It has been shown that equation 3.21 gives reliable results when applied to practical systems [39]. Equation 3.21 can be rearranged again to give a dimensionless expression for the cathodic sheath thickness, in a manner similar to that used in the derivation of equation 3.18,

$$\frac{d}{\lambda_D} = \left(\frac{e\,V_s}{k\,T_e}\right)^{\frac{3}{4}} \tag{3.22}$$

3.4 Characteristic frequencies

When dealing with the subject of ac excitation, it must be realized that plasmas have a large variety of characteristic frequencies and time constants. The electron plasma frequency ω_e and the electron-neutral collision frequency ν_e have been introduced already, see chapter 2. Other time-related parameters are the ion plasma frequency ω_i and the electrical plasma decay time τ. In this section, typical values for these parameters will be derived. In order to simplify the comparison with excitation frequencies, the calculated data are all expressed in Hz by dividing the respective angular frequencies by 2π: $\omega_e/2\pi$, $\omega_i/2\pi$, $\nu_e/2\pi$.

The decay time in the low-pressure, high area-to-volume type of discharge considered here, is determined mainly by wall recombination of ions and electrons. An estimate can be obtained by dividing the total charge in the system by the total loss current. For the cylindrical geometry of the standard system it follows that

$$\tau = \frac{e\,n_o\,\pi\,r^2\,l}{J_i\,(2\,\pi\,r^2 + 2\,\pi\,r\,l)} \tag{3.23}$$

where J_i is the ion current density at the plasma boundary as given by the Bohm criterion. In most of the practical etch systems $r \gg l$. If this is taken into account, and J_i is substituted for by equation 3.11, the following expression is obtained,

$$\tau \simeq \frac{e\,n_o\,l}{2\,J_i} \simeq l\sqrt{\frac{m_i}{k\,T_e}} \tag{3.24}$$

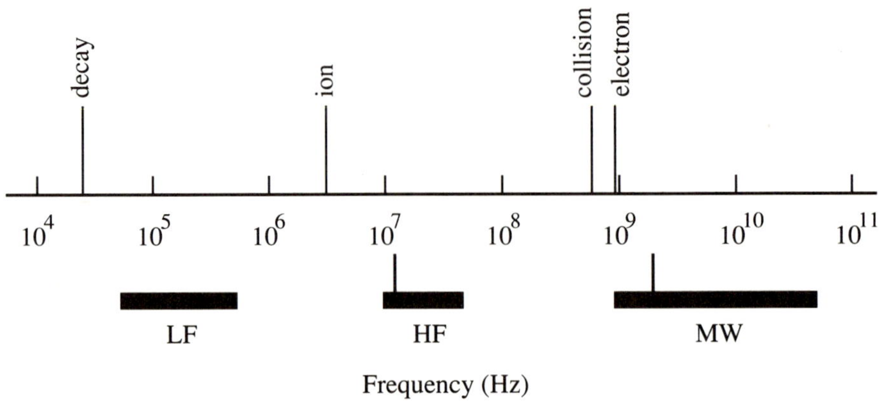

Figure 3.5. Characteristic discharge frequencies and operation ranges for plasma etching and deposition.

Inserting the appropriate values for the various parameters yields a decay time of 20 μs, which is the half-cycle time of a 25 kHz wave. Actual decay times reported for parallel-plate ac discharges are of the order of 100 μs [40,41]. This difference is caused by the fact that the plasma density in any real discharge is not uniform throughout the glow region, but falls off considerably from the center plane to the electrodes due to ambipolar diffusion (outdiffusion of electrons and ions at the same rate).

The electron plasma frequency as derived in chapter 2 is given by

$$\omega_e = \sqrt{\frac{e^2 \, n_o}{\epsilon \, m_e}} \tag{3.25}$$

Apparently, a plasma density of 10^{16} m^{-3}, as assumed for the standard system, corresponds to 0.9 GHz. The ion inertia gives rise to a similar expression for a characteristic ion plasma frequency if m_e in equation 3.25 is replaced by m_i. For Ar, this gives a transition frequency of 3.3 MHz, in agreement with published data on frequency effects in rf glow discharges [42]. The electron-neutral collision frequency has already been calculated in the introduction to this chapter; it amounts to 0.6 GHz.

Figure 3.5 shows the characteristic frequencies derived above, together with the excitation bands commonly used for dry etch (and deposition) plasmas. Systems not working in the indicated ranges are very rare. From this point on, the 50–500 kHz band will be denoted as LF (low frequency), the 10–50 MHz band as HF (high frequency), and the 1–50 GHz band as MW (microwave). Note that the use of MHz and GHz frequencies is in practice often limited to the industrial frequencies of 13.56 MHz and 2.45 GHz.

3.5 The ac sheath

From the discussion in the previous section, it will be clear that ac excitation must be treated in conjunction with the frequency range under consideration. This section will therefore differentiate between the LF, HF, and MW ranges from figure 3.5. In order not

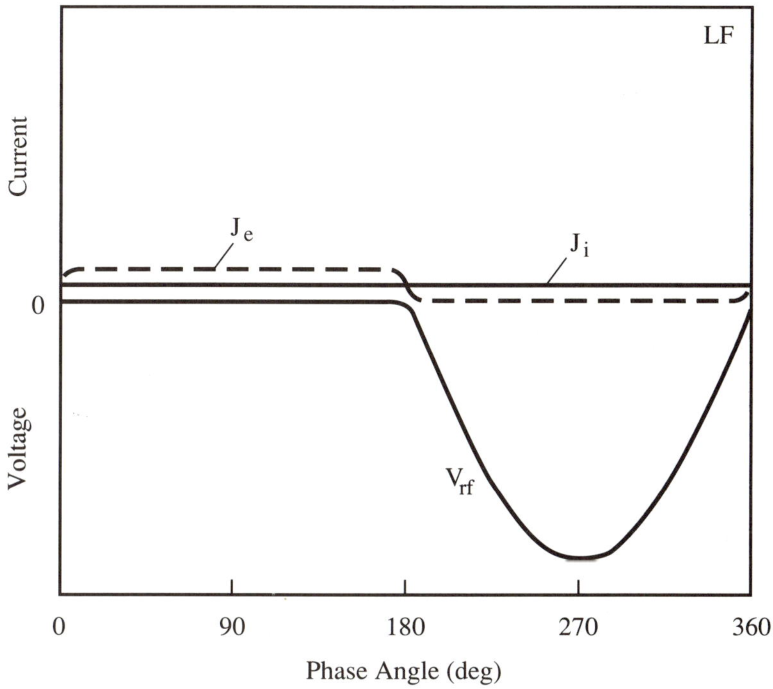

Figure 3.6. Potential (V_{rf}) and current density for electrons (J_e) and ions (J_i) at the electrode surface in LF operation.

to complicate matters too much, a stable, uniform plasma is considered, which is enclosed between two infinite parallel plate electrodes. The plates are spaced wide enough to neglect mutual influencing, but close enough to assume zero potential drop in the bulk of the plasma. There are no external dc currents.

At the low end of the frequency scale excitation will be slow relative to the plasma decay. Therefore, the discharge extinguishes and strikes again every half ac cycle, similar to the operation of an ordinary mains fluorescent lamp. Since dry etching is never done under these conditions such systems will not be treated here.

The lowest practical operation range is the LF band. This is above the decay limit; the discharge will remain "on" throughout the excitation cycle. The plasma will be pulsed, as both ions and electrons follow the electric field; the ion frequency limit is not reached yet. Even if plasma density fluctuations are neglected – a basic assumption throughout this chapter – each electrode will alternatively be going through cathodic and anodic cycles.

Figure 3.6 illustrates this phenomenon for one of the electrodes in the idealized system introduced above. The horizontal axis is a time scale in units of phase angle. Like in figure 3.3 the plasma is used as the reference potential; the voltage levels indicated in this and the next figure are that of the electrode relative to the plasma. Note that the second electrode has exactly the same I-V behavior but is shifted 180 degrees in phase angle. A situation is depicted where a sinusoidal voltage in the LF range is applied across the system; the electrode under consideration is positive for the first half cycle (0-180 deg). Because the ions have enough time to respond to the exciting field, an anode sheath will be formed similar to that in a dc discharge, see figure 3.2. This means that the potential across the sheath $V_s \simeq 0$ during this period, and that $J_e \simeq 2\ J_i$.

In the second half cycle (180-360 deg), the electrode changes polarity to become negative, i.e., the cathode. As a result of this, a cathode fall is build up between electrode surface and plasma. The sheath potential will follow the applied voltage, as there is only the negligible anodic voltage drop at the other electrode in this cycle. At the same time, the electron current density drops to zero. Under the assumption of a stable plasma with invariant density, the ion flux density J_i will be constant over the entire cycle.

Apparently, in LF systems the electrode is continuously bombarded by ions with an energy varying from almost zero to the peak value of the applied ac voltage. Also, net positive charge is accumulated at the electrode during the cathodic cycle, and negative charge during the anodic part, as there is no external dc path. Hence, the applied ac voltage and the resulting ac current are a measure of the ion bombardment energy and flux, respectively. Quantitative calculations on the behavior of a 100-kHz discharge confirm the qualitative description given here [43,44]. Because of the constant jumping of the plasma between the two electrodes, the visual appearance of the dark spaces on either side of the discharge will be like a somewhat blurred dc cathode fall with enhanced light emission. Experiments using a framing camera synchronized with a 1-MHz discharge have demonstrated that at a frequency of that magnitude, the LF system resembles the negative glow from a (switched) dc system, including the secondary electron beam from the cathodic electrode [45].

For excitation frequencies above the ion inertia limit, the ions cannot respond to the field anymore (HF range in figure 3.5). The ac current in the discharge will then be carried by the electrons only and have, contrary to the LF case, no direct relation with the ion flux. These are shortwave radio frequencies (meter waves) which are also used for applications such as radiographic control and citizen band private broadcasting. In literature it is common use to refer to this mode of operation as rf (radio-frequency) discharge etching. Most dry etching systems operate at the industrial frequency of 13.56 MHz.

The behavior of the cathodic sheath in an HF system is in principle similar to that in the LF system described above. The real difference lies in the anodic cycle. Although the electrons are moved back to the electrode surface in that period, the ions cannot follow because they do not "see" the rf field. Therefore, a true anode sheath is not formed, and a cathode fall remains in place over the complete ac cycle. This is illustrated in figure 3.7; applied voltage and further assumptions are the same as in the I-V curves for the LF case (figure 3.6).

Because of the presence of two "cathodes", the applied voltage will now result in smooth sinusoidal sheath potentials on both electrodes. Throughout almost the entire cycle only ions will reach the electrode and cause a net positive charge build-up. This is compensated for in the middle of the anodic period (around 90 deg) when the instantaneous sheath potential is close to zero and the electrons can break through. The electron current density in this pulse can be very high; it is only limited by the random current density in the plasma. Although the ion flux density will be constant, and not dependent on the immediate changes of the rf field, the ions certainly undergo acceleration on their way to the electrode surface. The ion energy is given by the time-averaged value of the sheath potential at that electrode, i.e., half the peak value of the voltage applied across the discharge. This potential is for obvious reasons known as the dc self-bias. In a perfectly symmetrical system, such as the one assumed here, it cannot be measured. However, for reactors with a sufficiently large difference in size

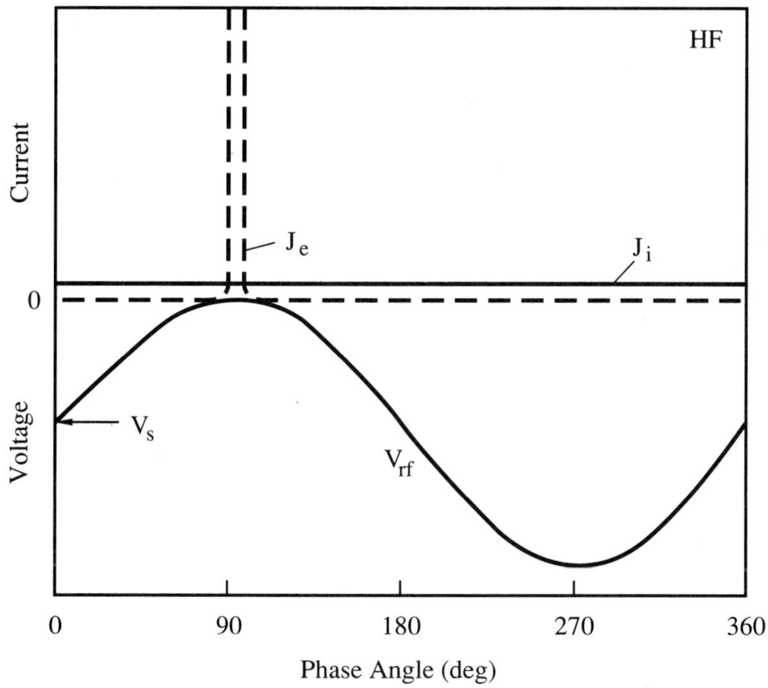

Figure 3.7. Potential (V_{rf}) and current density for ions (J_i) and electrons (J_e) at the electrode surface in HF operation. Arrow indicates ion energy / dc self-bias (V_s).

between the two electrodes, the dc self-bias can be observed as a dc off-set across the system, see chapter 5.

From the above discussion it appears that the HF discharge can be described in terms of a stationary background of ions through which a cloud of electrons rolls back and forth. The ion density gradient near the electrodes will be quite similar to that in a cathode fall from a dc discharge operating at a comparable voltage (figure 3.3). The electrons, while being swept to the electrodes for the surface charge neutralization pulses at 90 degrees phase angle, also follow this same density gradient, since the overall system is a plasma after all and will retain charge neutrality [46,47,48]. Since the maximum possible charge density near the electrodes is determined by the ions, the dark space in an HF discharge will be almost as clearly visible as in a dc cathode fall, even though the real plasma boundary oscillates.

The transition from LF to HF excitation in terms of the ion response to the electric ac field is not as abrupt as assumed in the description given above. At 13.56 MHz, modelling as well as experiments show a residual energy spread of ions arriving at the electrode surface, see figures 3.8 and 3.9, after references [44] and [49], respectively. The variation is caused by the fact that the transition time of the ions through the dark space is not infinitely long relative to the cycle time of the exciting field. Observed energy splits can in fact be used to calculate the dark space thickness and vice versa [37,50]. With an oscillating ion beam in the space charge region, one could once again doubt the validity of the Child-Langmuir theory in HF systems. From a rigorous solution of the fluid dynamics in rf discharges, it appears that the absolute magnitude of the ion current from equation 3.17 may be too high;

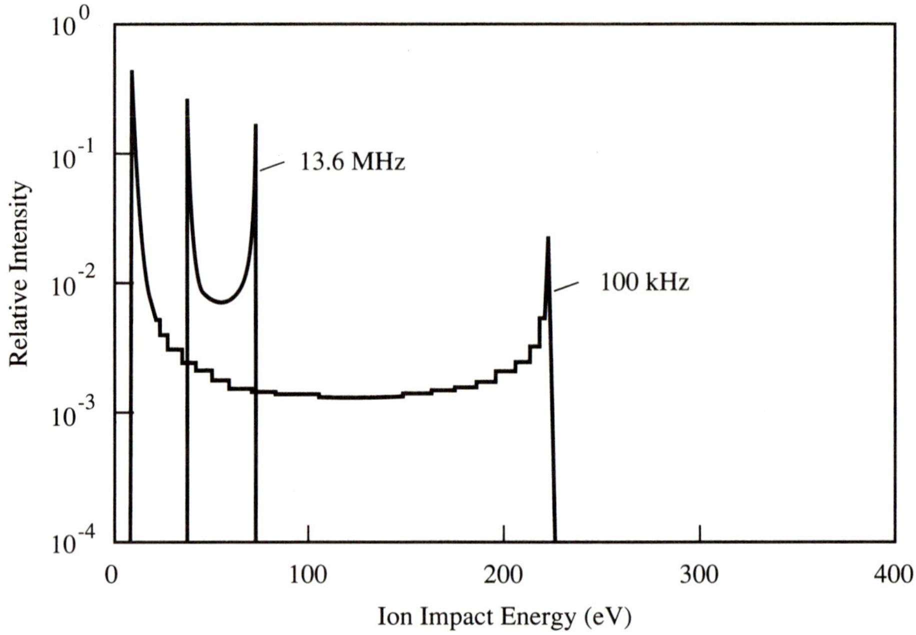

Figure 3.8. Calculated Ar^+ ion energy as a function of frequency.

the correction factor is, however, less than 25% at potentials typical for the excited electrode in etch reactors [46,47].

It is possible to calculate the amplitude of the oscillating plasma boundary in HF-excited systems. As an example, an approximation will be given for the maximum displacement in the cathodic period relative to the ion sheath edge. Assume a step function for the electron wave front. During the cathodic period the electrons will move away from the ion sheath, into the bulk plasma ($x < 0$). The ion density at the electron wave front is then constant, and equal to n_o. As a further approximation, the charge density in the actual ion sheath region is neglected ($0 < x < d$). Poisson's relation can then be solved directly as the only variable is

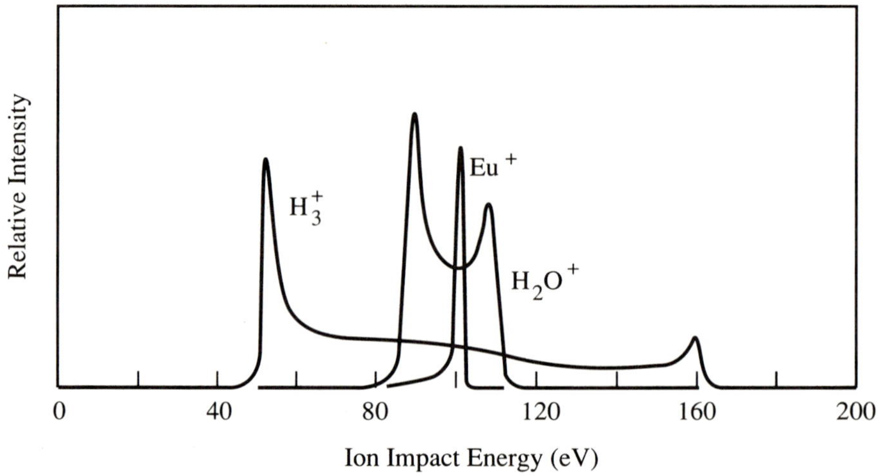

Figure 3.9. Energy spectrum measured at 13.56 MHz as a function of ion mass.

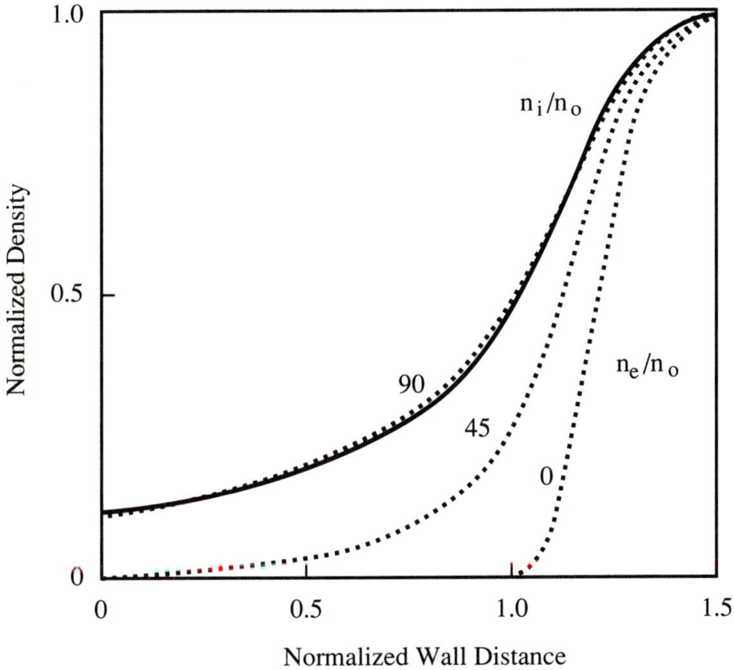

Figure 3.10. Normalized density gradients in the HF sheath. Solid line indicates the ions; dotted lines are electron densities for 0, 45, and 90 degrees phase angle.

x; the density is either n_o or 0. Let the maximum displacement be b. The maximum sheath potential is $2V_s$, which is reached when the boundary is at a position $x = -b$. Integration of equation 3.16 with boundary conditions $V = 0$ and $\frac{d}{dt}V = 0$ at $x = -b$, and $V = 2V_s$ at $x = d$ gives

$$2\,V_s \;=\; -\int_{-b}^{d}\int_{-b}^{0}\frac{e\,n_o}{\epsilon}\,dx\,dx \;=\; -\,\frac{e\,n_o\,(d\,+\,b)\,b}{\epsilon} \tag{3.26}$$

In the standard system the sheath potential is 500 V, the plasma density 10^{16} m^{-3}, and the ion sheath thickness according to the Child-Langmuir equation 7 mm. The maximum plasma boundary displacement is then found to be 0.1 mm, not much indeed. Note again that equation 3.26 is valid in the cathodic cycle only.

More elaborate calculations are needed to describe the position of the electron wave front in the anodic cycle, i.e., when it moves away from the steady plasma into the ion sheath. Figure 3.10 shows the results of a self-consistent solution of Poisson's equation, together with the continuity equations using the correct boundary conditions, after Biehler [47]. It shows the electron and ion gradients at several phase angles in the anodic part of the cycle. The existence of such substantial oscillation effects has been demonstrated experimentally by using rapid optical measurements [51].

It will be clear that the ac current through the HF discharge will be mostly electron displacement, with almost no net charge transport. Hence, the rf current density, which is carried by the plasma electrons, will be much higher than the flux of particles (ions)

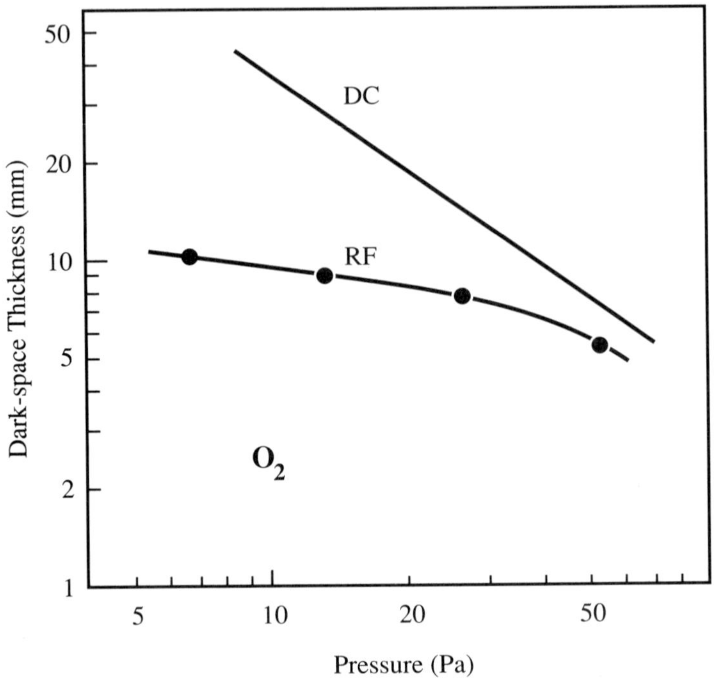

Figure 3.11. Optically determined thickness of a dc cathode fall (DC) and a 13.56-MHz electrode dark space (RF) in O_2 on an aluminum electrode.

bombarding the wall. Unlike the situation in the LF system, there is no direct relation between the two.

The reason why in the HF discharge cathodic sheaths are formed at all, and not floating sheaths, is because the ion space charge region forms an appreciable series impedance for the ac current. The wall sheath is almost free of electrons for most of the cycle time, such that it can be treated as a capacitor [48]. The ac current across this capacitive impedance will result in a voltage drop, which increases the sheath thickness (Child-Langmuir, see equation 3.17), which in its turn causes a higher impedance, etc., until a steady state is obtained. As this mechanism of sheath formation in the HF discharge is quite different from that in the dc case (see the next section), it will not come as a surprise that the variation with pressure is also not similar. Figure 3.11, after references [29,52], shows dark-space thicknesses measured at a 13.56-MHz electrode in comparison with the expected $1/p$ behavior of a dc cathode fall in the same atmosphere. It can be seen that the HF sheath thickness levels off at pressures below 50 Pa.

With microwaves (MW), the situation is entirely different from both the LF and HF cases. This range is far above the electron-neutral collision frequency, and the electrons can undergo many field-induced oscillations before hitting another particle. Another even more important observation is that the wavelength of MW radiation is of the order of centimeters. This implies that any practical reaction vessel is in fact a resonance cavity, where current and voltage levels are determined mainly by the geometry of the cavity, and the discharge itself acts as a lossy dielectric. In the MW band the frequency is high enough to neglect the capacitive impedance of the space charge regions at the plasma-wall interface altogether.

Therefore, all wall sheaths are floating and can be described with equation 3.18. The presence of the electron plasma frequency in the MW range will induce resonance effects.

3.6 Electrical properties

Impedances and I/V characteristics play an important role in the practical operation of dry etching systems. A simple and useful approach to describe the ac discharge is by means of an electrical equivalent circuit. Edge effects are neglected for the moment.

In the previous section, it was shown that the dark space region is virtually free of electrons. It can, therefore, be treated as a capacitor,

$$C_s = \frac{\epsilon A}{d} \tag{3.27}$$

where d is the time-averaged dark space thickness, and A the electrode surface area perpendicular to the ac current. In the standard argon system $A = \pi r^2 = 0.28$ m^2. This is strictly spoken true for HF excitation only [48]; for simplicity it is assumed to hold for all operating ranges.

The acceleration of ions in the dark space results in a resistive ion sheath impedance component,

$$R_s = \frac{V_s}{J_i A} \tag{3.28}$$

where V_s is the time-averaged sheath potential and J_i the ion flux density from the Bohm criterion.

Net transport of charge in the bulk plasma can, to a first approximation, be described with a mobility-limited electron drift [25],

$$J_{rf} = e\, n_o\, v_d = e\, n_o\, \mu_e\, E_{rf} \tag{3.29}$$

where v_d is the electron drift velocity, μ_e the electron mobility and J_{rf} the current density caused by an rf field with a root mean square strength of E_{rf}. The plasma resistance for a planar system is then

$$R_g = \frac{l'}{e\, n_o\, \mu_e\, A} \tag{3.30}$$

where l' is the plasma length in the direction of the rf current vector. At high frequencies or low pressures, the electrons can oscillate between collisions. Therefore, the plasma impedance has an inductive component which depends on the electron-neutral collision frequency [25]

$$L_g = \frac{R_g}{\nu_e} \tag{3.31}$$

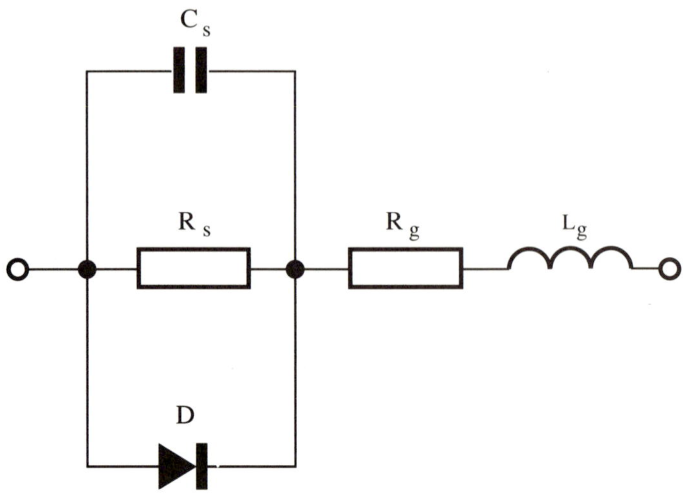

Figure 3.12. Electrical equivalent circuit for an rf discharge in contact with an electrode.

Figure 3.12 combines the above components in an electrical equivalent circuit for the sheath-plasma system. The rectifier D in the diagram denotes the transfer of electrons from the plasma to the electrode surface in the anodic rf cycle where the sheath potential approaches zero.

The relative importance of the various impedance components can be estimated by substituting some representative parameter values. Consider one half of the standard Ar system; $l' = \frac{1}{2}l - d = 1.8$ cm. It can be calculated that R_s = 820 Ω and R_g = 0.13 Ω. Let the excitation frequency be 13.56 MHz, such that the angular frequency $\omega = 8.5 \times 10^7\ \mathrm{s}^{-1}$. The reactive impedances are then $1/\omega C_s$ = 31 Ω for the dark space and ωL_g = 0.02 Ω for the bulk of the plasma. The root mean square current density across the dark space capacitor – which is the electron displacement current – follows from V_s and C_s, and is found to be 40 A m^{-2}. This is 20 times higher than the conduction current J_i, but still 8 times less than the electron random current, see equation 3.5.

For the 13.56-MHz example it appears that $R_g \ll 1/\omega C_s \ll R_s$ and that the admittance of L_g is negligible. The HF-band discharge is clearly almost completely capacitive [52]. In the LF band $1/\omega C_s > R_s$, while ωL_g has become even smaller. Obviously, both C_s and L_g can be neglected. There is now no low-impedance capacitive by-path through the dark space; the discharge is almost completely resistive and the rf current is equal to the ion current in the dark space. As indicated above, the ion current is 1-2 orders of magnitude less than the rf current in an HF system. Therefore, LF discharges require a much higher operating voltage than HF systems under otherwise similar conditions, see figure 3.13 after Donnelly et al. [53].

At the high end of the HF band $\omega L_g > R_g$, because the collision frequency is crossed. As R_g is small relative to the other components, the value of $1/\omega C_s$ is still sufficient for the discharge to stay capacitive. The transition to inductive behavior does not take place until near ω_e. If the excitation frequency reaches the electron plasma frequency, the system becomes resonant and the plasma effectively reflects all incoming waves (cut-off) [54]. Above ω_e the electrons are too slow to keep the plasma field free there. The discharge becomes electrically transparent and the reactor turns into a lossy wave guide (see for example reference [55]). No

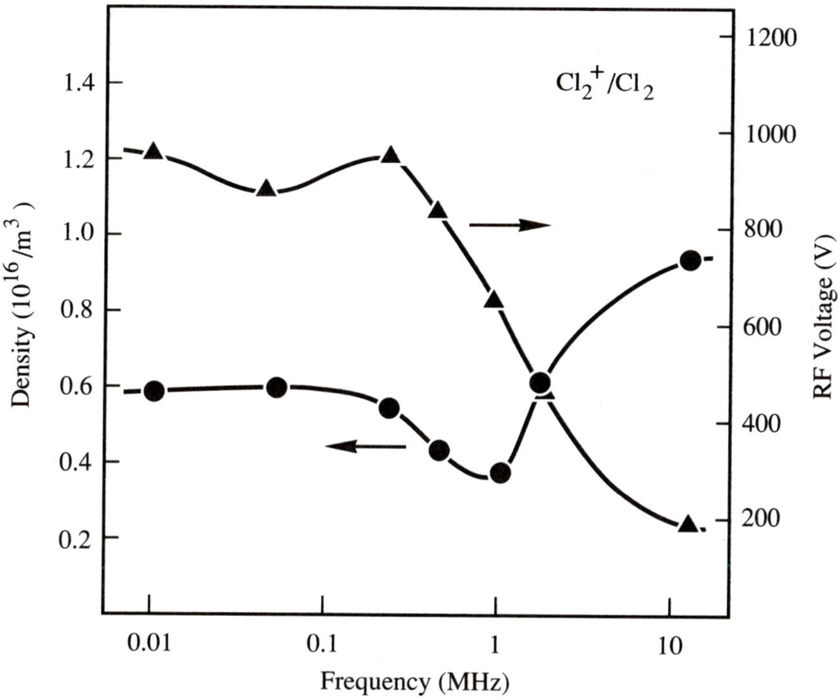

Figure 3.13. Density of Cl_2^+ ion, and reactor peak-to-peak rf potential as a function of frequency for a Cl_2 discharge.

rf potential drop is produced in the dark space; all surfaces in contact with the MW discharge will be at or near floating potential.

Circuit models similar to the one presented in this section have been used with success in the description of rf discharges. Generally, the agreement between expected and measured dark-space behavior is good [29,31,48,52,56,57]. The interpretation of the resistive part is more problematic. Although the assignment of a significant part of the electrical losses to the acceleration of ions in the dark space is in principle correct, certain inconsistencies in the experimental work done so far are indicative of resistive elements other than those given in the diagram shown in figure 3.12 after [52]. Some of the possible mechanisms will be touched on briefly in the next section.

3.7 Energy transfer

It is enlightening to look at the various energy loss and gain mechanisms in rf discharges. Calculations are again done for one half of the standard Ar discharge. Unless stated otherwise the excitation frequency is set to 13.56 MHz.

In the derivation of the discharge decay time it was mentioned that the major charged particle loss mechanism in the reactors under consideration, is recombination of electrons and ions at walls and electrodes. The power input needed to replace this loss by ionization is

$$P_1 = \eta \frac{J_i A}{e} \tag{3.32}$$

where η is the effective energy per ionizing collision and A the electrode surface area introduced in the previous section. Assuming that $\eta = 30$ eV, which is about 2 times the ionization energy of Ar (to compensate for non-ionizing inelastic collisions), and $J_i = 2$ A m^{-2} it appears that $P_1 = 18$ W.

The energy dissipation by the electron drift in the plasma is

$$P_2 = (J_{rf} A)^2 R_g \tag{3.33}$$

where R_g is the glow resistance and J_{rf} the root mean square rf current density. With 40 A m^{-2} for J_{rf} and 0.13 for R_g it is found that $P_2 = 17$ W. A much larger contribution to the energy dissipation in the discharge comes from the ion wall bombardment. The energy carried to the electrode by the ions is

$$P_3 = J_i V_s A \tag{3.34}$$

which in this example amounts to 300 W. Secondary electrons liberated by the ion bombardment are accelerated back into the plasma, thereby providing an additional high-energetic power input. A conservative estimate for the secondary electron yield is 10% [32]; the total power delivered back into the plasma is 30 W. The total of 47 W from primary and secondary electrons is more than the power P_1 needed to sustain the discharge.

Experimental work has shown that the amount of external rf power needed to achieve discharge conditions similar to those in the assumed standard system is at least 5 times more than that which can be accounted for by the mechanisms introduced so far [52]. This can be largely accounted for by power losses in the external circuit, see references [58,59,60], but certain discrepancies remain. Calculations using both analytical methods and Monte Carlo simulation have shown the existence of a dissipation mechanism known in the literature as "wave-riding" [61,62,63,64]. This mechanism can be envisaged as plasma electrons "surfing" on the "wave" created by the oscillating, electron-repelling plasma-sheath boundary. For a qualitative understanding of the wave-riding model it should be observed that the velocity of the sheath edge is considerable, even at lower frequencies. Evidently, the maximum velocity of the oscillating boundary is of the order of ωd, where ω is, as before, the excitation frequency. In the standard system this amounts to 6×10^5 m s^{-1}, or almost as high as the electron random velocity. There is experimental evidence for enhanced electron-impact during that part of the rf cycle where the sheath velocity is at its maximum, i.e., in the anodic period [61]. Another mechanism suggested in theoretical work involves ion-acoustic waves; it is probably important at low frequencies [65].

3.8 Macroscopic phenomena

Although much data is available on individual systems, the information needed to predict the behavior of practical rf glow discharges over a wide range of conditions in terms of rf power, gas pressure, and excitation frequency is quite scarce.

Table 3.1. Macroscopic relationships between external and internal discharge variables (+, – and □ indicate positive, negative and no relationship, respectively).

	Power	Pressure	Frequency
Ion current	+	□	□
Ion energy	+	–	–
Radicals	+	+	+

From the existing literature it can be concluded that the electron density (hence, the ion current to electrodes and walls; Bohm criterion) is about proportional to the rf power, although there are indications of upper and lower limits that may be reactor geometry dependent; frequency – see figure 3.13 – and pressure seem to have little effect [39,53,66].

For constant pressure, the ion bombardment energy has a power dependence lying between square root and linear, at least for HF systems [52,67,68]. Because of the rapidly increasing number of ion collisions in the dark space, the bombardment energy decreases with pressure.

A useful empirical observation is that the higher the frequency, the higher the degree of gas dissociation (not the ionization!) for a given rf power density. This can be judged from the experimentally observed formation of dissociation products (glow polymer and molecular chlorine) in a carbon tetrachloride glow discharge, this being negligible in LF systems, significant in the HF region, and almost complete in MW excitation [69,70].

The above data has been summarized in table 3.1. Of course, this table just indicates trends, and does not imply a quantitative relationship. An important conclusion is that gas pressure and excitation frequency appear to be interchangeable. This is confirmed with results of dry etch processes, see chapter 6.

A last remark is dedicated to negative ions. In all calculations and derivations done so far, it was assumed explicitly that all ions are positive. Many gases commonly used in dry etching are, however, known to have a high collision cross-section for electron attachment. Examples are Cl_2, BCl_3, and NF_3, but also CCl_4 and SF_6 [71,72,73]. Glow discharges in these gases will hold large amounts of negative ions. Optogalvanic measurements indicate that the density of negative ions in such systems can be up to 50 times larger than the number of electrons [72].

The impact of these observations is considerable. In the first place, the lifetime of the negative ions will typically be very high, as they cannot reach the walls of the reactor system. Their thermal energy is far too low to cross the positive space charge region of even a floating sheath, and neutralization by collision with positive ions is unlikely, because of the low degree of ionization. Since the main mechanism behind plasma sheath formation is the equalization of the negative and positive particle loss fluxes, and the density of negative particles that can be lost at all – the electrons – is so much smaller than that of the positive particles, the sheath potential in strong electro-negative gases will typically be significantly lower than in gases not exhibiting electron attachment. Also, the electron density and the ion flux density through the sheath are not related anymore.

Under certain conditions the addition of electronegative gases can cause the discharge to extinguish, because the electrons needed to sustain the plasma are captured into relatively inert, low-energetic negative ions. The dry etch gas SF_6, for example, is frequently used as "stopping gas" in high-voltage switches, where it suppresses the arc otherwise formed when the contact is broken.

Chapter 4

GAS AND SURFACE PROCESSES

4.1 Introduction

The ac discharge introduced in the previous chapter consists, apart from the parent gas, of a multitude of neutral and positively or negatively charged, ground-state, excited and fragmented particles. These species are generated mostly by collision processes between electrons and the neutral gas. Electrons have a dominant role, as they are by far the "hottest" particles in a cold plasma.

During an elastic collision, the kinetic energy of the colliding particles is conserved, and the internal energy levels of all particles are unchanged. If electrons are involved in the collision concerned, the amount of energy transferred will be low, as the masses of electrons and atomic or molecular species differ by some four orders of magnitude. With inelastic collisions, the kinetic energy of the colliding particles is used to change the internal energy of the colliding particles. Examples of inelastic processes are electronic, vibrational and rotational excitation, dissociation and ionization. Besides the generation of excited and ionized species, particles are lost by decay and recombination. In the discussion of the different inelastic collision processes not only the direct, but also some of the reverse processes will be highlighted. For extensive discussions about electron collision processes, the reader is referred to references [74,75,76]; for spectroscopic data in relation to molecular structure, to references [77,78].

The second part of this chapter, starting in the section on plasma-wall interactions, deals with the effects that occur when the particles generated in the plasma reach the wall. The accelerated positive ions are especially important because of the directionality they can induce in etching. Many processes have to be viewed simultaneously; the synergetic effects of having fluxes of reactive neutrals and ions at the same time, is the key factor determining many dry etching processes.

4.2 Collision cross section, free path, and mobility

Let particles with diameter a move through a background of uniformly distributed, stationary, similar particles. If the center of a beam particle comes within a distance a of the mass center of a background particle, the two will collide. In other words, each beam particle sweeps out a cylindrical volume with diameter $2a$. This process is usually described in terms of the cross-sectional area of the cylinder σ, which is called the collision cross section,

$$\sigma = \pi a^2 \tag{4.1}$$

Once the collision cross section for a certain process is known, the average distance between two successive collisions can be determined, that is, the mean free path λ. If the collision cylinder has a length equal to the mean free path, it defines a volume in which exactly one collision takes place. This is equal to the space occupied by one particle of background gas, so

$$\sigma \lambda = \frac{1}{n} \tag{4.2}$$

where n is the background gas particle density. Hence, it appears that λ is inversely proportional to the gas pressure.

An alternative way of obtaining the mean free path is as follows. Consider a cube of background gas which is penetrated by N beam particles. After having travelled a distance x into the cube, the number of beam particles not having collided yet is $N(x)$. Within a distance dx, the number of particles that will collide dN is given by

$$dN = n \sigma N(x) dx \tag{4.3}$$

With the boundary condition $N(x) = N$ at $x = 0$, this differential equation can be solved. The result is

$$N(x) = N \exp(- n \sigma x) = N \exp\left(\frac{-x}{\lambda}\right) \tag{4.4}$$

The probability $P(x)$ of having a collision over a distance x is

$$P(x) = 1 - \exp\left(\frac{-x}{\lambda}\right) \tag{4.5}$$

For the distance $x = \lambda$ the probability is 63%, for $x = 3\lambda$ the probability is 95%. This shows that λ is a statistical property. Following a statistically more sound derivation, which takes into account a moving background instead of a "frozen" gas, the mean free path can be calculated to be a factor $1/\sqrt{2}$ smaller than is given in equation 4.2.

The collision cross section for elastic interaction in simple gases can be estimated with reasonable accuracy from the dimension of the atoms or molecules. Argon, for example, has a geometrical diameter of about 0.3 nm; the cross section according to equation 4.1 would then be 2.8×10^{-19} m^2. The mean free paths for many atoms, including argon, in their parent gas is known [79]. Converting the available data to the standard conditions from the previous chapter (20 Pa and 400 K), the mean free path for argon atoms is 4.5×10^{-4} m. Entering this value in equation 4.2, together with the gas density for the standard conditions (4×10^{21} m^{-3}), the "real" cross section is found to be 5.6×10^{-19} m^2. This shows that the calculations according to the above equations can be correct to within a factor of 2.

It is to be realized, however, that atomic particles are no billiard balls. In reality, the cross section is a parameter describing the interaction area between electron wave functions.

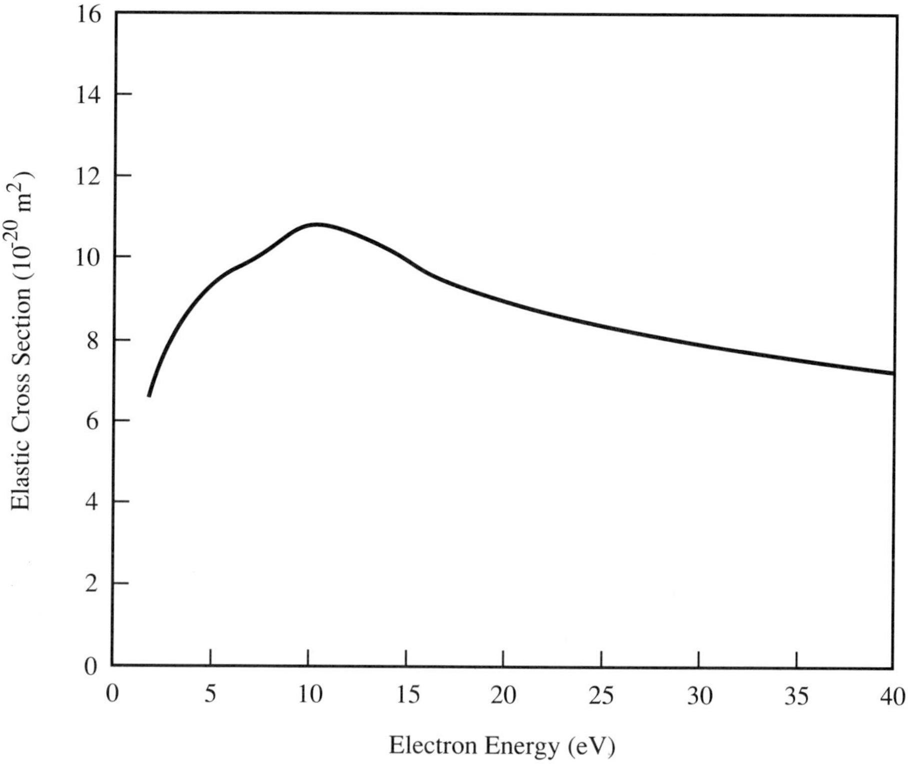

Figure 4.1. Cross section for elastic collision between electrons and O_2 molecules as a function of the electron energy.

Other than in the simple ball model, it depends on the kinetic and potential energy of both beam and background particles. Excited species, for example, have larger cross sections. Usually, the cross section rises at low energy levels with the velocity of the impinging particles because of threshold effects; it decreases at high energy as a result of the fact that any interaction needs a certain minimum contact time. For details about the calculations of collision cross sections, the reader is referred to reference [80].

Figure 4.1 after reference [74], shows the cross section for elastic collisions between electrons and oxygen molecules as a function of the electron energy. The cross section can be seen to be almost independent of the electron energy in the range under consideration. It amounts to about 8×10^{-20} m^2 between 3 and 40 eV [22,81,82]. Note that this is not the same as the total collision cross section; the latter can be much larger.

The angle at which an incoming particle is deflected is, like the (angle-independent) average magnitude of the cross section, a function of energy. For electrons, scattering is more or less random at low energy. With increasing energy, the deviation from the original direction decreases rapidly. At high energy, the scattering angle approaches zero, even though the amount of inelastic energy transfer may still be considerable. The situation is complicated further because each collision process has its own cross section. Summation of the cross sections for all possible processes in a given system leads to what is called the total collision cross section σ_{tot}. It is the latter parameter which determines the mean free path, which explains why a calculation of λ from purely geometrical parameters will always give too high

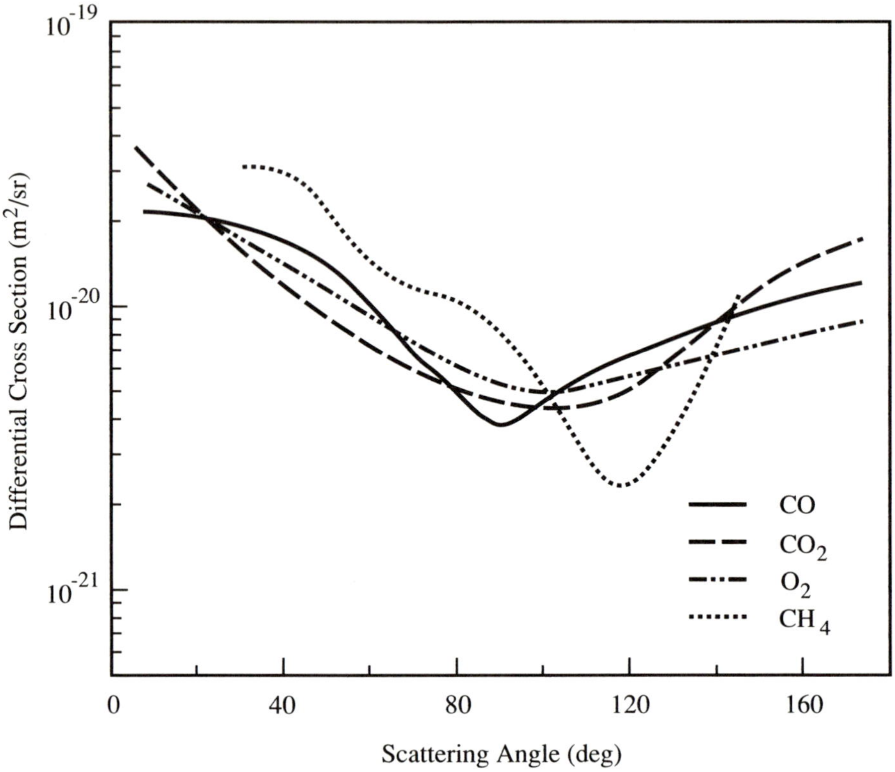

Figure 4.2. Scattering of a mono-energetic beam of electrons at 10 eV in CO_2, CO, O_2 and CH_4.

a value. In figure 4.2 after reference [74], the angle-dependent (differential) cross section for elastic electron scattering is shown for a number of molecular gases. All graphs clearly show a dip near 90 degrees, implying that the majority of the electrons is either reflected back into the source, or scattered forward without deflection. These minima may be associated with the electronic shell structure of the target particles.

At electron energies below about 1 eV, many gases show a pronounced "window" where the elastic cross section drops significantly. This window is known as the Ramsauer minimum; it is a quantum-mechanical effect which can be explained from the Heisenberg uncertainty principle [79]. In molecular gases, the Ramsauer minimum is masked effectively by low-energy inelastic processes, such as vibrational and rotational excitation. The latter two processes are not possible in the noble gases; there, the minimum can also be traced back in the total collision cross section. Argon, for example, has a window between 0.1 and 1 eV where the total collision cross section is lowered by two orders of magnitude. It will be clear, that this can have a considerable influence on the electron energy distribution.

Charged particles

For collisions between two charged particles, the short-range van der Waals force valid for neutral-neutral and neutral-charged interactions, must be replaced by the long-range Coulomb force. Let two ions approach each other on a line connecting their centers, each with the same mass m, velocity v, and charge $+e$. Since the two particles repel each other, a point of closest approach exists where the velocities are zero and the mutual potential energy is at maximum. From energy conservation and the Coulomb equation

$$2 \frac{m\, v^2}{2} = \frac{e^2}{4\, \pi\, \epsilon\, a} \tag{4.6}$$

where ϵ is the dielectric constant and a the distance at closest approach. The parameter a can be treated as a collision diameter. From the above equation is seen to be inversely proportional to the energy of the particles. Assuming argon ions moving at thermal velocities equivalent to 400 K, the mass $m = 6.6 \times 10^{-26}$ kg, and the mean thermal velocity $\overline{v} = 400$ m s^{-1}, it appears that the minimum distance is about 20 nm. That is, indeed, much more than the argon ion diameter. Because a is inversely proportional to the kinetic energy, faster particles will approach each other closer. This does not imply that the interaction is not felt at larger distances; in fact, all particles not following a head-on approach will be deflected at distances as far as the Debye length (0.1 mm in the standard argon system).

Drift is defined as the motion of charged particles in a neutral background as a result of an external electric field. For the drift of ions in moderate fields, the drift velocity v_d can be assumed to be negligible relative to the thermal velocity $\overline{v}$. This is equivalent to the assumption of zero drift velocity after collision. In that case, the time t between collisions, which is equal to the inverse of the collision frequency ν, does not depend on the electric field strength, and can be written as

$$t = \frac{1}{\nu} = \frac{\lambda}{\overline{v}} \tag{4.7}$$

The acceleration $\frac{d}{dt}v$ as a result of the electric field E is given by

$$\frac{dv}{dt} = \frac{e\, E}{m} \tag{4.8}$$

Thus, the length s of the free path between two collisions

$$s = \tfrac{1}{2} \frac{dv}{dt}\, t^2 \tag{4.9}$$

The drift velocity is the average velocity over this path. Combining the above equations it appears that

$$v_d = \frac{s}{t} = \frac{e\, \lambda}{2\, m\, \overline{v}}\, E \tag{4.10}$$

The proportionality factor between v_d and E is the mobility. In the above derivations, use has been made of statistical properties without proper averaging. Although this is not completely correct, it is customary to use the following approximation for the mobility μ,

$$\mu = \frac{e\, \lambda}{m\, \overline{v}} = \frac{e}{m\, \nu} \tag{4.11}$$

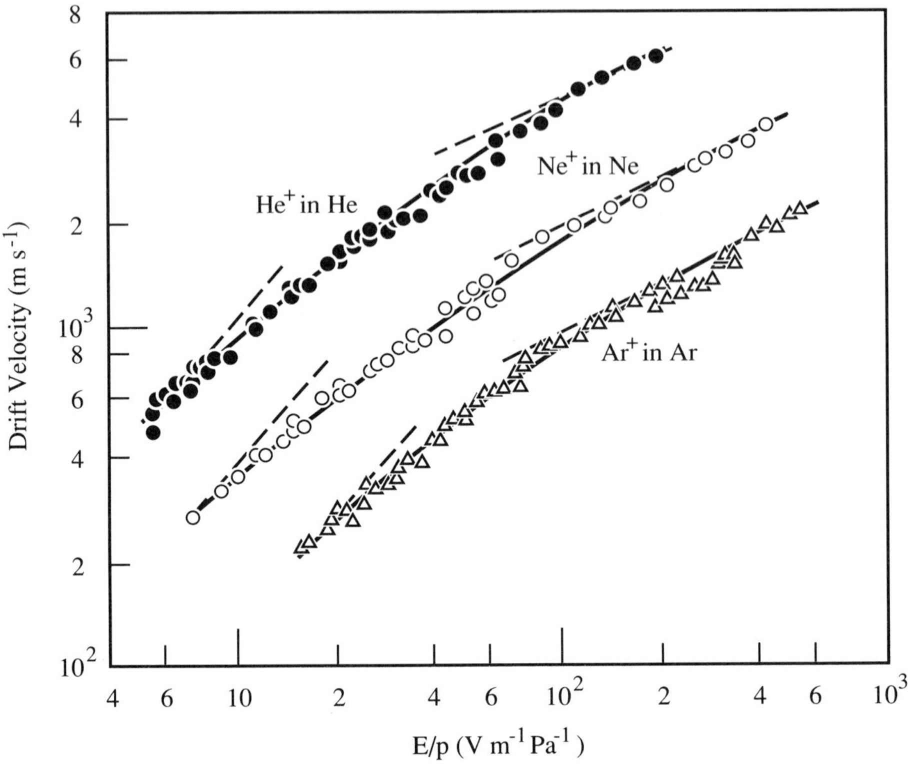

Figure 4.3. Drift of noble gas ions in their parent gas.

Because λ is inversely proportional to the gas pressure, μp is a constant. Hence, the drift velocity is proportional to the reduced field E/p. The mobility for argon ions in argon at 20 Pa and 400 K according to equation 4.11 can be calculated to be 2.7 m^2 V^{-1} s^{-1}, if use is made of experimental values for the free path of the neutral gas atoms. From experimental drift data extrapolated to 20 Pa and zero reduced field, the mobility is found to be 0.8 m^2 V^{-1} s^{-1} [34]. This is still of the correct order of magnitude, but the deviation is significant. In fact, equation 4.11 always overestimates the ion mobility by a factor of 3–5 for both positive and negative species. This can be explained from polarization of the parent gas. Dipoles induced by the ions in the surrounding gas will distract passing ions from their original path without actually colliding. As a result, the path between two "real" collisions will not be linear, which is interpreted on a macroscopic scale as a lower effective drift velocity.

The above derivations are valid for low field strengths only. In high fields, the drift velocity cannot be neglected relative to the thermal velocity, which leads to a reduction in measured mobility. If the drift becomes of the same order of magnitude as the mean thermal velocity or above, it can be derived that $v_d \propto \sqrt{E/p}$. In gases at room temperature, this point is reached when E/p = 50–100 V m^{-1} Pa^{-1}, see figure 4.3 after reference [34].

Drift of electrons in gases is a situation quite different from that of ions. In the first place, electrons have a much higher temperature than the neutrals; even a small drift velocity is large relative to the thermal velocity of the gas. Secondly, the momentum transfer in elastic collisions with the gas is extremely inefficient. The assumptions used above are, therefore, incorrect even for low fields. For atomic gases, the drift velocity can be shown to be proportional to the square root of the reduced field. In moderate fields, which in practice is above 1 V m^{-1} Pa, the electrons can gain enough kinetic energy from the field to reach the threshold for inelastic collision with the parent gas. The situation then once again becomes more comparable with the situation as described for ions, and the mobility can be approximated by equation 4.11. In high fields, the drift component starts to affect the overall electron energy distribution. A general prediction of the drift behavior of the electrons cannot be given anymore in this situation. In plasmas of the type used for dry etching, it appears that the electron drift is proportional to $\sqrt{E/p}$ if the field strength is more than 1000 V m^{-1} [68], as is the case with ions. Under such conditions, the drift energy is sufficient to induce ionizing collisions. In other words, a chain reaction leading to breakdown ionization can occur.

Reaction rate and the electron energy distribution function

If the collision cross section for a certain gas process is known, the reaction rate of that process can be calculated. Consider a medium containing electrons with a density n_e and atoms A with a concentration n_{A}. Let inelastic collisions between the electrons and the atoms occur, through which the atoms A are excited to a species with a higher potential energy A^*,

$$e + \mathrm{A} \rightarrow \mathrm{A}^* + e \tag{4.12}$$

If the concentration of excited species is small relative to that of the parent gas, reverse reactions can be neglected, and the reaction rate is given by

$$\frac{dn_{\mathrm{A}^*}}{dt} = k_r\, n_{\mathrm{A}}\, n_e \tag{4.13}$$

The reaction rate constant k_r in this equation is proportional to the average product of the inelastic collision cross section σ and the electron energy ε. This can be written as

$$k_r = \int_0^\infty \sigma(\varepsilon)\, v\, f(\varepsilon)\, d\varepsilon \tag{4.14}$$

where $f(\varepsilon)$ is the electron energy distribution function. Evidently,

$$\int_0^\infty f(\varepsilon)\, d\varepsilon = 1 \tag{4.15}$$

since all electrons must be found in the energy range between zero and infinity. A similar relation is obtained if the electron velocity is used as the variable rather than the electron energy. Assuming that all electrons are in thermodynamic equilibrium, the distribution function can be approximated by a Maxwell-Boltzmann relation:

$$f(\varepsilon) = \frac{1}{\sqrt{\pi}} \frac{1}{k\,T_e} \left(\frac{\varepsilon}{k\,T_e}\right)^{\frac{1}{2}} \exp\left(\frac{-\varepsilon}{k\,T_e}\right) \tag{4.16}$$

where k and T_e are the Boltzmann constant and the electron temperature, respectively.

With the total collision cross section $\sigma_{tot} = 1/n_A\lambda$, see equation 4.2, and the electron collision frequency $\nu = \overline{v}/\lambda$, see equation 4.7, the reaction rate can be expressed in electron-related properties only,

$$\frac{dn_{A^*}}{dt} = n_e\,\nu \int_0^\infty \frac{\sigma}{\sigma_{tot}}\, f(\varepsilon)\, d\varepsilon \tag{4.17}$$

The term σ/σ_{tot} is the reduced cross section; it is a measure of the probability for the process to occur during a random collision. The collision frequency is proportional to the gas pressure; an estimate for this can be obtained from known mobility data, see equation 4.11. Equation 4.17 may be used in situations where the electron velocity is a major parameter, such as drift-enhanced electron-impact ionization.

It is, at least in theory, possible to determine the electron temperature by entering known collision cross sections in equation 4.17, together with data for the electron density and the reaction rates. The big disadvantage of such a method is that electrons with a non-thermal distribution are neglected. Figure 4.4 shows the electron density for a Maxwellian distribution (see equation 4.16). Note the difference in shape between this graph and the corresponding velocity plot in chapter 2, which results from the fact that the horizontal axis is in units of energy (ε) now, with $\varepsilon = \frac{1}{2}mv^2$. Typical for this thermal distribution is the high amount of low-energy (bulk) electrons and low amount of high-energy (tail) electrons.

Many collision processes have a threshold value of several eV. In argon, for example, the threshold energy for excitation is 11.6 eV, for ionization it is 15.8 eV. The fraction of electrons in a 2-eV Maxwellian electron gas, having an energy higher than these thresholds, can be calculated to be about 0.01 for excitation and about 0.001 for ionization. This would mean that sustaining a discharge with the low electron density and temperature usual for dry etching would be extremely difficult. However, in any practical dry etching system, the "bulk" and "tail" parts of the electron distribution are not in thermal equilibrium. Energy is brought into the discharge mainly by "hot" electrons originating from the ion sheath zones. These electrons are created partly by wave-riding of bulk electrons on the oscillating sheath edge, and partly by acceleration through the dark space of secondary electrons from bombardment processes on walls and electrodes. After many inelastic collisions, the "hot" electrons cool down and finally equilibrate with the rest of the "thermal" electrons in the bulk of the energy distribution. Therefore, the discharge holds many more high-energy electrons than what would be expected from a Maxwell-Boltzmann model. Thus, reactions with a high threshold, such as ionization and dissociation (of molecules), are dominated by the non-thermal tail electrons, while low-threshold processes, such as excitation, are sustained by the thermal bulk electrons. Assuming a straightforward thermal model in the interpretation of equation 4.17 will, therefore, result in unrealistically high values for the electron temperature.

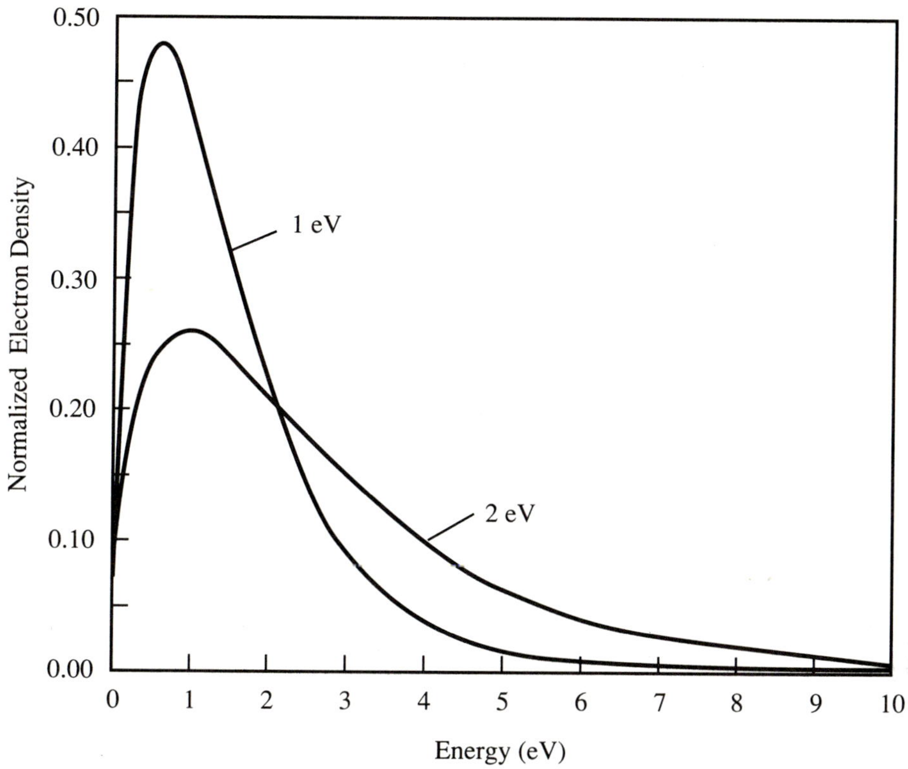

Figure 4.4. Normalized electron density in a Maxwell-Boltzmann distribution for electron temperatures of 1 and 2 eV.

4.3 Excitation

Electronic excitation

The excitation of a ground-state atom A to an excited-state particle A^* by collision with an electron can be expressed as

$$A + e \xrightarrow{k_r} A^* + e \tag{4.18}$$

where k_r is the reaction rate coefficient. This is the most straightforward example of an inelastic collision process. Kinetic energy from the colliding electron is used simply to raise an electron from a low-energy state to one with a higher potential energy. The excited state may decay by emitting a quantum $h\nu$,

$$A^* \rightarrow A + h\,\nu \tag{4.19}$$

The lowest electronically excited states in the noble gases lie at least 10 eV above ground level. Hence, the threshold value for the corresponding collision cross section curve is also above 10 eV. Other atomic gases have a much lower minimum energy distance, say 1–2 eV. Wavelengths of radiation corresponding with decay of such states are between 100 and 1000 nm,

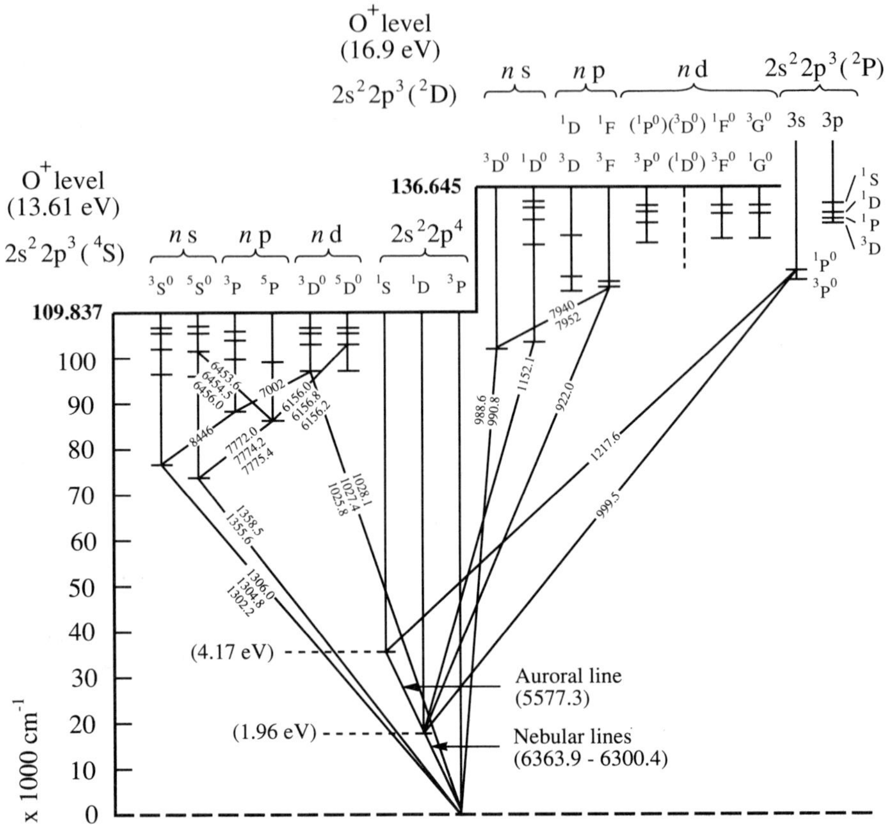

Figure 4.5. Energy diagram and characteristic transition wavelengths for the O atom.

hence, from near ultraviolet to near infrared. Figure 4.5 after reference [83], shows some of the energy levels of the oxygen atom, together with the characteristic wavelengths of the allowed transitions. The value of the collision cross section after threshold rises either rapidly to a sharp maximum a few eV higher, or to a broad and ill-defined maximum 10–20 eV above threshold. Which shape is actually observed depends on the multiplicity of the corresponding energy levels. Transitions with like multiplicity (singlet-singlet) are broad; for unlike multiplicity (singlet-triplet) these are sharp.

The reaction rate of the decay process described in equation 4.19 is

$$\frac{dn_{A^*}}{dt} = \frac{n_{A^*}}{\tau} \tag{4.20}$$

where τ is the characteristic lifetime of the excited state. The solution of this equation is an exponential decay

$$n_{A^*} = n_{A_0^*} \exp\left(\frac{-t}{\tau}\right) \tag{4.21}$$

where $n_{A_0^*}$ is the density of A* at $t = 0$. In steady state, the direct collision process (equation 4.18) is in equilibrium with the reverse reaction (equation 4.19). In other words, the rate of production is equal to the rate of decay,

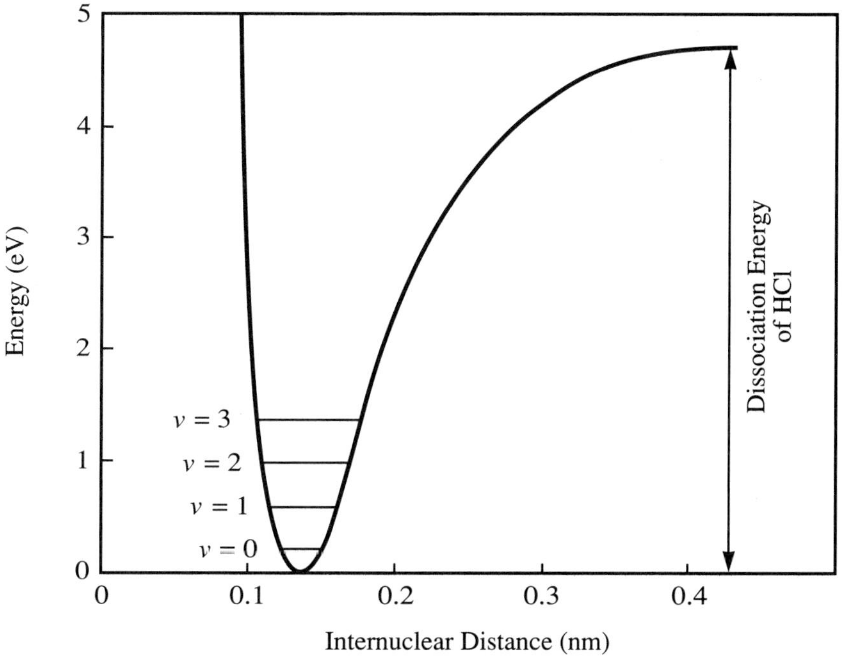

Figure 4.6. Potential energy of ground-state HCl as a function of the interatomic distance. Also shown are the lower vibrational sub-levels.

$$k_r \, n_{\rm A} \, n_e \;=\; \frac{n_{\rm A^*}}{\tau} \tag{4.22}$$

see also equation 4.13. If the electron density n_e, the parent gas density and the constants k_r and τ are known, the excited state density $n_{\rm A^*}$ can be calculated.

Generally, the lifetimes of excited atoms are short compared to the time between collisions. Typically, τ is in the range 10^{-8}–10^{-11} s for the above decay mechanism. If direct radiative decay is quantum-mechanically forbidden, life times can be much longer. This results in so-called metastable states. A metastable atom such as He* 2^1S_0 has a life time of about 10^{-2} s, but much longer life times are also possible. The low-energy singlet state of atomic oxygen ^{1}D, for example, can live as long as 2–3 minutes [79]. Obviously, metastable states decay mostly by collisions with other particles, for example in electron-impact excitation to a next higher spectroscopically allowed state, or through super-elastic electron collision. The latter means that the potential energy is given back as kinetic energy to the colliding electron. In low-pressure systems, de-excitation will probably take place at the chamber walls.

With molecules, the mechanism of excitation is more complicated than with atoms, since vibrational and rotational sub-states have to be taken into account too. For diatomic molecules, the potential energy diagram can be depicted as shown in figure 4.6 after Barrow [84]. Variations in the interatomic distance in molecules give rise to vibrational sub-levels. The corresponding energy levels can be written as

$$\varepsilon_{vib} \;=\; (v \,+\, \tfrac{1}{2})\; \varepsilon_{vib,o} \tag{4.23}$$

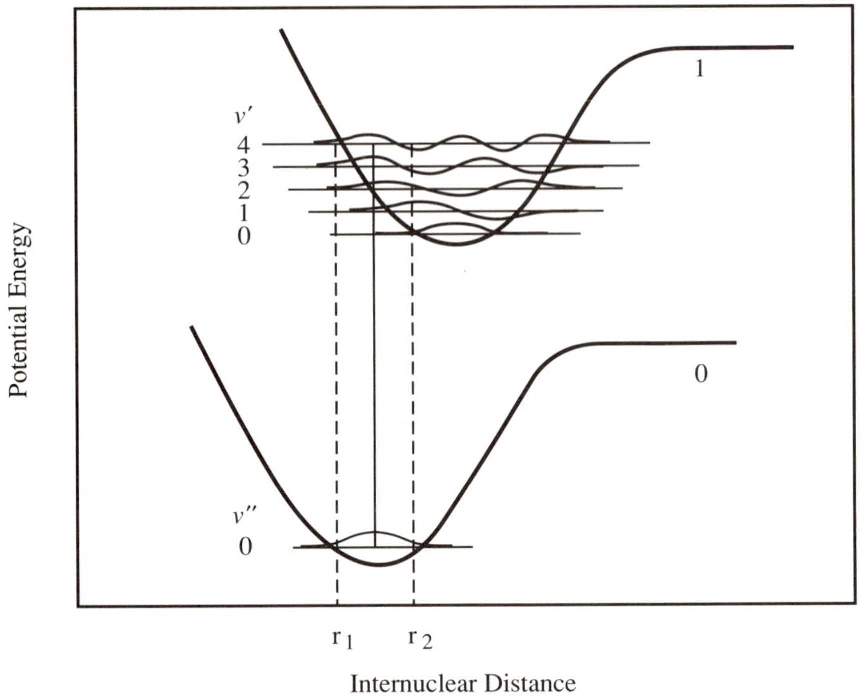

Figure 4.7. Illustration of the Franck-Condon principle for excitation of a molecule from the ground state 0 to a stable electronically excited state 1. The most likely transition is for quantum number $v'' = 0 \rightarrow v' = 2$.

where $\varepsilon_{vib,o}$ is the energy associated with the natural vibration frequency of the molecule, and v is the quantum number ($v = 0, 1, 2, ...$). Moving from one level to another is governed by the selection rule $\Delta v = \pm 1$. Each of the vibrational levels has a fine-structure caused by rotational excitation given by

$$\varepsilon_{rot} = J\,(J + 1)\,\varepsilon_{rot,o} \tag{4.24}$$

with J the quantum number for rotation ($J = 0, 1, 2, ...$), and $\varepsilon_{rot,o}$ the angular momentum energy. If the molecule has a permanent dipole, the selection rule is $\Delta J = \pm 1$. Symmetrical molecules such as O_2 do not have a dipole moment; in that case, rotational transitions are dictated by the quadruple moment, and $\Delta J = \pm 2$.

The energy gap from one electronic state to the next is large compared to the energy difference between the vibrational and rotational quantum levels within one state. Electronic excitation in molecules involves a transition between a multitude of vibrationally and rotationally resolved levels. This is shown in figure 4.7 after [78]; the lower curve depicts the ground state, the upper one a stable excited state. Electronic transitions are related to electronic energy diagrams such as figure 4.7 on the basis of the Franck-Condon principle. This principle stems from the idea that electrons move and rearrange much faster than the nuclei of molecules. For example, the time for an electron to circle a hydrogen nucleus can be calculated from Bohr's model to be about 10^{-16} s, whereas a typical period of vibration of a molecule is about 10^{-13} s. Spectral transitions must, therefore, be drawn vertically and not, as one might otherwise be tempted to do, from the potential minimum of the lower curve to

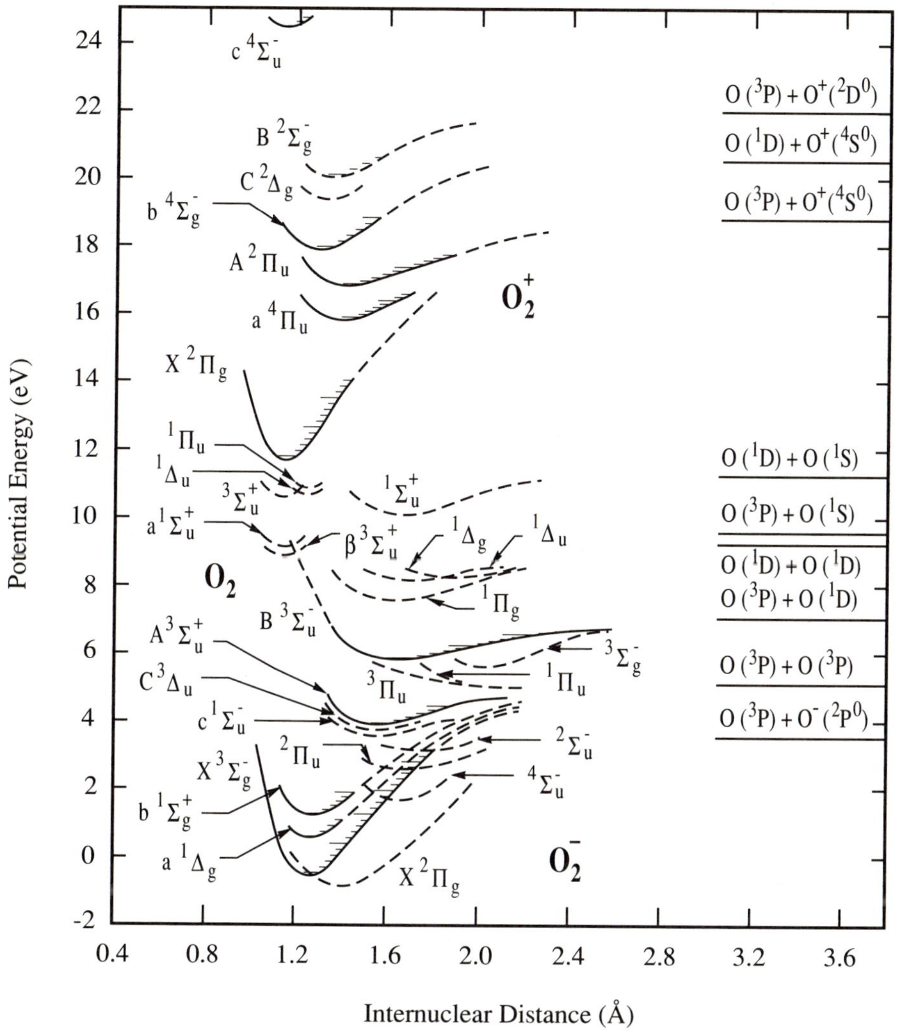

Figure 4.8. Potential energy curves for O_2 molecules and O_2^+ and O_2^- ions. Ground state level for O_2 is $X^3\Sigma_g^-$, for O_2^+ $X^2\Pi_g$ and for O_2^- $X^2\Pi_g$.

that of the upper curve. If the molecule in figure 4.7 is in the ground state and in the lowest vibrational level, where $v = 0$, the nuclear separation will lie between the extremes r_1 and r_2. According to the Franck-Condon principle, transitions from the ground state to the excited state will occur only to sub-levels where the nuclear separation still lies between these two extreme values. The transition probability between two states is the Franck-Condon factor; it is proportional to the integrated product of the respective wave functions.

A molecule in an excited state differs from one in ground state both in total internal energy and in electron wave function. Significant differences in chemical behavior are not unusual; sometimes the ground state and the excited state can be considered as two entirely different components. In figure 4.8 after reference [83], the potential energy curves of a number of electronic states of the O_2 molecule, and the O_2^+ and O_2^- ions are given. The molecular ground state of oxygen is $X^3\Sigma_g^-$. The electronic states $a^1\Delta_g$, with an energy of

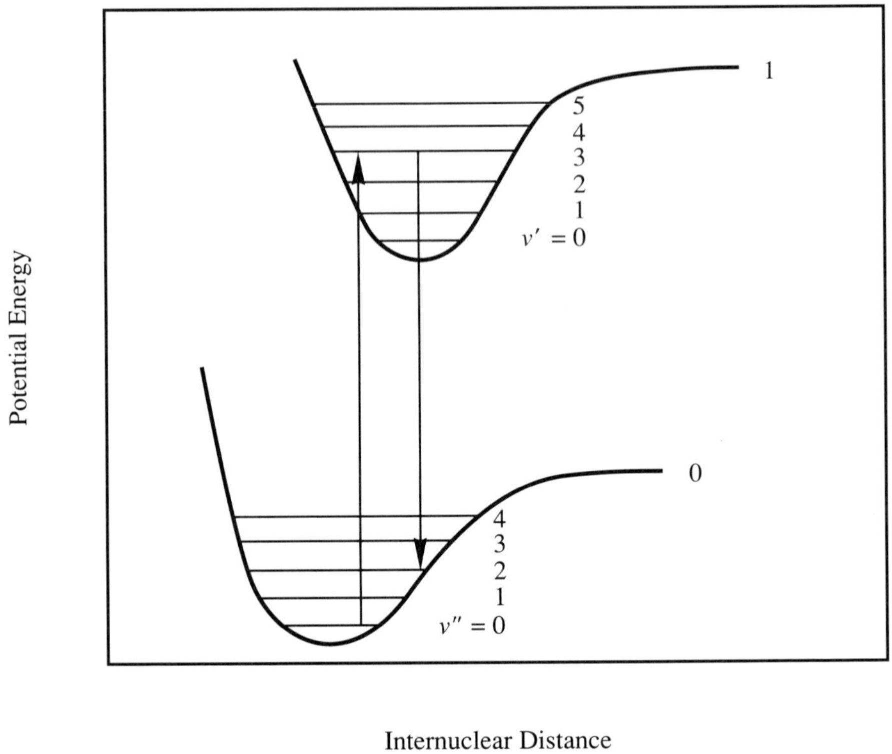

Figure 4.9. Schematic illustration of the potential energy curves of a ground state (0) and a nearby electronic state (1). Vibrational excitation of a ground state molecule occurs if the final vibrational level lies above the original one.

0.977 eV, and $b^1\Sigma_g^+$, with an energy of 1.627 eV, are both metastable, with lifetimes of 3900 and 12 sec, respectively.

In gases with a permanent dipole, such as CO, the cross section for electron impact excitation reaches maxima up to two orders of magnitude higher than for symmetric species, such as O_2. The cross section rises monotonously with energy from threshold to a maximum at about ten times the threshold electron energy value. Above the maximum, the cross section decreases slowly. Spin-forbidden processes occur readily through electron exchange, which results in metastable atoms and molecules. The collision cross section curve for such processes rises rapidly to a sharp maximum within a few eV from threshold [82]. These effects are similar to the singlet-singlet and singlet-multiplet transitions mentioned in the discussion on electronic excitation of atoms.

Vibrational excitation

The separation between two vibrational levels in an electronic state of a molecule is of the order of 0.1 eV, corresponding with radiation in the mid infrared (wavelength around 10 μm). Thus, it appears that in molecular gases even very low energy electrons can loose energy through inelastic collisions. Nevertheless, electron-impact induced vibrational excitation can also involve electrons of much higher energy. This is illustrated in figure 4.9.

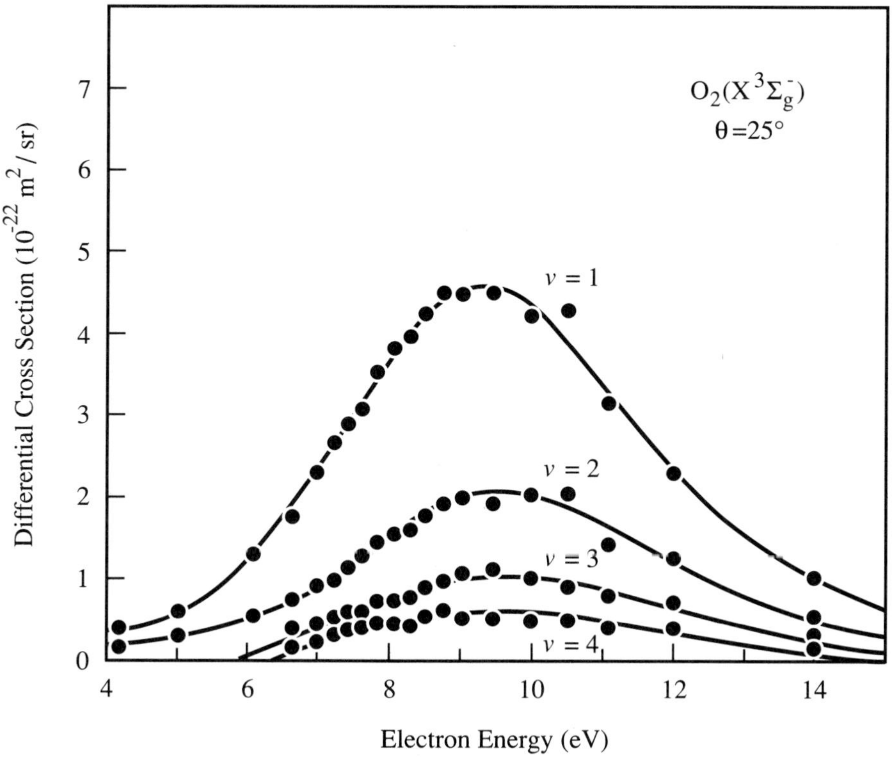

Figure 4.10. Cross section for vibrational excitation of ground state O_2 ($v = 0$) by electron impact.

Here, an electron from the electronic ground state (0) and with vibrational quantum number v"= 0, is excited to a nearby electronic state (1) with vibrational quantum number v'= 3. Note, that this is equivalent to the formation of an unstable negative ion through temporary electron capture. The excitation is followed by a relaxation to the ground state with a vibrational level with v" = 2. The final vibrational level is determined by the transition probability of vibrational level in the excited state to the one in the ground state (the Franck-Condon factor). A high-energetic vibrational transition as described above is possible in oxygen [85]:

$$O_2\ (v'' = 0)\ +\ e\ \rightarrow\ O_2^-\ \rightarrow\ O_2\ (v'' \geq 1)\ +\ e \qquad (4.25)$$

The maximum of the cross section for these vibrational transitions lies at about 9.5 eV, and has a value of the order of 10^{-22} m^2, see figure 4.10 after reference [83].

Radiative decay of vibrationally excited ground state molecules is possible, but often quantum-mechanically forbidden (selection rule $\Delta v = \pm 1$). A dominant mechanism for de-excitation of vibrationally excited molecules is, therefore, superelastic electron collision, where the internal energy of the particle is transformed into kinetic electron energy. These collisions between excited molecules and electrons are important at energies close to the thermal energy of the electrons; they have a significant influence on the bulk part of the electron distribution function [86,87]. The wall also plays a role. For an rf-driven nitrogen discharges under conditions typical for dry etching, it was found that de-excitation was at the reactor wall; gas-phase decay could be neglected [40].

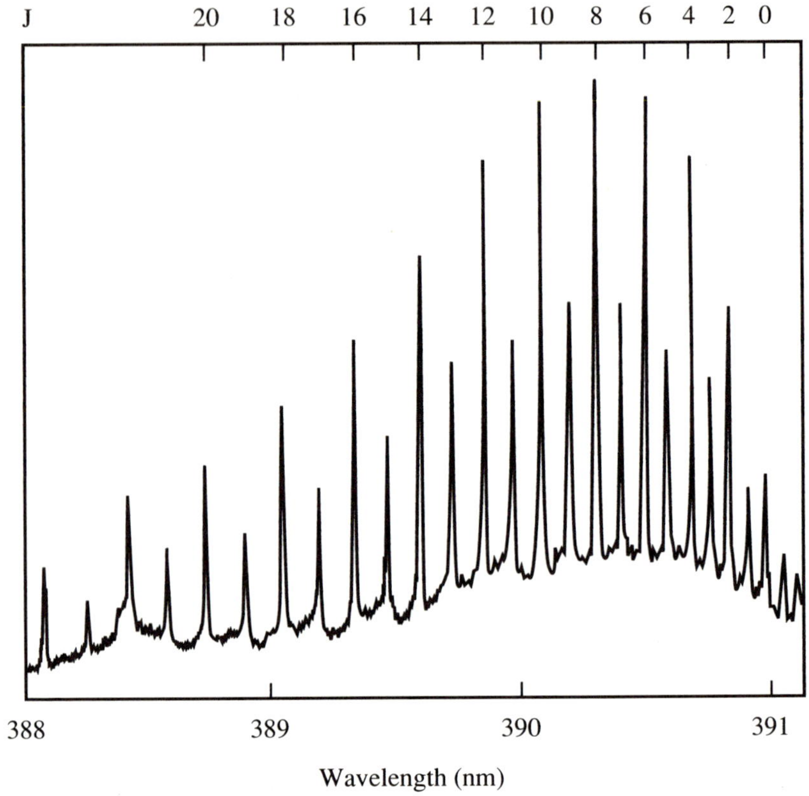

Figure 4.11. LIFS emission spectrum of ($B^2\Sigma_u^+$, ν' = 0 → $X^2\Sigma_g^+$, ν'' = 0) N_2^+ at 500 K showing rotational fine-structure.

Rotational excitation

The energy level spacing in rotational transitions is of the order of 0.01 eV, which is a factor 10 less than for vibrational states. As this is of the same magnitude as kT_g, the excited levels are heavily populated even at room temperature. In electron-molecule collisions at low electron energies, rotational excitation of molecular gases such as N_2, O_2 and CO, is a main loss mechanism for low energy electrons. Collision cross sections for rotational excitation are low and difficult to measure. Often, values can be obtained by theoretical calculations only. Summation of all rotational states for the oxygen molecule yields a collision cross section of the order of 10^{-21} m^2 [80].

The distribution over the different rotational states in a given gas can be described by a Maxwell-Boltzmann distribution function. The rotational distribution of ground state N_2^+ in a low-pressure rf-driven plasma is given in figure 4.11. From such data, a figure for the gas temperature can be obtained because the exchange rate between rotation and translation is large. The N_2^+ ion has both even and odd rotational levels. For quantum-mechanical reasons, these groups cannot interconvert by collisions with other molecules. This implies that two separate energy distributions will develop.

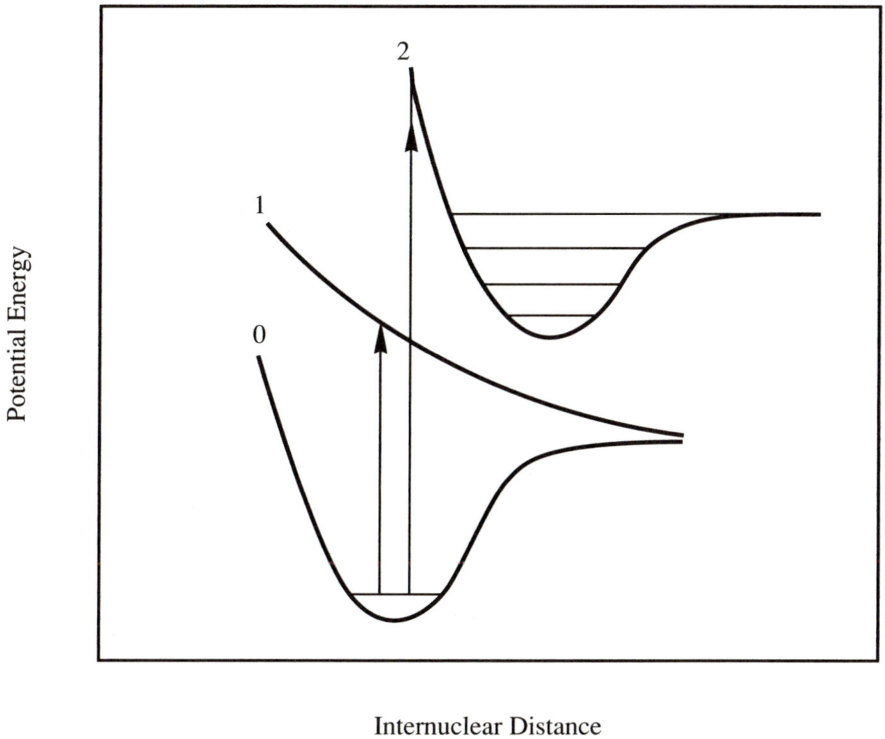

Figure 4.12. Dissociative excitation from the ground state (0) of a diatomic molecule to a repulsive state (1) and above the dissociation limit of a higher electronic state (2).

Dissociation

If a molecule is electronically excited to a level in a repulsive state, or a level in an attractive state above the binding energy, the molecule will dissociate, see figure 4.12. The resulting fragments can be atoms, molecules, and ions in either the ground or some excited state.

A dissociation reaction for molecular oxygen is the excitation of the ground state O_2 by electron impact to the repulsive $B^3\Sigma_u^-$ state. As the repulsive state is highly unstable, this is followed by very fast dissociation into a ground state atom $O(^3P)$ and an excited atom $O(^1D)$,

$$O_2 + e \rightarrow O_2(B^3\Sigma_u^-) \rightarrow O(^3P) + O(^1D) + e \qquad (4.26)$$

Figure 4.13 after reference [83], gives the collision cross section for dissociation of O_2 by electron impact as described in equation 4.26. The collision cross section is relatively high. Threshold is at about 7 eV, which is somewhat higher than the value for electronic excitation.

The reverse reaction to dissociation is recombination. Recombination reactions of metastable fragments can be studied conveniently in a decaying rf discharge. If the power to the system is switched off, the electrons cool down rapidly since the accelerating high-frequency field is gone. Soon, no further formation of radicals by electron impact occurs, and an undisturbed study of atomic and molecular reactions is possible. An example is the

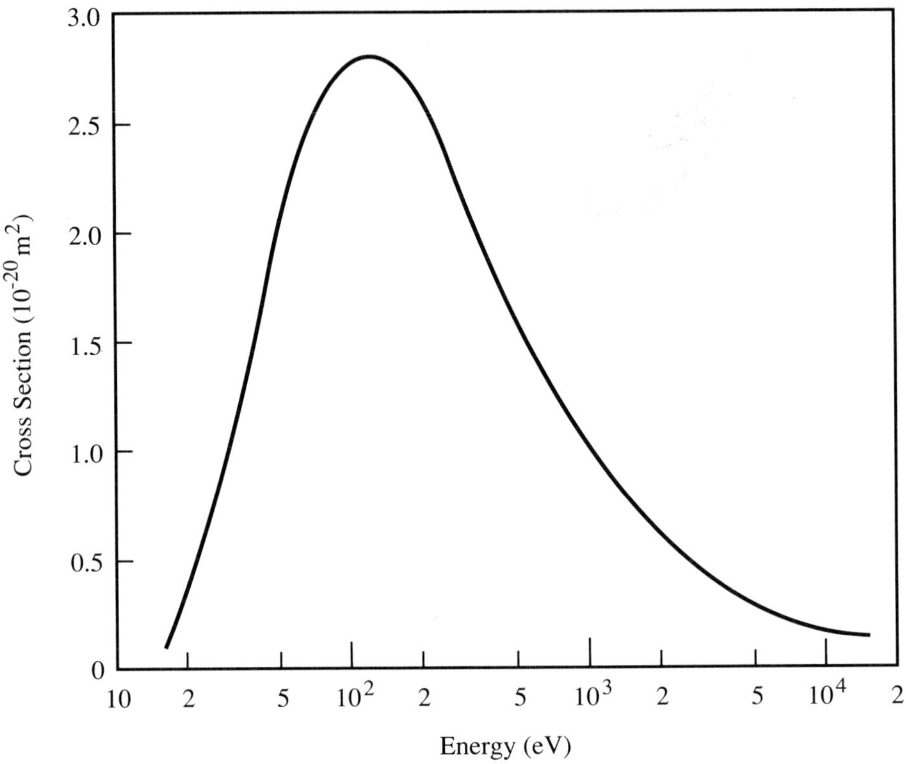

Figure 4.13. The collision cross section for dissociation of O_2 as a function of the electron energy.

recombination of Cl in a chlorine plasma in a parallel-plate reactor. Two mechanisms play a role here: gas and wall recombination. In the latter process, radical diffusion to the wall is an essential step. The total decay rate after the power has been switched off, can be written as a differential equation in t and x,

$$\frac{dn_{\mathrm{Cl}}}{dt} = -2\, k_3\, n_{\mathrm{Cl}}^2\, n_{\mathrm{Cl}_2} + D\, \frac{d^2}{dx^2} n_{\mathrm{Cl}} \tag{4.27}$$

where k_3 is the three body recombination rate and D the diffusion coefficient. The boundary condition is set by the surface reaction ($x = 0$),

$$D\, \frac{dn_{\mathrm{Cl}}}{dx} = \gamma\, \frac{n_{\mathrm{Cl}}\, \overline{v}}{4} \tag{4.28}$$

where γ denotes the surface recombination coefficient and $\overline{v}$ is the mean chlorine atom velocity. In a reactor with an electrode spacing of a few cm, a pressure above about 150 Pa, and a sufficiently high degree of dissociation, gas recombination dominates. In that case, two Cl-radicals can recombine in a three-body collision with Cl_2. The chlorine molecule is needed since the impulse must be conserved. The process has a rate coefficient k_3 of 7×10^{-44} m^6 s^{-1} [88,89]. At lower gas pressures, wall recombination is the main process. Chlorine radicals recombine on SiO_2 and Al_2O_3 surfaces by first order kinetics as per equation 4.28, with a surface recombination coefficient γ of approximately 0.01 [88]. For oxygen radicals, the surface recombination coefficient is much lower; values reported for silicon and aluminum oxide are 1.6×10^{-4} and 2.1×10^{-3}, respectively [90].

4.4 Ionization

Electrons can excite atoms and molecules into positive ions if they have a kinetic energy higher than the first ionization potential of the particle hit. This usually means a value above 10 eV. At higher electron energy, say 100 eV, ionization to multiple charged states becomes possible. A parameter often used to describe ionization by electrons accelerated in an electric field is the ionization coefficient α_i. This parameter should not be confused with the degree of ionization α, as introduced in chapter 2. By definition, the product of the electron drift velocity and α_i is equal to the ionization rate per electron,

$$\alpha_i \, v_d = \frac{1}{n_e} \frac{dn_{A^*}}{dt} \tag{4.29}$$

Combining the above relation with equation 4.17, which is the general equation for the rate of excitation by electron impact, it can be derived that

$$\frac{\alpha_i}{p} = \frac{\nu}{v_d \, p} \int_0^\infty \frac{\sigma}{\sigma_{tot}} f(v) \, dv \tag{4.30}$$

Dividing by the gas pressure p is useful because it takes out the variation in the collision frequency ν, which is linearly proportional to p. Equation 4.30 can be used as input to calculations concerning the onset of breakdown ionization in electric fields. Because of the highly anisotropic energy distribution in rf discharges the applicability in steady-state dry etching plasmas is limited, however.

In general, the collision cross sections for ionization by electron impact are high and broad. Figure 4.14 after reference [83], shows the cross section for single ionization of an oxygen molecule by electron impact,

$$O_2 + e \rightarrow O_2^+ + 2\,e \tag{4.31}$$

The collision cross section peaks at 120 eV with a value of a few times 10^{-20} m^2. At 1000 eV, the value of the cross section is still around 10^{-20} m^2. It will be obvious that secondary electrons coming from the sheath regions in rf discharges, having kinetic energies in the range of 100–1000 eV, are very important for the gas ionization in the plasma.

In the case of dissociative ionization, a combination of ionization and dissociation occurs. In oxygen,

$$O_2 + e \rightarrow O^+ + O + 2\,e \tag{4.32}$$

For dissociative ionization, the collision cross section is of the order of 10^{-22} m^2 at 23 eV. At high electron energies (150–200 eV), the collision cross section increases to about 10^{-20} m^2.

A neutralization reaction for positive molecular ions is dissociative recombination. For an oxygen ion, recombination with an electron can take place only if conservation of momentum is fulfilled. The only way to achieve this is by dissociation of the molecule itself into two atoms,

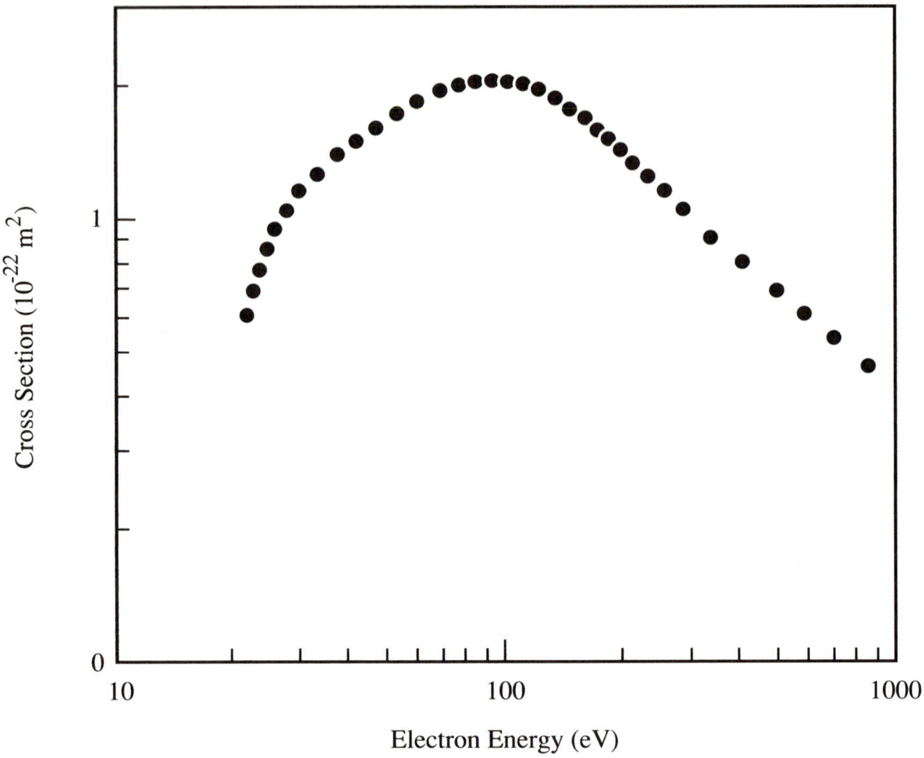

Figure 4.14. The collision cross section for ionization of O_2 by electron impact.

$$e + O_2^+ \rightarrow O + O \tag{4.33}$$

The reaction requires the densities of both ions and electrons to be sufficiently high; the rate coefficient for the above process is only 1.7×10^{-13} m^3 s^{-1} [75]. Obviously, a similar process is not possible with positive atomic ions; a third body is needed during the collision of an electron and an ion. At moderate pressures, the probability of such a process is low, unless the third body is the reactor wall.

The decay of N_2^+ in a dry etching system after switching off the rf power off is shown in figure 4.15 after Visser et al. [40]. Initially, a fast decay with second order kinetics is observed, followed by a much slower exponential deçay on a millisecond time scale. The fast decay is associated with dissociative recombination as was illustrated in equation 4.33; the second-order kinetics is brought about by the requirement that electron and ion density be equal. Dominant at low ion density is the diffusion of electrons and ions to the electrodes, which results in an exponential relation [23]. Summarizing, the decay in figure 4.15 can be written as

$$\frac{dn_{N_2^+}}{dt} = -k_d \, n_{N_2^+}^2 + D \, \frac{d^2 n_{N_2^+}}{dx^2} \tag{4.34}$$

where k_d and D are the reaction rate coefficient for dissociative recombination and the diffusion coefficient, respectively. The recombination rate coefficient has been measured to be 1.6×10^{-13} m^3 s^{-1} [75].

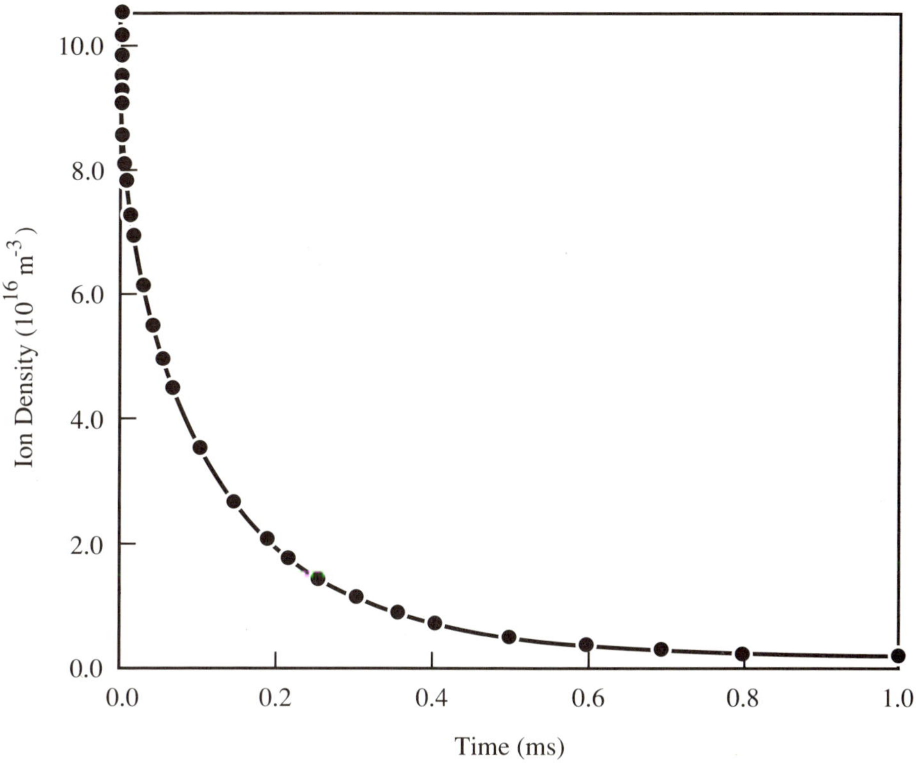

Figure 4.15. Decay of N_2^+ in an dry etching system at 30 Pa.

Figure 4.16 after von Engel [79], summarizes the electron collision cross sections in hydrogen for the most important inelastic processes: vibrational and rotational excitation, dissociation, ionization, and dissociative ionization. The difference between the curves for these sub-processes and the total cross section for momentum transfer is equal to the elastic cross section. From the figure, it appears that elastic processes are important at low electron energy, while inelastic processes dominate at high energy. Also, the only important process in the high energy range, say 100–1000 eV, is straightforward ionization. At still higher electron energy, the ionization cross section approaches the extrapolated σ_{tot} curve. The absolute value decreases slightly less than linear with energy. This means that electrons with an energy above 1000 eV have a rapidly decreasing chance of interacting with the target gas, simply because they have insufficient time to do so. In a dry etching system they will pass straight on from one electrode to the opposing one without colliding with the gas on their way. Other gases show very much the same behavior.

Negative ions

Significant amounts of negative ions can be created in electronegative gases. The simplest process is direct attachment. In this case, a third particle is needed to meet the requirements for impulse conservation. For oxygen,

$$O_2 + O_2 + e \rightarrow O_2 + O_2^- \qquad (4.35)$$

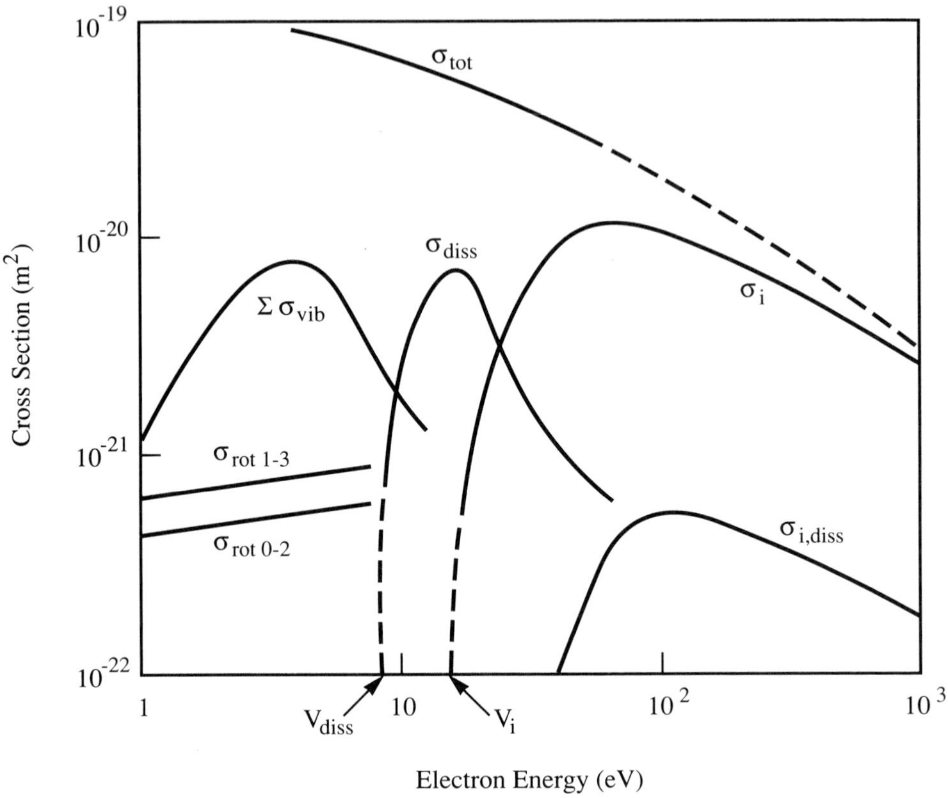

Figure 4.16. Inelastic and total electron collision cross sections in hydrogen.

the reaction rate coefficient is 2×10^{-42} m^6 s^{-1} at 300 K and 0.2 eV [91]. Dissociative attachment of oxygen is the next possibility. What fragments are generated depends on the electron energy. Between 4.4 and 10 eV, attachment to a ground state oxygen molecule leads to a negative ion according to

$$e + O_2(X^3\Sigma_g^-) \rightarrow O_2^{-*}(^2\Pi_u) \rightarrow O^-(^2P) + O(^3P) \tag{4.36}$$

The cross section for this process lies between 5 and 10 eV and peaks at 6.7 eV with a value of 2×10^{-22} m^2 [83]. Ion pair formation is a mechanism which dominates at higher electron energy,

$$e + O_2 \rightarrow O^- + O^+ + e \tag{4.37}$$

The electron must have sufficient kinetic energy to excite the molecule to a dissociative state, resulting in positive and negative charged fragments.

The opposite to attaching an electron to a neutral particle, is detachment from a negative ion. Detachment processes occur mainly at the instant the negative ion collides with other particles. A subdivision can be made between direct and associative detachment, the former often being the main process. Direct detachment of the negative oxygen ion is

$$O^- + e \rightarrow O + 2e \tag{4.38}$$

with a rate coefficient of 1.3×10^{-14} m^3 s^{-1}. At collision energies below 0.1 eV, associative detachment can occur,

$$O^- + O \rightarrow O_2 + e \tag{4.39}$$

with a rate coefficient of 2×10^{-16} m^3 s^{-1}. Provided that sufficient amounts of charged and neutral particles are present, three-body associative detachment is possible. This ion-ion recombination is basically the reverse process of ion-pair formation,

$$O^+ + O^- + O_2 \rightarrow 2\,O_2 \tag{4.40}$$

The reaction rate coefficient for the latter process is 2×10^{-37} m^6 s^{-1} [83].

Charge transfer

During the collision between an ion and a neutral, the charge of the ion can be transferred to the neutral. These charge transfer collisions are highly probable for ions moving in their parent gas; the collision cross section is usually significantly larger than the geometric diameter of the particles involved. A possible reaction in oxygen is

$$O^+ + O_2 \rightarrow O + O_2^+ \tag{4.41}$$

with a reaction rate of 1.2×10^{-17} m^3 s^{-1} [92].

The importance of charge transfer becomes apparent when taking a closer look at the situation in the dark space region between plasma and wall. The ions are accelerated by the field and get high kinetic energies. When charge exchange occurs the "ex-ion" becomes a high-energetic neutral, while the "ex-neutral" becomes a low-energetic ion. Thus, charge transfer results in low energy ions and high energy neutrals, while the total ion density remains constant. In practice, this means that the ions finally impinging on the electrode and the substrates are less energetic and less directional than would be expected. On the other hand, neutrals can be both very hot and quite directional.

4.5 Plasma-wall interactions

Most particles generated in the plasma will eventually reach the walls of the plasma container and the substrates to be etched. An impression of the order of magnitude of the effects involved can be obtained by once again considering the standard argon discharge from chapter 3. At 20 Pa and 400 K the density of the gas is 4×10^{21} m^{-3}. The mean thermal velocity corresponding to 400 K (0.025 eV) is 400 m s^{-1}. The thermal flux density on the walls is given by $\frac{1}{4}n\bar{v}$, which amounts to 4×10^{23} m^{-2} s^{-1} in this case. From the particle flux, the time needed to form a monolayer of neutrals on a clean surface can be calculated. Not all particles will adhere to the bombarded surface; the relevant factor here is the sticking coefficient – the chance that a particle will stay for some time on the surface instead of being reflected immediately. With a sticking coefficient of 0.02, it appears that the surface will be covered within one millisecond. In other words, all surfaces are continuously covered by neutrals under normal dry etching conditions.

Neutrals may adsorb, dissociate, and react on the surface. This is true especially for electronically excited molecules and atoms. It should be noted, however, that the surface residence time is rather limited. For CF_4, the residence time is approximately given by

$$\tau = 5 \times 10^{-13} \exp\left(\frac{\varepsilon}{k\,T}\right) \tag{4.42}$$

With an activation energy ε of 0.25 eV, and a surface temperature of 400 K, the residence time is only 2 ns. If particles from the surface layer can diffuse into the substrate, either spontaneously, or assisted by other bombardment mechanisms, reaction with the substrate can occur. This is eventually followed by etching, provided that the reaction products are volatile.

Flux densities of charged particles at the walls of a plasma container are much lower. Because of ion sheath formation, the positive ions determine the wall current. In the previous chapter, the equilibrium current in the standard argon system was calculated from the Bohm criterion to be 2 A m^{-2}, equivalent to 10^{19} m^{-2} s^{-1}. This is more than 10,000 times less than the neutral flux. The energies of all charged particles impinging on the surface, except for negative ions, can be of the order of the peak rf potential across the system; a typical value is 500 V. Note, that negative ions are too slow and of the wrong polarity to pass the potential barrier of the dark space.

In principle, electrons can also induce dissociation and desorption of reactants on the surface. The collision cross sections for electron-stimulated desorption is, however, often much smaller than for the corresponding gas phase processes of excitation, dissociation and ionization. Also, the required energy is quite high, around 1 keV, a value not commonly found in dry etching. Because positive ions have both a high mass and a high velocity, they are the most important species in stimulating surface processes. Low energy ions can change the surface morphology, high energy ions can knock atoms out of the surface lattice. The latter is what is known as sputtering.

Many additional processes can occur as a result of fast particle impact. Ions and electrons, but also neutrals and photons, will cause secondary electrons to escape from the surface of the bombarded substrate. The origin of this secondary electron emission can be either the kinetic energy or the internal energy of the incident particle. The emission coefficient, which is defined as the number of electrons created per impinging particle, depends on both substrate composition and energy [75,79]. For ions hitting metallic surfaces with an energy between 100 and 1000 eV, the emission coefficient lies between 0.001 and 0.1, raising about linearly with the impact energy. For electron bombardment, the emission coefficient for similar energies is much larger, and almost constant. It can range from about 1 on most metals to more than 20 on certain oxides. The secondary electrons have high kinetic energies when entering the plasma, as they are accelerated across the sheath.

From the decay of excited species, photons with energies of many eV can reach the surface too. The adsorption cross sections are low, however, similar to the situation with electrons. In practice, photon fluxes required to obtain etching are of a density requiring focussed laser illumination [93,94]. Nevertheless, the effect should not be neglected since it can result in lattice damage to VLSI devices. Energetic photons can have significant effects on the density of traps and other undesirable semiconductor states.

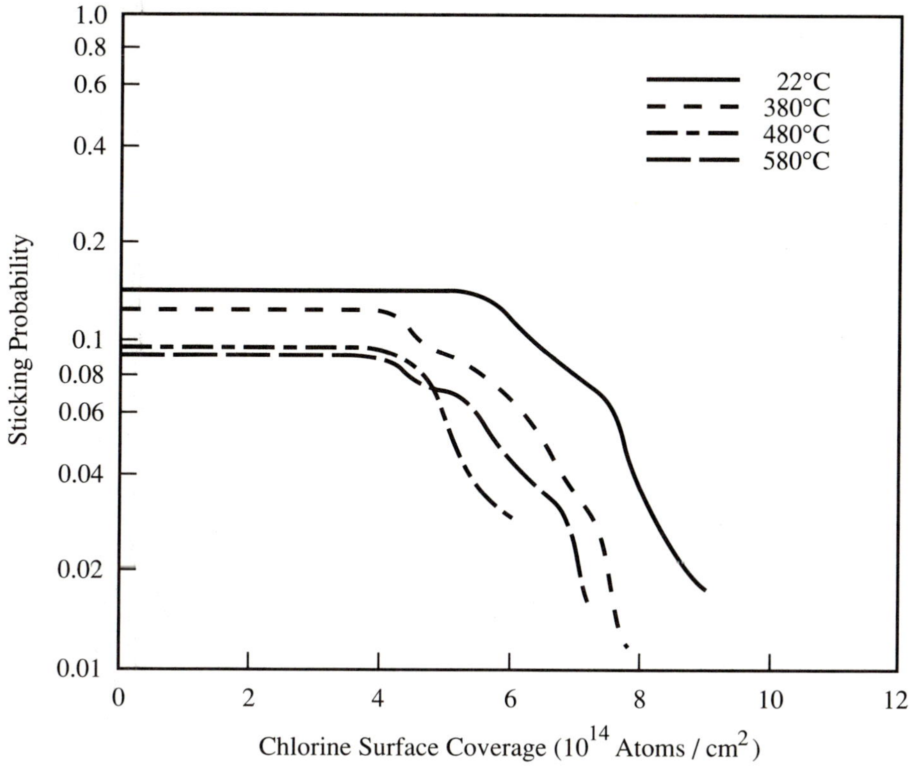

Figure 4.17. The sticking coefficient of Cl_2 on Si as a function of surface coverage and substrate temperature.

4.6 Thermal etching

Thermal etching is etching by chemical reaction only. It consists of chemisorption of the active species, diffusion into the lattice, formation of a reaction product and the desorption of that (volatile) product.

An example of chemisorption is shown in figure 4.17 after Florio and Robertson [95]. Displayed is the sticking coefficient as a function of surface coverage and temperature in the Cl_2 on Si(111) system. The initial sticking coefficient can be seen to be of the order of 0.1, and fairly constant. It drops rapidly when the saturation coverage of the surface is reached. In this case it corresponds to the formation of a Cl atom monolayer, and no further reaction occurs.

A different reaction sequence is found in the reaction of XeF_2 molecules with Si. When incident on a solid surface, the XeF_2 molecule dissociates in Xe, which desorbs instantaneously, and F_2. The latter is then chemisorbed. As the saturation coverage is approached, the reaction probability (or sticking coefficient) of XeF_2 decreases too, but soon reaches a steady state. In the steady state situation, adsorbed F diffuses into the lattice, and chemical etching of the silicon occurs by the desorption of SiF_x species [96,97].

Apparently, silicon is etched chemically by F_2, and not by Cl_2, at least at moderate temperatures. This can be understood by calculating the potential energy barrier for F and Cl to penetrate into the silicon [98]. For F the potential barrier is about 1 eV, whereas for Cl the value is much higher, of the order of 13 eV. The potential barrier can be decreased by surface contamination and by dislocations in the substrate lattice.

A reaction between Cl_2 and silicon can be brought about if the temperature is raised considerably. At temperatures between 800 and 1500 K the main etch product is $SiCl_2$; below 800 K $SiCl_4$ is formed [99]. The reaction products have a Maxwell-Boltzmann kinetic energy distribution [100],

$$f(\varepsilon) \propto \exp\left(\frac{-\varepsilon}{k\,T}\right) \tag{4.43}$$

where T is the temperature and ε the energy of the emitted particles. Contrary to silicon, aluminum reacts rapidly with chlorine molecules at room temperature. Below 450 K, the main reaction product is Al_2Cl_6. In the range between 500 and 650 K $AlCl_3$ is formed mainly, whereas above 700 K the reaction product is AlCl [101,102,103].

4.7 Sputtering

Sputtering is the term reserved for ion-bombardment induced erosion of a solid substrate. A difference is usually made between physical sputtering, and chemical sputtering. In the latter case, the ions undergo a chemical reaction at the surface, in addition to causing mechanical disruption [104,105].

Many processes of sputtering of polycrystalline or amorphous elemental targets by atomic ions can be described by a linear model, as shown in figure 4.18. After entering the matrix of atoms, the incoming particle is gradually decelerated. At ion energies above some 100 keV, the incoming ion loses its energy by interaction with the electrons of the target material. Below 100 keV, the energy loss is mainly by nuclear stopping. Here, the incoming ion slows down by consecutive elastic collisions with target atoms which are initially at rest. During these collisions, fast recoil particles are created, which lose energy by setting other target atoms in motion, resulting in an increasing number of moving atoms with decreasing energy. Thus, an area of moving target atoms, the collision cascade, is developed. The number of recoil atoms in the collision cascade remains low and the main process is knock-on collisions. About $1\text{–}5 \times 10^{-13}$ s after penetration of the ion into the substrate, the transferable energy has decreased below the displacement threshold of a few tens of eV. On a much longer time scale (10^{-11}–10^{-9} s) the collision cascade is damped by energy dissipation, resulting in local heating of the target material [104]. If the impulse imposed on a surface atom within the collision cascade is above the surface barrier energy, it will be ejected out of the target material. Note, that the projectile itself will normally be implanted in the lattice; only target material is eroded. If erosion proceeds, the previously implanted atoms will also be affected by subsequent cascades, and eventually ejected.

The sputter yield Y is defined as the number of atoms that are sputtered away by one incoming particle. The incoming particle is usually an energetic ion, but energetic neutrals can cause the same effect as well, since sputtering is caused by impulse transfer. The actual yield is a function of the target material and its morphology, as well as the characteristics of the bombarding particle.

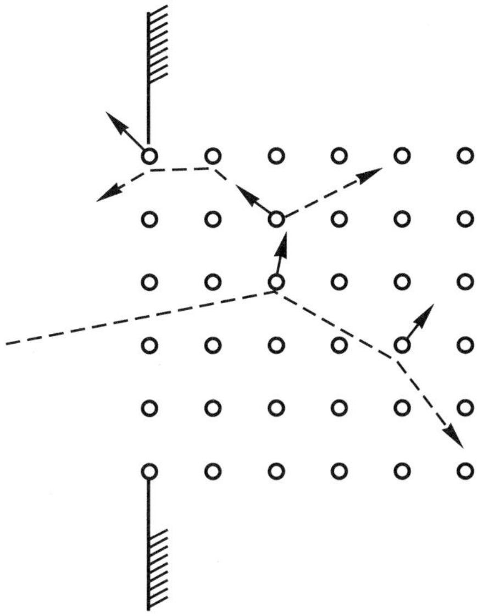

Figure 4.18. Schematic illustration of the linear cascade regime. The recoil atoms receive enough energy to generate recoil cascades in the surface layer. However the density of recoil atoms remains low, so knock-on collisions mainly occur.

In the regime of the linear collision cascade, the theory of Sigmund can be applied to predict sputter yields. For a detailed treatment of the theory, see references [104,106,107,108]. According to Sigmund's theory, the sputter yield Y for perpendicular incidence, expressed in number of atoms per incoming projectile, is given by

$$Y(\varepsilon_p) = C_{pt}\ S_n(\varepsilon)\ \frac{\varepsilon_p}{\varepsilon_{pt}} \tag{4.44}$$

where ε_p is the projectile energy and S_n the energy-dependent nuclear stopping cross section or stopping power. The subscripts p and t denote projectile and target, respectively. Two characteristic constants C_{pt} and ε_{pt} appear in the equation, both being a function of the atom numbers $\mathcal{Z}_t$ and $\mathcal{Z}_p$ of the target and projectile species, respectively. Furthermore, the characteristic energy ε_{pt} is proportional to $1+(m_p/m_t)$, where m_p and m_t are the respective atomic masses; C_{pt} is inversely proportional to the surface escape energy barrier U_0 of the target material. The barrier energy is for many materials close to the sublimation enthalpy; an exception is silicon, as shown in figure 4.19 after Baller and Dieleman [109,110].

If the projectile energy is below about 3 keV, which will in practice always be true for dry etching, and the ratio of projectile and target atom number $1/5 < \mathcal{Z}_p/\mathcal{Z}_t < 5$, then the Sigmund yield relation has an approximate solution given by [108,111]

$$Y(\varepsilon) \approx \frac{5}{3U_0}\sqrt{\frac{\mathcal{Z}_t\ \varepsilon_p}{1000}} \tag{4.45}$$

where ε_p and U_0 are both in eV. Equation 4.45 neglects the fact that all sputter processes have a threshold energy below which no target erosion takes place. The threshold energy for

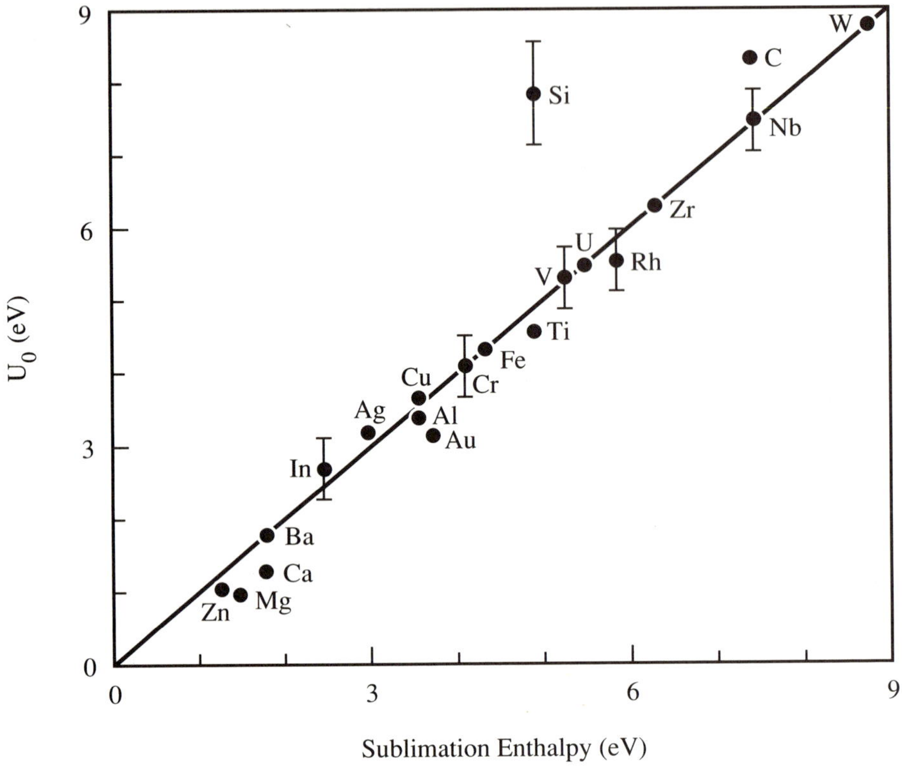

Figure 4.19. The surface escape barrier energy U_0 versus the sublimation enthalpy for several elemental targets.

the atom number ratio range under consideration may be estimated to be

$$\varepsilon_{p,th} \approx 8\, U_0 \tag{4.46}$$

This is an empirical relation, rather than the result of a sound theory. Nevertheless, if the value of ε_p in equation 4.45 is corrected for the threshold value from equation 4.46, an estimate of the yield can be obtained which is valid to within 25%.

As an example, consider a Si substrate being bombarded by 500 eV Ar^+ ions at normal incidence. The surface escape barrier U_0 for silicon is 7.8 eV; thus, according to equation 4.46, the threshold energy should be 62 eV. From equation 4.45, the sputter yield can be calculated to be 0.5. These values are in good agreement with experimentally obtained data.

For incidence at an angle θ relative to the surface normal, the linear cascade theory predicts a rise in yield proportional to $\cos^{-n}\theta$, because the energy deposited in the outermost atomic layers increases. Experimentally, such a dependence is observed to hold up to about 50 deg. At higher angles of incidence, a competitive effect occurs; reflection of ions from the surface. Sputter yield curves, therefore, tend to flatten off to a maximum near 70 deg, after which they rapidly drop to zero as θ approaches 90 deg. The angle dependency of the sputter yield of Si in Ar^+ is plotted in figure 4.20 after Yamamura and Itoh [105].

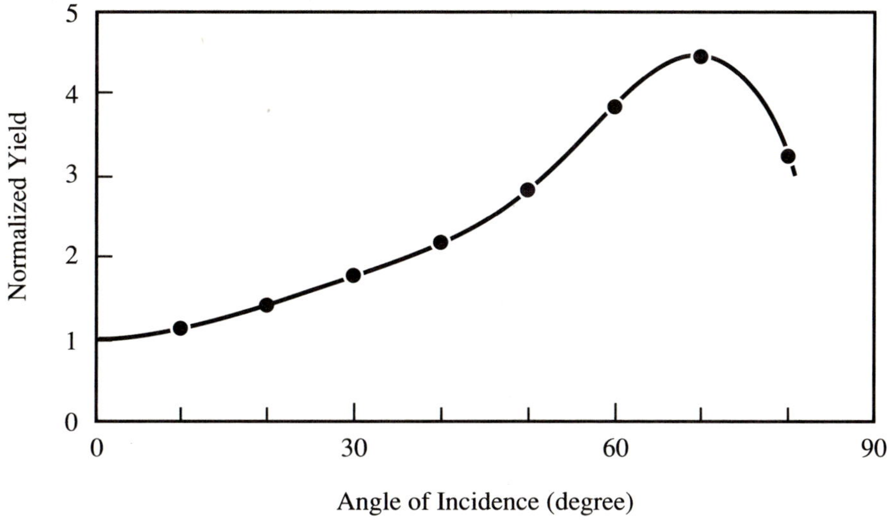

Figure 4.20. Angle of incidence dependence of the sputter yield for silicon bombarded with Ar^+ ions.

The kinetic energy distribution perpendicular to the surface of target atoms sputtered as the result of a collision cascade, can be derived theoretically to be [112]

$$f(\varepsilon_p) \approx \frac{\varepsilon}{(9\,\varepsilon + U_0)^3} \tag{4.47}$$

Figure 4.21 after Zalm [108], shows the kinetic energy distribution of atoms sputtered from a Si target bombarded by 3 keV Ar^+ ions at normal incidence. The dashed line is the prediction

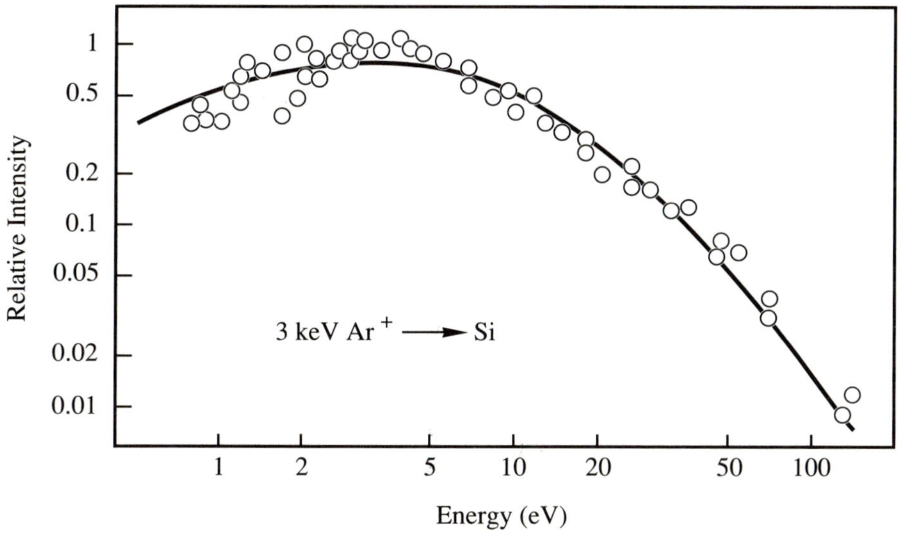

Figure 4.21. Kinetic energy distribution of Si atoms sputtered from a silicon target by 3-keV Ar^+ ions.

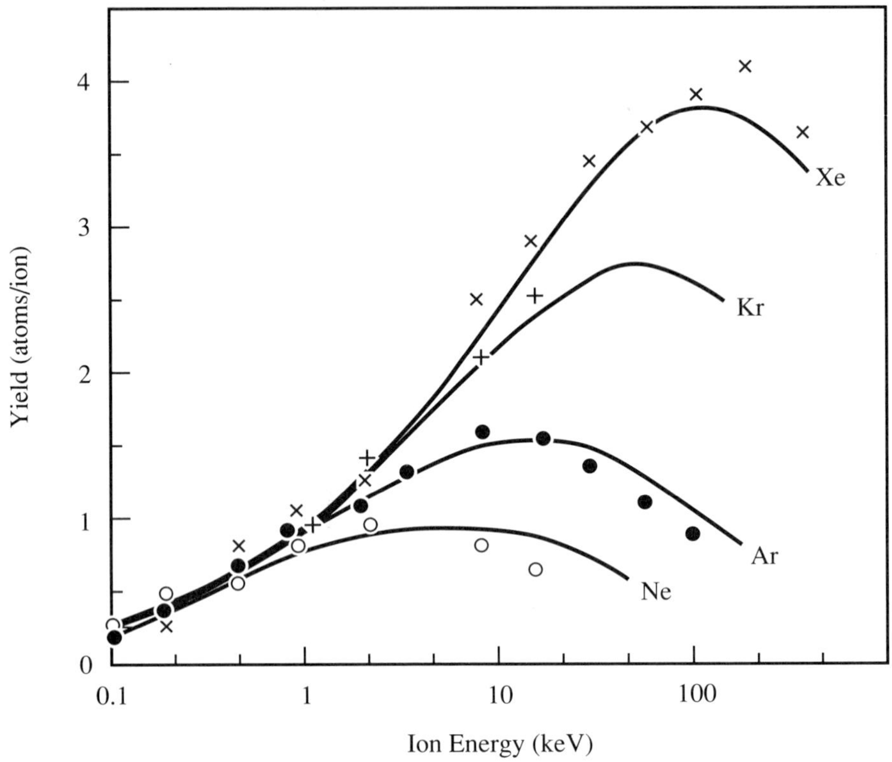

Figure 4.22. The sputter yield of Si for different noble gas ions as function of the ion energy.

from the linear cascade theory. Hence, it appears that the surface barrier energy U_0 can be determined from measuring the energy distribution of the sputtered particles [113].

Figure 4.22 after Zalm [108], shows the sputter yield of Si as a function of the energy of different noble gas ions; the drawn lines are predictions from the linear cascade theory. Note, that the sputter yield at low energies (< 1 keV) becomes independent of the mass of the bombarding ion, in agreement with the approximate solution in equation 4.45.

Sigmund's theory applies to the linear cascade regime only. For very light, as well as very heavy projectiles, the cascade model fails. Instead, two other mechanism are found, known as the single knock-on and the spike model [104,107], as depicted in figure 4.23. In the single knock-on regime, the incoming particle transfers its energy directly to the target atoms, of which some are emitted (sputtered) from the surface, and some diffuse into the substrate without causing other displacements. The sputtering results from primary recoils instead of recoil atoms generated in a collision cascade. Single knock-on holds for very light projectiles, like H^+ and He^+, and for energies from near-threshold up to about ten times the threshold energy; such particles have an inefficient energy transfer. At incident energies below 100 eV, the penetration depth is of the order of 1 nm and no collision cascade is created either. In this case, ion bombardment results in deposition.

The spike model is to be used if the density of the incident atoms is so large that all atoms in the local interaction volume ("spike volume") are set in motion instantaneously. Spiking is found for heavy, high-energetic projectiles. The amount of energy deposited in the spike volume may be interpreted as a temperature. Note, that a regular collision cascade

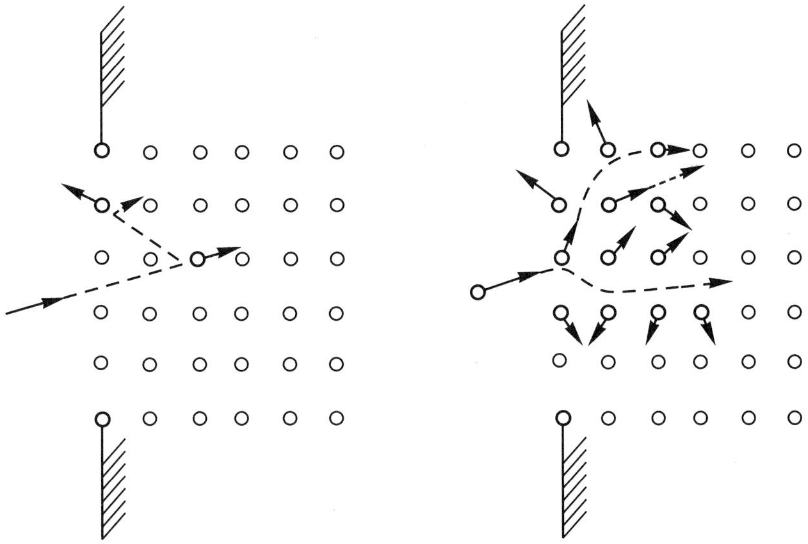

Figure 4.23. Schematic illustration of the single knock-on regime (left) and the (thermal) spike regime (right).

can, after some time, also be interpreted as a spike. The difference lies in the fact that "real" spiking occurs immediately below the surface, which implies that the energy distribution of the emitted particles is directly related to the situation in the spike volume.

Higher sputter yields are obtained with inert molecular ions rather than with noble gas ions of comparable mass. This can be explained from the fact that if a molecular ion bombards a surface with sufficient energy ($>$ 50 eV), it will break up into its constituent atoms. Each individual fragment creates its own collision cascade; when these collision cascades overlap, a more than linear effect in the sputter yield is observed [104,114,115].

In bombardment of surfaces with chemically active ions, the predictions of the linear cascade do not hold. Implantation of the reactive ions can alter the binding energy as well as the concentration of surface atoms. Formation of volatile reaction products between incoming ions and a target atom enhances the sputter yield relative to the model predictions. On the other hand, a decrease is observed if the reaction product is non-volatile. An example of enhanced etching is the bombardment of silicon with F^+, giving gaseous SiF_4; bombardment of silicon with N^+, on the other hand, leads to reduced yields as a result of formation of the non-volatile Si_3N_4 [116]. In comparing Cl^+ and Ar^+ for silicon sputtering, both having about the same mass, it is found that the chemically reactive ion has the highest sputter yield, as shown in figure 4.24 after Tachi et al. [117]. Interestingly, the reaction product silicon chloride is not formed in this case; the halogen is implanted only.

4.8 Synergetic effects

Simultaneous bombardment of a substrate by both energetic ions and chemically reactive molecules or atoms can lead to etching which is substantially faster than would be predicted from simply adding together thermal etching and sputtering. This synergetic effect was shown

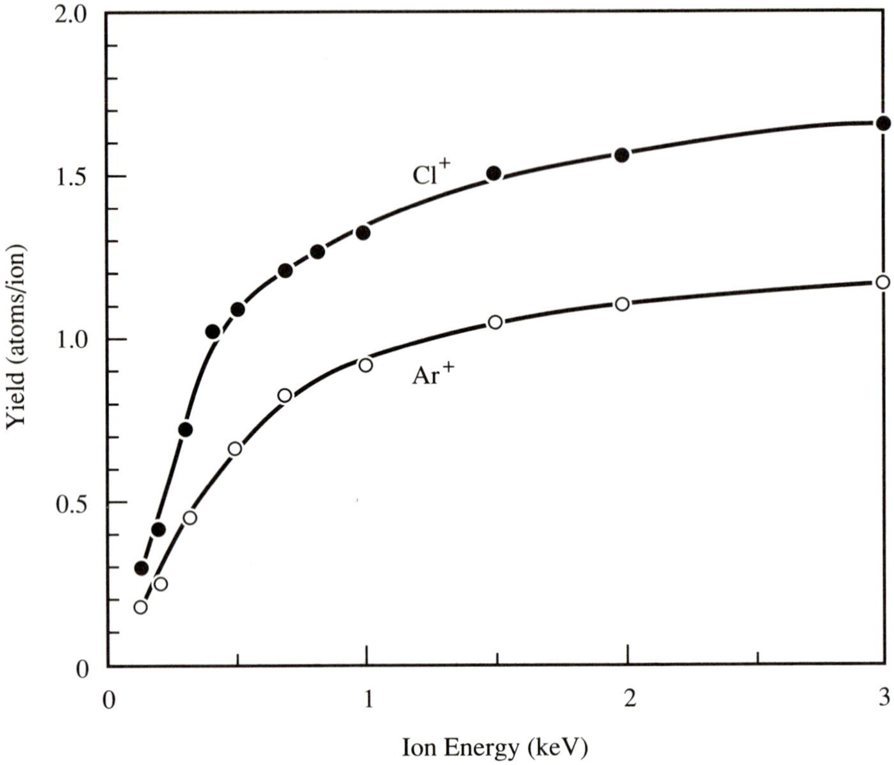

Figure 4.24. The sputter yield of Si by noble gas ions (Ar^+) compared with chemically reactive ions of similar mass (Cl^+).

for the first time by Coburn and Winters [118] in the etching of silicon with separated and combined beams of XeF_2 molecules and Ar^+ ions, see figure 4.25. The enhancement in etch rate cannot be ascribed to heating of the target by the ions; the rate is independent of the ion current density as long as a continuous coverage of reactive molecules is guaranteed. If electronic excitation was to be responsible, electrons would be more effective than ions; this was not observed. Enhanced etching occurred only for target materials with volatile halides. Simultaneous bombardment of aluminum by XeF_2 and Ar^+, for instance, did not lead to higher rates, because the non-volatile AlF_3 is formed. Synergetic effects of ions and neutrals are considered to be the most relevant surface processes involved in anisotropic dry etching. Later work has shown, that the interpretation of experimental results is often quite complicated, especially if the energy and composition of the ejected particles must be explained [119,120].

The mass spectrum of the particles emitted from a silicon surface as a result of simultaneous exposure to Cl_2 molecules and 1-keV Ar^+ ions is shown in the top part of figure 4.26 after Kolfschoten et al. [121]. This spectrum deviates substantially from that of gaseous and solid $SiCl_4$, which are plotted in the middle and bottom parts of the figure. At ion energies above 100 eV and simultaneous exposure to Cl_2 molecules and Ar^+ ions, the species ejected from the Si substrate are Si, Cl, SiCl, $SiCl_2$ and resputtered implanted Ar. Below 100 eV, the products Si and Cl are no longer observed. The reaction products found in this process differ from those observed during thermal etching of silicon with molecular chlorine; in the latter case, $SiCl_2$ and $SiCl_4$ are formed.

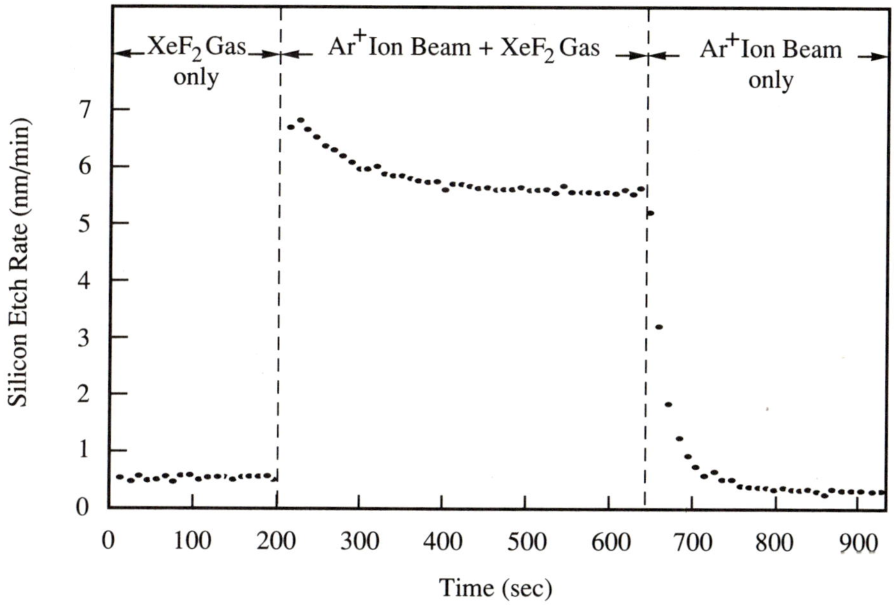

Figure 4.25. Etch rate of silicon as a function of separated and combined beams of XeF_2 and Ar^+.

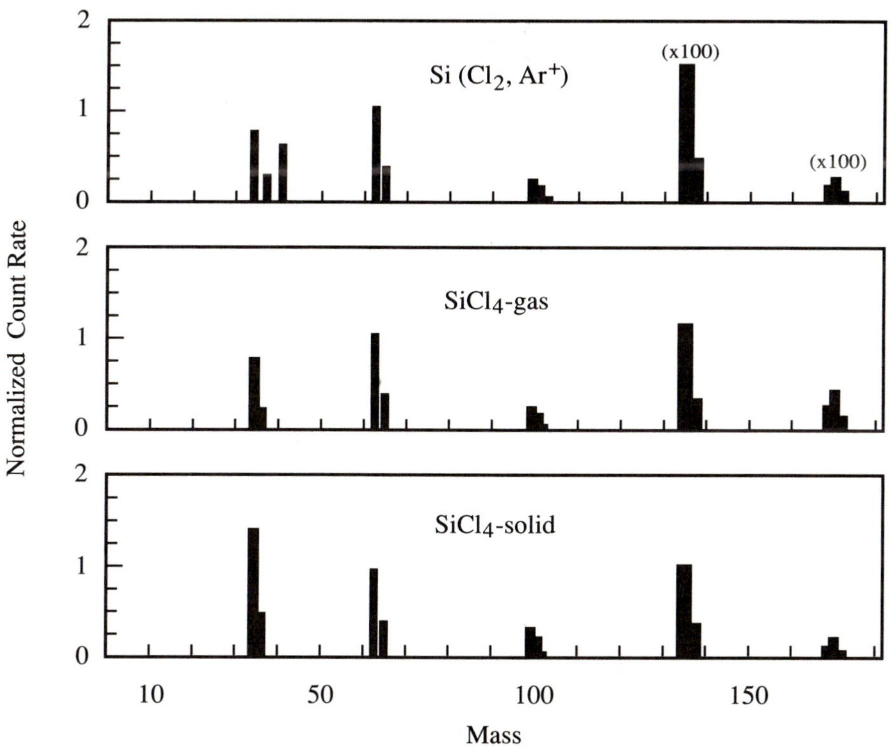

Figure 4.26. Mass spectra of (top) Si simultaneously bombarded by 1 keV Ar^+ ions and Cl_2 molecules at 300 K; (middle) $SiCl_4$ gas and (bottom) $SiCl_4$ solid, bombarded by Ar^+ ions. Spectra have been normalized to $SiCl^+$.

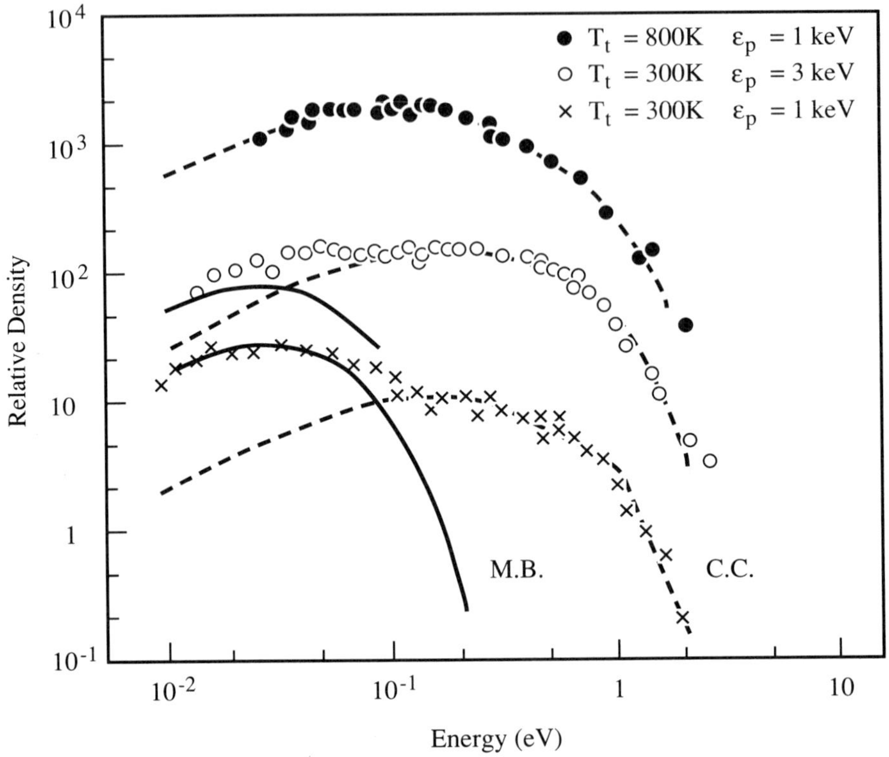

Figure 4.27. Kinetic energy of $SiCl_2$ ejected from a Si target bombarded with Ar^+ and Cl_2, with varying substrate temperature T and ion energy ε. The solid lines represent low temperature Maxwell-Boltzmann distributions.

A similar effect has been seen in sputtering of silicon by simultaneous bombardment with XeF_2 molecules and Ar^+ ions. In the reactive sputtering process, SiF and SiF_2 were dominant, and only small amounts of SiF_4 were detected. During thermal etching in XeF_2, SiF_4 is the main reaction product [122].

Figure 4.27 after Zalm [108], shows the kinetic energy of $SiCl_2$ emitting from silicon exposed to Ar^+ and Cl_2 for a variety of beam energies and surface temperatures. These curves can be resolved into two energy contributions. Firstly, a collision cascade distribution according to equation 4.47 at the high energy end, comprising more than 80% of the total emission. By fitting equation 4.47 to the curves, the surface barrier energy for the reaction products was found to be in the range of 0.2–0.35 eV. This is at least 10 times lower than values for elemental surfaces (see figure 4.19). At the low energy end, a second distribution with Maxwell-Bolzmann behavior according to equation 4.43 is found, with a temperature equal to that of the substrate. It is probably caused by ion bombardment enhanced diffusion to the surface, followed by evaporation at substrate temperature. The Maxwellian part contributes less than 20% of the total particle emission. Apparently, the dominant ejection mechanism at ion energies above 500 eV for chemically assisted ion etching is physical sputtering.

At ion energies below 250 eV, a third contribution is observed in addition to the two mentioned above, namely Maxwell-Boltzmann emission with temperatures near 2500 K (0.22 eV) [123]. With decreasing ion energy, the relative importance of this distribution increases at the cost of the collision cascade. Below 100 eV the cascade contribution is not observed anymore. This third distribution can be ascribed to ion-enhanced evaporation of the reaction products. Note the similarity between the temperature of this distribution and the surface escape barrier for reaction products deduced from the collision cascade emission.

In the simultaneous bombardment of Si by Cl_2 molecules and Ar^+ ions, reduction of the reactive gas flow, increase of the energy of the ions, and increase of the substrate temperature all have the same net effects, namely [113,119]

- reduction of the Maxwell-Boltzmann part of the energy distribution
- a shift to less chlorinated reaction products
- an increased surface barrier energy.

A notable phenomenon is the low surface barrier energy U_0 for the reaction products SiCl and $SiCl_2$. Since no substrate temperature dependence is observed, the mechanism cannot be straight thermal evaporation of the reaction products. Various models have been proposed to explain the low values of U_0. They agree in the observation that bond breaking and mixing of chlorine into the lattice are important factors, but a complete image has not yet been reached [110,113,120,124,125,126].

The fundamental studies on particle interaction with surfaces discussed so far are all conducted in more or less high vacuum. From the results, general guidelines for dry etching processes can be deduced, but caution should be taken with the details. In the simultaneous bombardment with Cl_2 and Ar^+ beams, for example, SiO_2 is found to erode about 5–10 times slower than Si. Hence, there is selectivity, but it is limited. Common dry etch systems based on rf discharges can do much better, however; in pure chlorine plasmas, the selectivity can be almost infinite, as discussed in chapter 6. The difference in operating pressure is probably the major factor disturbing the straight extrapolation of beam data to discharges. At the relatively high pressure where most reactors operate, backscattering of low-volatile reaction products against gas molecules and polymerization will be much more pronounced than can be simulated in the low background pressure regime of the beam studies.

Chapter 5

REACTOR TECHNOLOGY

5.1 Introduction

The previous chapters covered the chemical and physical principles of dry etching discharges, as well as the interaction of these plasmas with their surroundings. The hardware to produce such discharges will be dealt with in this chapter. A large variety of reactors is in use. The frequency of operation of dry etch systems which are or have been available commercially ranges from 50 kHz in the LF band to 50 GHz in the MW band, but HF, more specifically 13.56 MHz, is used mostly. Because the dimensions in present VLSI devices require highly directional pattern transfer, bombardment of the wafers with ions is a necessity. This is accomplished by mounting the electrodes inside the reactor, and bringing the wafer in direct contact with the plasma, as in sputter deposition. Very common in this respect is the parallel-plate reactor, with electrode diameters between 150 and 750 mm. The early so-called barrel reactors have external electrodes, and are suitable for non-directional etching only. Recent additions to the family of dry etchers use magnetic fields to improve system performance, or have special electrode designs.

Though there are many differences, all dry etching reactors have in common the requirement of a properly matched connection between electrode and rf generator, and the need to measure and control the amount of energy delivered to the plasma. In this chapter, such general reactor engineering aspects will be explained first. Then, the various reactor concepts will be presented more or less chronologically, the newest systems being treated last.

5.2 Electrode construction

A typical rf-powered electrode assembly is shown in figure 5.1 after de Vries and van den Hoek [127]. At the backside of the electrode a metal shield is fitted which is electrically grounded, thus suppressing the occurrence of discharges at the backside of the electrode. The distance between the shield and the electrode must be less than the ion sheath (dark-space) thickness for the conditions used, to prevent striking of a discharge. For this reason the shield is usually called the dark-space shield. In chapter 3, the dark-space thickness was shown to depend on gas type, pressure, power input, and frequency. It would be nice to use a very small gap in order to be sure to suppress parasitic discharge formation. Unfortunately, the reactor parasitic capacitance will then often become too large, thereby inducing large blind currents and power losses. Moreover, dust particles in a small gap cause arcing.

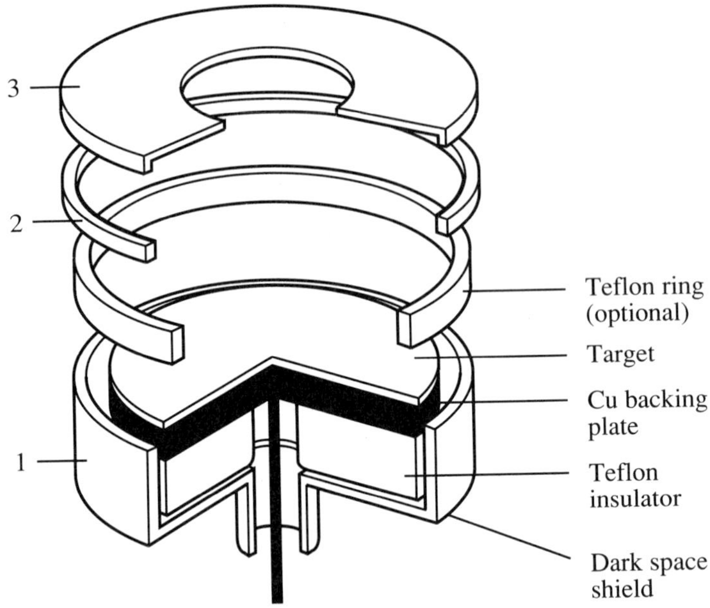

Figure 5.1. Schematic drawing of an rf-powered electrode assembly.

The shape of the electrodes and the dark-space shields can have a significant influence on discharge behavior. Sharp edges cause high electric fields, which can lead to locally intensified plasmas, instabilities and sputtering of electrode materials. Holes, ports, and bolts in walls and electrodes induce hollow cathode type behavior. The latter means, that electrons are caught between the high voltage sheaths on either side of the hole, and start bouncing back and forth. Thereby increasing the local degree of excitation, dissociation and ionization. Gas inlets and pumping ports are also suspect.

Material use in a reactor is not trivial. The reactor, for example, must withstand the aggressive environment, but should also be compatible with process, vacuum, and rf engineering demands. Aluminum is readily attacked by chlorine and must be protected by a layer of, for example, aluminum oxide deposited by anodization or plasma spraying. Stainless steel, on the other hand, can cause problems because metals like iron, nickel, and chromium can be sputtered, and are known to degrade silicon device performance, see chapter 6. The electrode material, especially that of the powered electrode, has a large influence on plasma behavior, both physically and chemically. Due to the large variety of interactions between gas molecules and the wall, the electrode material can have a large chemical influence if it reacts with the gas radicals; it can also act as a recombination center [128]. Physically, it affects the plasma because of variations in the secondary electron emission coefficient with ion bombardment [129].

A noticeable side-effect of covering the electrode surface with an insulator (quartz, anodization) is that measuring the dc self-bias across the discharge can become a difficult job. To do this, a conductive channel is needed. This path can be either around the electrode edge, or a leakage current straight through the insulator. Having an insulating plate on top of the electrode is a limitation in this respect, but it certainly does not preclude dc measurement. In the strong electric fields that will often be present at the electrode edge, the surface of many isolators becomes conductive. Ion bombardment can also give rise to a significant decrease in the surface resistance. Whether a conductive path is formed or not depends critically on the construction of electrode and dark-space shield [127]. If a dark-space shield such as plate 3 in figure 5.1 is fitted, the dc bias voltage present on the electrode-plasma interface cannot be measured at the back stem, because the shield plate covers a sufficiently large part of the electrode. In case plate 2 and the Teflon ring are present, and plate 3 is omitted, the plasma still does not "see" the isolator-backing plate interface, but now the electric field at the edge induces enough parasitic conductance.

5.3 Matching networks

Most rf generators have a fully resistive output impedance which matches the usual 50 or 75 Ω coaxial transmission lines. However, in chapter 4 it was shown that discharges operating in the HF band (10–50 MHz) behave mainly capacitive with a small resistive component, while LF systems (50–500 kHz) are mainly resistive with a small capacitive component. The goal of a matching network is to adapt the variable impedance of reactor and discharge to the fixed cable resistance. In unmatched systems, large reactive currents can flow that do not contribute to the net power transfer. Another reason for using matching networks, is that in high frequency systems, standing waves can occur on the transmission line, which cause losses and rf interference. Under extreme conditions this can lead to arcing in the connectors and cables [130,131].

Consider a load with an impedance

$$Z_2 = R_2 + j\,X_2 \tag{5.1}$$

which has to be matched to a system with an open circuit voltage V and an impedance

$$Z_1 = R_1 + j\,X_1 \tag{5.2}$$

The power absorbed by the load is

$$P = \frac{V^2\,R_2}{(R_1 + R_2)^2 + (X_1 + X_2)^2} \tag{5.3}$$

From this equation it can be derived that maximum power transfer is achieved when $R_1 = R_2$ and $X_2 = -X_1$. In other words, the network output impedance has to be the complex conjugate of the load input impedance.

Different types of matching networks are used for the different frequency regimes. For systems operating in the HF band (13.56 MHz), the L-type and the Π-type networks are used mostly (see figure 5.2). Matching is achieved by simultaneously adjusting the capacitors C_1 and C_2. This can be done automatically with servomotors and a circuit that measures the phase angle and magnitude of the matching network input impedance [132]. Consider a load impedance Z_{load}, which can be written as

$$Z_{load} = R + jX \tag{5.4}$$

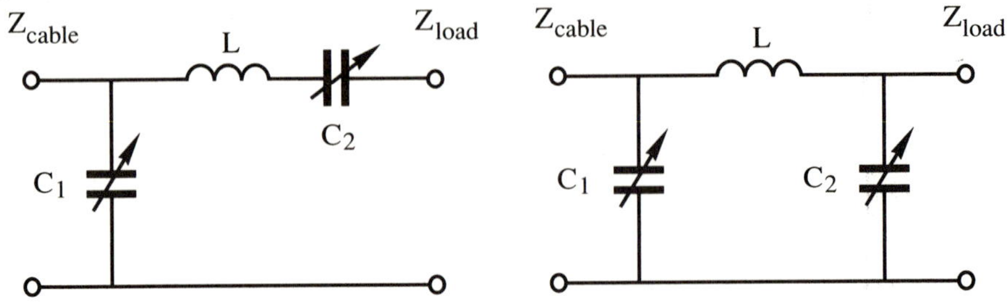

Figure 5.2. Schematic diagrams of L-type (left) and Π-type (right) matching networks.

To transform this complex impedance to a real cable impedance Z_{cable} with the aid of an L-type matching network, the capacitance and impedance values in the network, see figure 5.2, have to be adjusted to satisfy

$$R = \frac{Z_{cable}}{1 + (\omega\, C_1\, Z_{cable})^2} \tag{5.5}$$

and

$$X = \omega\, L - \frac{1}{\omega\, C_2} - R\sqrt{\frac{R}{Z_{cable} - R}} \tag{5.6}$$

where ω is the excitation angular frequency.

The Π-type matching network has a wider matching range than the L-type, but invokes more reactive currents since it has a closed loop within the circuit [131]. Reactive blind currents can be the source of significant power losses. In certain matching networks and reactor set-ups, up to 80% of the transferred power can be lost [52,60]. Moreover, the Π-network does not have a capacitor in series with the load. Hence, a net dc current can flow through the plasma and the substrates to annihilate the reactor dc self-bias which would otherwise have developed. In general, dc current through silicon wafers should be prevented to reduce reliability problems in sensitive devices.

In microwave systems the excitation frequency is in the low GHz band (2-50 GHz); the corresponding wavelengths are of the order of centimeters. Because of this, the discharge chamber is transmission line, matching and reactor in one. Tuning, if at all possible, is done by moving metal pieces inside the waveguide connecting reactor and generator.

LF systems are mostly matched with a step-up transformer, such as shown in figure 5.3. Reactive components are thus neglected and the output resistance is varied in steps. In this case, no series capacitor is used either, and dc currents can flow in the system. It is, of course, always possible to include a blocking capacitor in the reactor connection in case of the above two matching systems. Especially in the LF system, care should be taken to use a large enough value (say 5 nF) to prevent losses by voltage division.

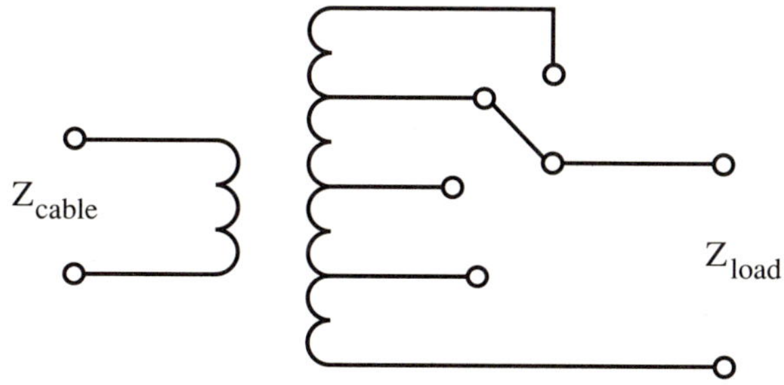

Figure 5.3. Diagram of a step-up transformer matching network.

5.4 Power, voltage, and current control

In high-frequency systems, rf power is usually measured with a transmission line power meter [130]. For proper operation, this type of meter must be connected on both sides to a constant and resistive impedance. This will normally be the cable connecting generator and matching network, hence, the term transmission line. The system is shown schematically in figure 5.4; it is simply a pick-up wire fitted parallel to the rf-carrying feedthrough in a coaxial system. One end is connected to earth via a resistance R; this can very well be the wire itself. The floating other end has a capacitance C and a mutual inductance L relative to the feedthrough; the voltage V_M on this floating end is rectified and measured. If $R \ll 1/j\omega C$, it can be derived that

$$V_M = j\,\omega\,R\,C\,V_{rf} + j\,\omega\,L\,I_{rf} \tag{5.7}$$

where V_{rf} and I_{rf} are the local values of voltage and current on the feedthrough. Note, that L has either a positive or a negative sign, depending on the direction in which it is fitted in the system (left and right situation in figure 5.4). Choosing the components such, that

$$R\,C = \frac{L}{Z_{cable}} \tag{5.8}$$

with Z_{cable} the cable (input and output) impedance, it appears that

$$V_M = j\,\omega\,L\left(\frac{V_{rf}}{Z_{cable}} + I_{rf}\right) \tag{5.9}$$

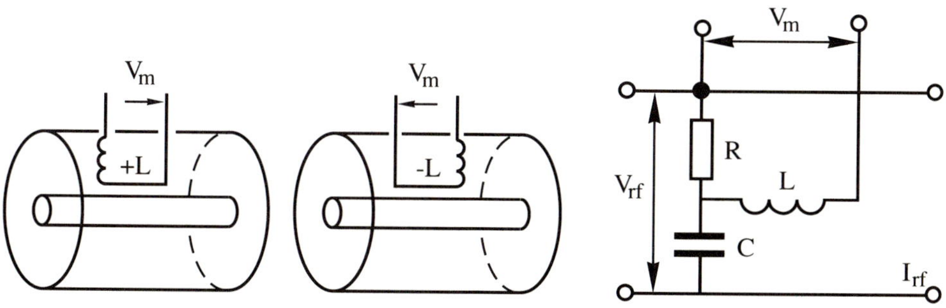

Figure 5.4. Basic circuit of a transmission line power meter.

If the cable is not perfectly matched at one end, the relation between the peak values of the rf voltage and current at any point on the transmission line (cable) can be written in terms of a forward (in-phase) and reflected (90-deg out of phase) wave,

$$V_{rf} = V_F + V_R \qquad \text{and} \qquad I_{rf} = \frac{V_F - V_R}{Z_{cable}} \tag{5.10}$$

where V_F and V_R denote forward and reflected voltage, respectively. Depending on the direction of the pick-up wire, and thus the sign of L, the output voltage of the power meter is either

$$V_M = j\,\omega\,\frac{L}{Z_{cable}}\,2\,V_F \qquad \text{or} \qquad V_M = j\,\omega\,\frac{L}{Z_{cable}}\,2\,V_R \tag{5.11}$$

This shows that the system from figure 5.4 responds either to the forward or the reflected wave; for this reason, it is also sometimes called a directional power meter. By adding some extra circuitry, the output is calibrated to read $\frac{1}{2}V_F^2/Z_{cable}$ ("forward" power) and $\frac{1}{2}V_R^2/Z_{cable}$ ("reflected" power). This makes the transmission line meter suitable for both rf power and matching control. It should be noted, that the net power dissipated in the load is not simply the difference between forward and reflected power. Similarly, the reflected power is not a real dissipated energy; it is only a measure for the presence of standing waves on the cable. Since the in-line meter has to be placed between generator and matching network, resistive losses in the matching network itself cannot be taken into account.

When attempting to measure rf voltages in high-frequency (13.56 MHz) systems directly, the first problem is to avoid unwanted rf interference through capacitive and inductive coupling. An even more severe problem is, that it is difficult to find a good reference ground, since every piece of wire and every part of the reactor will have some impedance relative to the "real" ground. Finally, the reactor construction itself may effectively preclude direct measurement of the rf voltage across the discharge, as the latter requires an undisturbed contact with the electrode surface in direct contact with the discharge. These problems are less important at low frequencies.

The rf current can be measured with either a current transformer or by probing the voltage across a small resistor in series with the cable. As mentioned earlier in this chapter, the reactive part of the current can be high as a result of matching network and reactor stray impedances, which does not add to the power dissipation in the plasma. This makes current measurements system dependent and, in a general sense, not representative for the properties of the plasma. An exception has to be made for low-frequency systems (100 kHz). In that case, the effect of system capacitances is low, and the rf current is roughly proportional to the ion current, as discussed in chapter 3. Under these conditions, it is sensible to use rf current feedback to control the power generator.

In principle, power dissipation can be determined directly by real-time multiplication of rf current and voltage and integration of the result,

$$P = \frac{\omega}{2\,\pi}\int_0^{2\pi/\omega} V(t)\;I(t)\;dt \tag{5.12}$$

with V and I the real-time voltage and current, respectively, and $2\pi/\omega$ the time of one rf cycle. Input and output impedance need not be resistive in this case. The applicability of this algorithm is limited to low frequency systems (LF) with frequencies up to about 1 MHz, by the availability of fast multipliers.

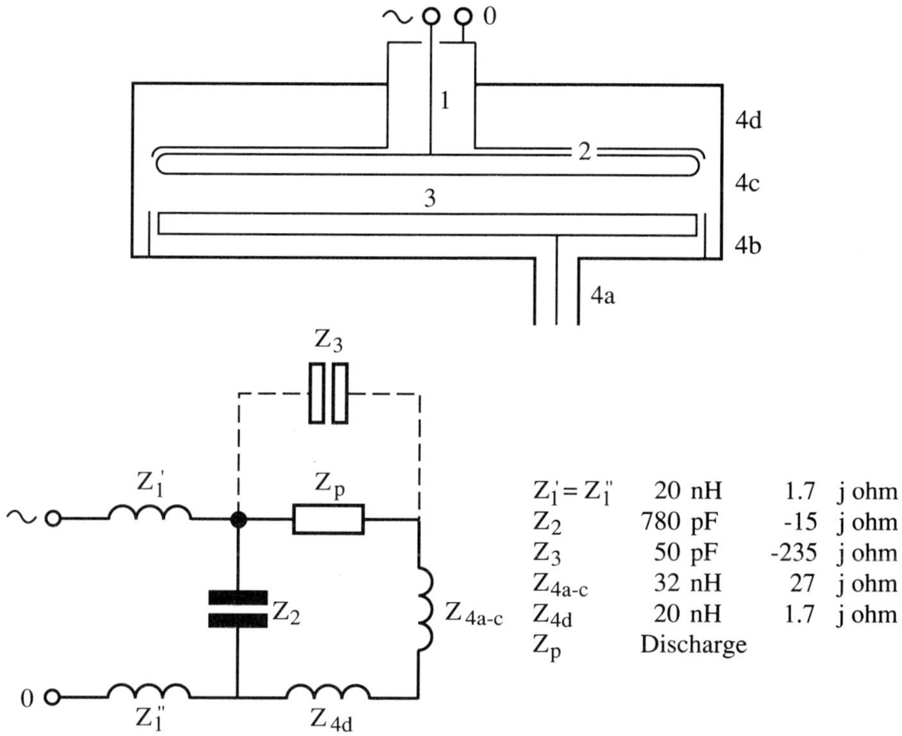

Figure 5.5. Schematic drawing of a planar reactor and its electrical equivalent circuit.

5.5 Parasitic impedances

The matching network and the reactor contain many electrical components – resistors, capacitances, inductances – which are not directly visible. In the MHz range and above, feedthroughs, shieldings, pipes, bolts and nuts, and even platework, represent impedances with values that cannot be ignored [130]. This is caused to a large extent by the skin effect – the phenomenon that high-frequency electrical currents in conductors are forced to the surface of the conductor. The nominal penetration depth δ in a plane conductor is given by

$$\delta = \sqrt{\frac{2\,\rho}{\omega\,\mu_M}} \tag{5.13}$$

where ρ is the resistivity and μ_M the magnetic permeability. For a copper conductor at 13.56 MHz, the value of δ is only 5 μm, which implies that in any practical piece of metal the current path follows the contours only.

As an example, a practical parallel-plate reactor will be analyzed; it is shown schematically in figure 5.5 after van Roosmalen et al. [52]. A 13.56-MHz power supply is connected to the top electrode via a matching network; a grounded metal shield is mounted to suppress parasitic discharges at the back of the electrode. The bottom electrode is grounded via a stub and an external connection. For the rf current flow, it has to be realized that the skin effect allows surface conduction only. The inductances and capacitances in the diagram from figure 5.5 were calculated from the equations for coaxial cables

$$L = \frac{\mu_M}{2\pi} \frac{z}{\ln(b/a)} \qquad \text{and} \qquad C = 2\pi\epsilon \frac{z}{\ln(b/a)} \tag{5.14}$$

and for parallel plates

$$C = \epsilon \frac{A}{l} \tag{5.15}$$

In these equations, ϵ is the dielectric constant, z the cable length, a and b the inner and outer conductor diameter, A the plate area, and l the distance between the plates. Note that the interelectrode capacitance vanishes from the circuit as soon as the discharge strikes.

The importance of these calculations becomes evident when rf power is applied [39]. Assume a capacitive current of 10 A (rms) in the discharge and an interelectrode peak voltage of 500 V, corresponding to 350 V (rms). These values are taken from the discussion on high-frequency sheath formation in chapter 3, under assumption that the reactor is asymmetrical. Network analysis then immediately shows that the average rf potential difference between the (grounded) bottom electrode and the (grounded) reactor side wall is 27 V; between the side wall and the edge of the top-electrode shield, another 17 V is found. The top-electrode shield as such draws twice as much current as the discharge itself. Combined with the high top stem inductance, the latter results in 200 V and 31 A at the reactor input clamps. This shows why in an HF system, parasitic discharges can strike between dc-grounded surfaces. Also, it is understandable why the measured dc self-bias voltage between powered electrode and ground – in an asymmetric system proportional to the discharge peak potential – can be much higher than the external rf peak voltage across the reactor. In this example, the dc potential is 500 V, since the internal rf peak voltage across the discharge was assumed to be 500 V, while the external rf potential is only 200 V (rms), see above.

With LF excitation, parasitic impedances are much less important, although it has to be realized that certain protective electrode coatings can act as series impedances. A major concern in LF reactors is the high potential that develops between the electrodes, even under moderate power operation. Only few rf feedthroughs can withstand the combination of amps and kilovolts. Moreover, the high potentials can cause arcing inside the reactor.

5.6 Afterglow etching

The oldest dry etching system still in abundant use is the barrel reactor. A traditional barrel reactor consists of a quartz cylinder (the barrel) with connections for gas lines and a vacuum pump; it is shown schematically in figure 5.6. Plate electrodes are strapped to the outside of the cylinder, usually in axial direction. Application of 13.56 MHz power to these electrodes provides a capacitively coupled energy input, which for moderate to high gas pressure is sufficient to maintain a discharge in the reactor. The wafers to be etched are placed on a quartz or metal boat inside the quartz tube. A metal Faraday cage is usually placed around the wafer boat to prevent charged particles from reaching the wafers. By doing so, the etching is limited to the neutral particles diffusing from the discharge zone between cage and reactor wall to the wafers. Although the etching zone is obviously not a plasma, it still emits light because of photon decay of excited molecules. The resulting dimly lit medium is known as the afterglow, hence the term afterglow etching. As pattern transfer

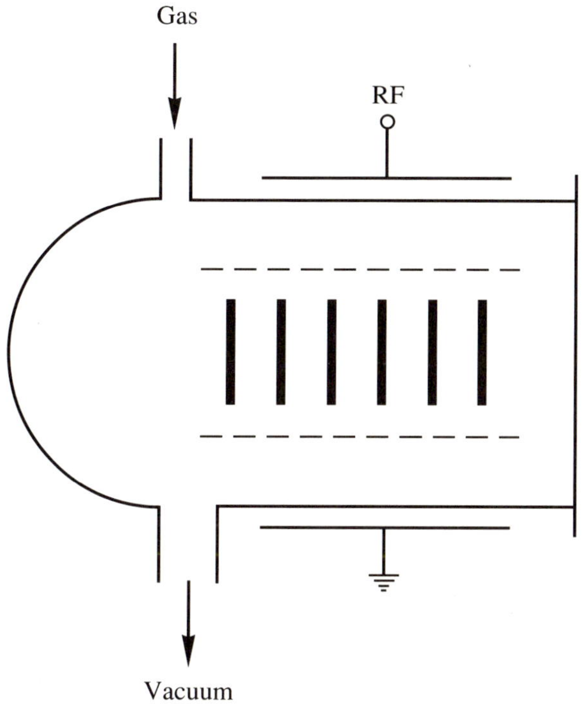

Figure 5.6. Barrel-type afterglow reactor. Typical conditions: $p = 100$ Pa; $P = 10$ kW m^{-3}.

is by chemical reaction between the excited gas and the substrate only, isotropic etching is obtained, similar to the action of wet chemical agents.

Because the wafers are floating inside the reactor without an effective heat sink, and relatively high rf power levels are needed to achieve sufficiently fast etching, the wafers will become quite hot; 200 °C is not unusual. Temperature control is hardly possible. Another problem with the barrel system is the sputtering of material from the metal cage, causing wafer contamination. Therefore, the cage is often made of hard-anodized aluminum; Al_2O_3 is a material with a low sputter yield. A third limitation is the inevitable high non-uniformity of etching. The non-uniformity results from the fact that the wafers must be relatively tightly packed in order to make the process economically interesting, while the etching species can reach the wafer surface by diffusion only. Because of this, the concentration of etching species – hence, the etch rate – rapidly decreases from the edge to the center of the wafers, an effect that becomes more severe with shorter wafer distance. Nowadays, the relatively cheap barrel etchers are mostly used for stripping photoresist and other non-critical applications.

A more recent modification to the idea of a barrel reactor is the vertical rf downstream etcher, as depicted in figure 5.7 after Kalnitsky and Chung [133]. Similar to the system in the previous figure, the discharge is induced inside a quartz tube by external electrodes driven at 13.56 MHz. The major difference lies in the position of the substrate. Only one single wafer is exposed to the afterglow emerging from the central discharge zone. It is protected from charged species by either a metal or quartz grid with sufficiently small holes. This construction allows good uniformity as well as good temperature control. The absolute etch rate needs to be quite high, however, since only one wafer is etched at the time. In systems in practical use, this is done by using very high gas flow rates, by increasing the wafer

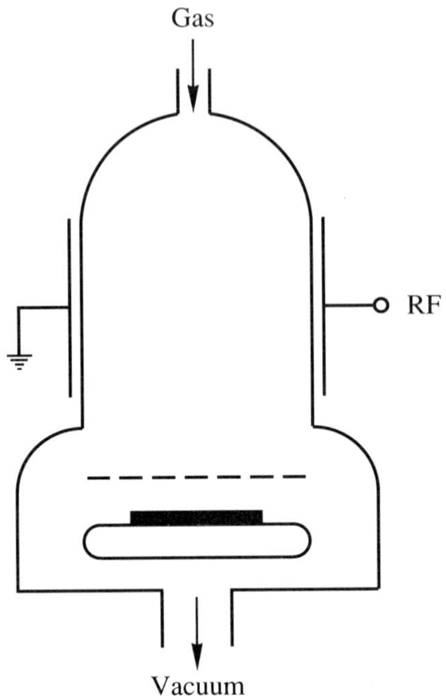

Figure 5.7. Vertical downstream reactor.

temperature, or by modifying the chemistry. Resist stripping, for example, can be helped by adding fluorocarbons to the usual oxygen discharge. This is discussed in more detail in chapter 6.

An alternative way to obtain fast etching by creating a high concentration of reactive neutrals is replacing the 13.56 MHz high-frequency source by microwave excitation [134]. This uses the rule of thumb that gas dissociation increases with the excitation frequency, as discussed in chapter 3. A microwave discharge can have a degree of dissociation up to 80%, but sheath voltages are approximately equal to the floating potential (10–20 V). This means that the microwave discharge is not suitable as such for directional etch processes; the latter usually require a high ion bombardment energy. On the other hand, it can be a good replacement for the vertical high-frequency tube discharge in figure 5.7. The principal advantage of a microwave etching system is its high efficiency and low ion-bombardment-induced damage. A disadvantage is the difficulty to produce uniform etching over large areas; the size of the resonance cavity is limited to the wavelength of the radiation used. Since the reactive species have to travel from the place where they are produced to the wafer, recombination reactions in the gas and especially at reactor walls are very important. A special case of microwave etching is the electron cyclotron resonance reactor, which will be discussed at the end of this chapter.

5.7 Reactive ion etching and the area theorem

It has already been emphasized several times that practical dry etch processes often require a substantial amount of high-energy ion bombardment for anisotropic pattern transfer. This can be achieved only in reactors with internal electrodes. Electrode geometry and

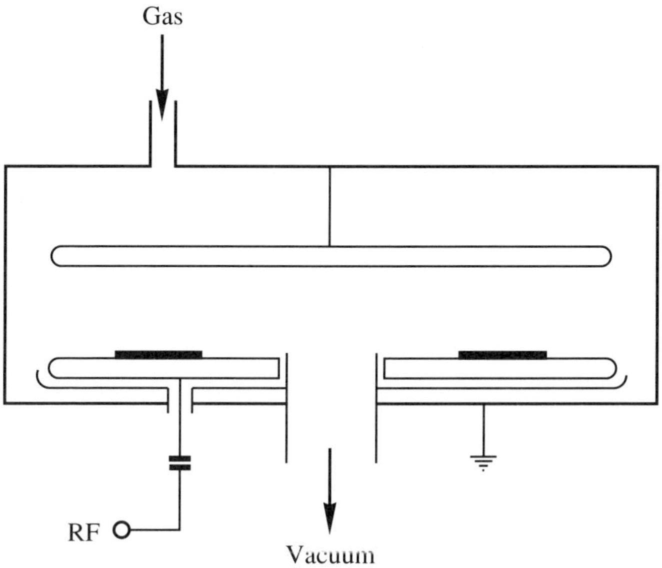

Figure 5.8. Batch-type cathode coupled (RIE) reactor. Typical conditions: p = 5 Pa; P = 5 kW m^{-2}.

symmetry aspects play an important role in such systems, especially if the excitation frequency is in the HF band. The traditional system which meets these demands is the parallel-plate reactor depicted in figure 5.8. Basically, it consists of a high-vacuum chamber, in which two parallel electrodes are positioned. The top electrode and the chamber walls are electrically grounded; the lower electrode and substrate holder are connected through a dc-blocking capacitor and matching network to an rf generator, mostly 13.56 MHz. The system shown in figure 5.8 is usually referred to as a reactive-ion etch (RIE) or cathode-coupled reactor. Because the ion sheaths in a 13.56-MHz discharge are capacitive, capacitive voltage division takes place if an rf current is drawn through the system,

$$\frac{V_{rf}}{V_g} = \frac{C_g}{C_{rf}} \tag{5.16}$$

where V is the sheath rf peak voltage, C the sheath capacitance, and rf and g are indices for the powered and the grounded electrode, respectively. The sheath rf peak voltage will be approximately equal to the sheath dc self-bias potential, as discussed in chapter 3. Writing the capacitance in terms of sheath thickness d and electrode area A, the following expression is obtained,

$$\frac{C_g}{C_{rf}} = \frac{A_g}{A_{rf}} \frac{d_{rf}}{d_g} \tag{5.17}$$

Since the powered electrode area A_{rf} in this reactor is small relative to the sum of all grounded surface areas A_g in contact with the plasma, the powered electrode capacitance will be less than that at the grounded electrode.

The positive ion current J_i density in the sheath can be written in terms of the sheath thickness d and the sheath dc self-bias potential V. Assuming collision-free movement of the ions in the sheath, the Child-Langmuir equation for space-charge limited current is applicable,

$$J_i = 0.6\,\epsilon\sqrt{\frac{e}{m_i}}\,\frac{V^{\frac{3}{2}}}{d^2} \tag{5.18}$$

where m_i is the ion mass (see also chapter 3). For a perfectly uniform plasma, the ion sheath current density is constant on all surfaces. Writing equation 5.18 twice, once for the grounded, and once for the powered electrode, the ion current density can be eliminated, and the result expressed as

$$\frac{V_{rf}}{V_g} = \left(\frac{d_{rf}}{d_g}\right)^{x/(x-1)} \tag{5.19}$$

with $x = 4$.

Combining the above equations 5.16 and 5.17 to eliminate C, and applying equation 5.19 on the result in order to eliminate d, it appears that

$$\frac{V_{rf}}{V_g} = \left(\frac{A_g}{A_{rf}}\right)^{x} \tag{5.20}$$

where x takes the same value as in equation 5.19. This relation is known as the area theorem; it was first proposed by Koenig and Maissel to explain the voltage division in low-pressure sputtering systems [56]. Equation 5.20 shows that the highest sheath potential is found at the smallest electrode, which in general will be the powered one. Also, small asymmetries in electrode area are predicted to cause large potential differences. According to equation 5.20, an area ratio of 1.5 leads, with $x = 4$, to a voltage ratio of 5. Obviously, the applied rf voltage is found almost completely across the ion sheath at the smaller electrode in that case. The external reactor dc self-bias voltage V_{dc}, which is given by

$$V_{dc} = V_{rf} - V_g \tag{5.21}$$

is about equal to the rf peak voltage at the powered electrode, hence, the maximum ion bombardment energy.

Caution should be taken in making quantitative use of equation 5.20. It is questionable whether the ion current density is really equal at both electrodes. It is also not easy to decide what surfaces should be assigned to what electrode in calculating the exact area ratio [32]. Measurements of the dc-bias developed in practical systems indicate that the area ratio is in principle valid. The exponent x, however, is much less than the value 4 resulting from straightforward application of the Child-Langmuir relation. Especially if the area ratio is more than 1.5 (or less than 0.7), a value between 2 and 3 appears to fit the experimental figures much better [135]. There are indications that the Child-Langmuir relation in collision-free ion sheaths does not give a correct description of the sheath behavior if the sheath potential is small – less than 100 kT_e. From a more accurate model for the sheath behavior in this situation, it has been derived that the exponent x should be about 2.3 [46,47]. One can doubt whether the idea of a collision-free sheath is valid at all under the usual dry-etch conditions.

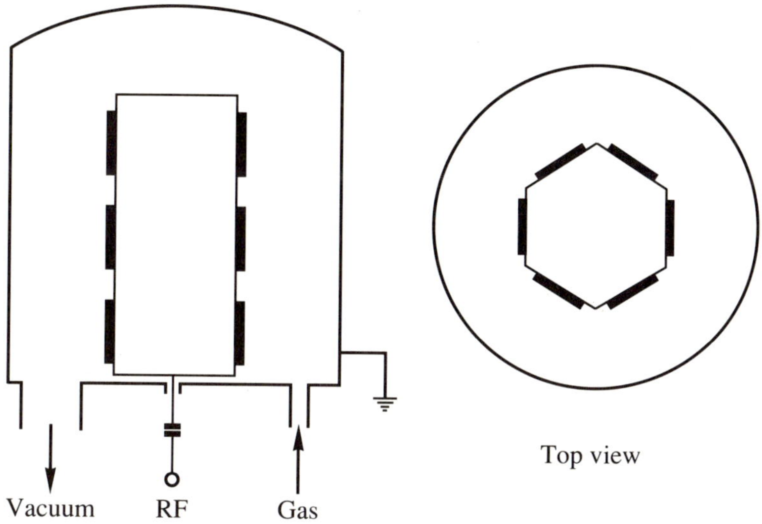

Figure 5.9. The hexode reactor. Typical conditions: p = 10 Pa; P = 5 kW m^{-2}.

In chapter 3, it was shown that the relation between ion sheath current, voltage, and thickness must be modified slightly if a mobility-limited current is assumed rather than free-falling ions. The relevant relation is

$$J_i = 1.4\, \epsilon\, \kappa\, \frac{V^{\frac{3}{2}}}{d^{\frac{5}{2}}} \tag{5.22}$$

where κ is a mobility-like proportionality factor. Using the line of work as has been used above for the Child-Langmuir relation, it now appears that for mobility limited ion current $x = 2.5$ in equation 5.19 [136], in agreement with experimental data.

Practical application of the area theorem is the coaxial or hexode reactor depicted in figure 5.9. Special feature of the hexode reactor is its coaxial electrode configuration. The inner cylinder is rf-powered; the outer is the counter electrode and therefore grounded. Here, a large area ratio is combined with a uniform concentric field distribution, giving the system the potential to etch uniformly, and with high ion energy. Since the wafers need to be clamped, the electrode needs flat surfaces. Therefore, the powered electrode is in fact hexagonal. With increasing wafer sizes, pentode shaped electrodes are also being used to hold the wafer. A weak point of the hexode reactor is the need for vertical wafer loading, which is intrinsically more difficult to automate, and plasma instabilities at the top end of the powered electrode.

When using higher gas pressures in the parallel-plate system from figure 5.8, the discharge contracts and the plasma loses contact with the side walls of the reactor [52,137]. This immediately influences the voltage distribution, as the effective area ratio drops (see equation 5.20). At sufficiently high pressure, i.e, 20–200 Pa, the voltages at powered and grounded electrode can be of comparable magnitude, implying that high energetic ion bombardment is now also encountered at the grounded electrode. Hence, the difference between placing the substrates on either the powered or the grounded counter electrode has disappeared. Fi-

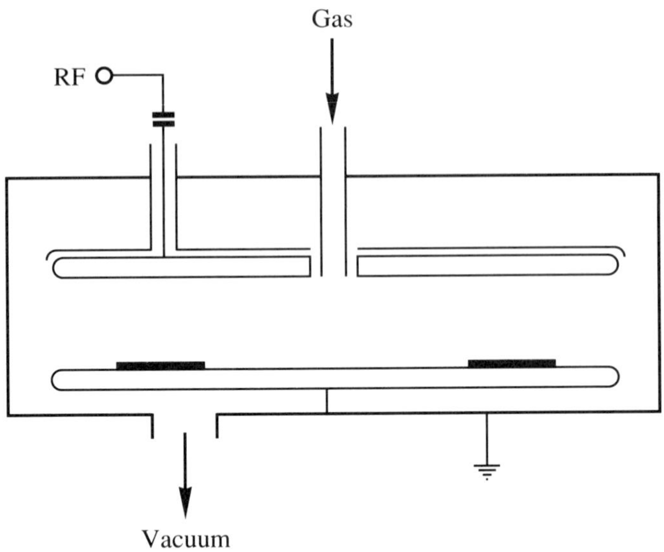

Figure 5.10. Batch-type anode-coupled (PE) reactor. Typical conditions: p = 30 Pa; P = 5 kW m^{-2}.

gure 5.10 shows such an anode-coupled or "plasma etch" (PE) reactor. The fact, that the substrates can be placed on a grounded surface, has advantages in handling and loading the system. Measuring the dc self-bias voltage across the reactor in figures 5.8 and 5.9 will give a value close to the ion-bombardment energy at the substrate electrode, since the systems are asymmetrical. As the PE system in figure 5.10 is almost symmetrical, the potentials at both electrodes cancel, and no net voltage will be found. This does not mean, of course, that the actual ion bombardment energy is low. From the above discussion, it can be concluded that there is no principal distinction between RIE and PE systems with respect to energetic ion bombardment. It would, actually, be better to speak of low-pressure cathode-coupled, and high-pressure anode-coupled systems, instead.

For reactors operating in the LF band, some conclusions have to be modified slightly. Symmetry aspects do play a role, but the effect is much less pronounced since there is no capacitive voltage division. Most systems are of the type shown in figure 5.10, with an operating pressure between 10 and 100 Pa. Generally speaking, it can be said that such systems are symmetrical. The full rf potential appears on each electrode during its cathodic cycle. This enhances the fact that the sheath voltages for an LF discharge are already substantially higher than in a corresponding HF system. Combined with the relative low degree of dissociation, it is now conceivable that high-pressure low-frequency anode-coupled discharges often exhibit a behavior quite similar to low-pressure high-frequency cathode-coupled systems [138,139]. Both have high energy ion bombardment and low concentrations of reactive neutrals.

Etch rates in the reactors with internal electrodes, as discussed so far, are between 50 and 100 nm min^{-1}. When used in production, a number of wafers therefore needs to be processed simultaneously, typically 6–24. For this reason such systems are called "batch" reactors. By increasing the power density in the symmetrical, high-pressure system, and lowering the frequency, if necessary, it is found that the etch rate can be increased up to 1 μm min^{-1}. With careful tuning, etching can still be directional. This observation has opened the door to a class of reactors where only one sample is etched at the time, as shown

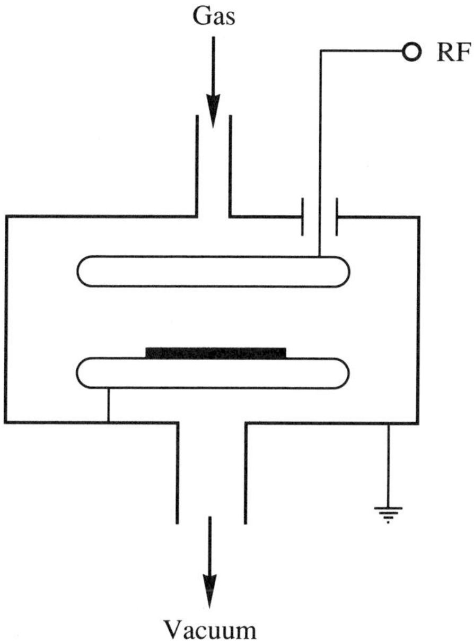

Figure 5.11. Single-wafer or in-line reactor. Typical conditions: p = 200 Pa; P = 50 kW m^{-2}.

in figure 5.11. Individual sample handling and the ease of automation are the main advantages of these so-called "single wafer" etch reactors; the throughput in terms of samples per hour can be the same as in batch-loaded reactors.

5.8 Ion-beam etching

In the reactors discussed up till now, no separate control of the generation of charged particles and their flux towards the substrate to be etched was available. An ion beam etcher does offer this opportunity [140,141]; a schematic plot of such a reactor is shown in figure 5.12. In this system, a plasma is generated in a separate source. This plasma source can be a conventional rf discharge of the type discussed previously. More frequently, a dc field is used (Penning source). Ionization can be enhanced by adding a hot wire (Kaufman source) or a hollow cathode. Usually, the efficiency is further improved by magnetic confinement with either permanent or electromagnets. The use of magnets is discussed in detail in the last section of this chapter. The result is a low-pressure, but highly ionized plasma. Ions generated in the plasma are accelerated from the source to the substrate to be etched through a set of dc-biased extraction grids. The substrate itself is placed in a remote chamber; the voltage difference between substrate table and the last extraction grid determines the bombardment energy. By careful adjustment of the voltage levels in the system and, optionally, using a neutralization wire in the ion beam exit, it is possible to make the resulting beam neutral by the addition of electrons. Achievable ion energies range from about 30 eV to several keV. Since the gas pressure is low, few charge transfer collisions take place, and a narrow ion energy distribution can be created. Nevertheless, beam divergence through collision with residual gas molecules in the substrate chamber, and energy variation from non-uniformities in the source plasma, usually limit the lowest practical ion energy to a few hundred eV.

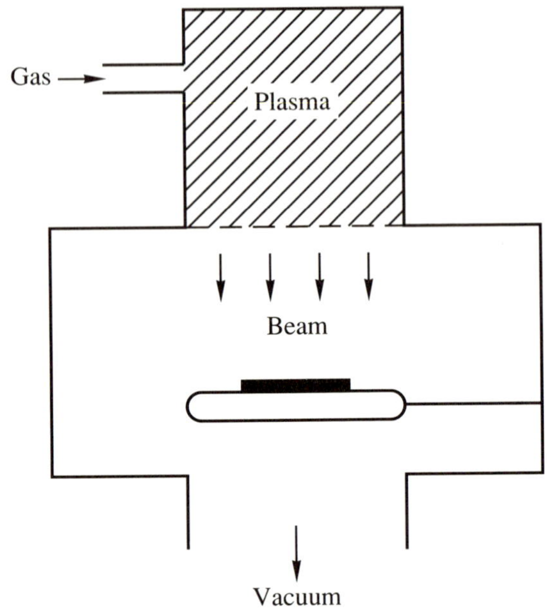

Figure 5.12. Ion beam reactor. Typical conditions: p = 0.05 Pa; J = 5 A m^{-2}; V = 500 V.

The oldest type ion-beam etcher is the ion miller [142]. In this reactor, the substrate is sputtered away by means of a beam of noble gas ions. High accuracy is possible, but selectivities and rates are low. In reactive ion beam etching (RIBE), a reactive gas is used [143]. This can help to increase both etch rates and selectivities. One step further is CAIBE – chemically assisted ion beam etching – which combines ion bombardment from a plasma source with an independent supply of reactive gas into the main chamber. In this way, it is possible to control ion and neutral fluxes independently, both in density and in energy.

A specific feature of ion-beam etching is, that the substrate can be tilted relative to the direction of ion bombardment. As the ion energy at the substrate to be etched and the angle of incidence are known, fundamental information, such as sputter yields, can be determined. A disadvantage of the – necessary – use of acceleration grids is the potential risk of contaminating the ion beam with sputtered grid material. Systems using hot filaments to produce the source plasma have a relatively short lifetime in the presence of reactive gases, as many conventional filament materials simply etch away. An example is tungsten, which is not very compatible with gases like fluorine and chlorine – tungsten halides being quite volatile. More recent solutions for the ion source problem involve the use of microwave excitation and the addition of magnetic plasma confinement [144]. Such systems come very close to what is known as ECR etching; this type of reactor will be discussed later on in this chapter.

A real restriction of ion beam etching for processing is, that the synergetic effects of ions and neutrals at the surface are not used as extensively as in regular dry etching. Experimental results show relatively low values for both etch rates and selectivities [145,146,147,148]. Therefore, ion beam etching is used for fundamental mechanistic studies such as described in chapter 4, rather than for wafer patterning.

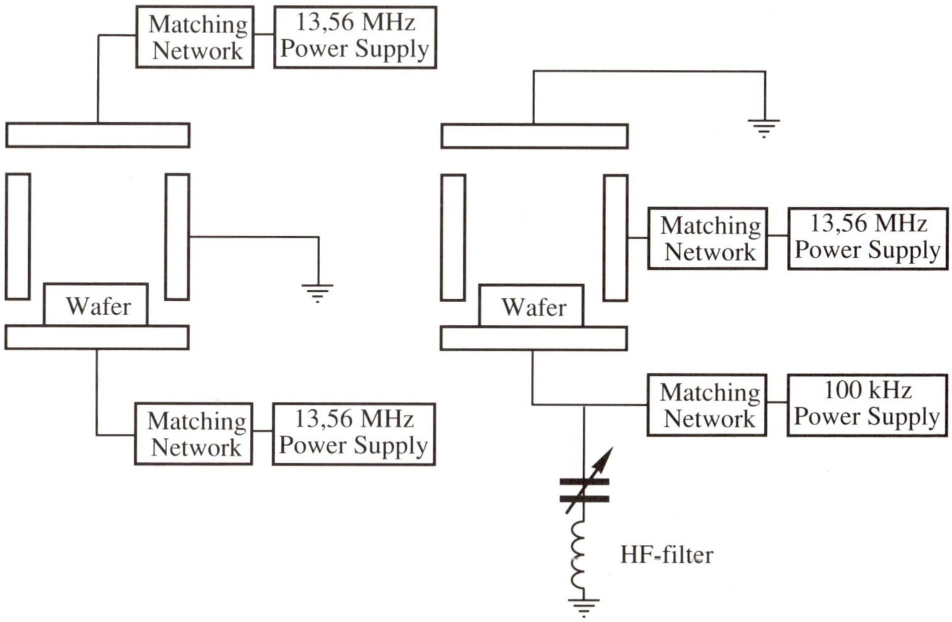

Figure 5.13. Triode reactors; the flexible diode system (left) and a dual-frequency etcher (right).

5.9 Multi-electrode and multi-frequency reactors

Fitting more electrodes and more frequencies than the usual single excited rf-powered electrode opens possibilities to decouple the production of charged and neutral particles, as well as influencing energy and density of the ion flux on the substrate surface separately. This would lead to a reactor type concurrent in flexibility with the ion-beam etcher.

If one frequency is used, several possibilities exist. In a conventional parallel-plate system, the non-rf-powered counter electrode can be connected to ground through an adjustable LC network. This induces an rf bias voltage on that electrode. A similar set-up is obtained if the rf power from the generator is divided between the two electrodes in a special matching network [149]. It is also possible to use two generators and two matching networks; this is known as the "flexible diode" system [150], as depicted on the left hand side of figure 5.13. In the latter case, special precautions need to be taken to prevent interaction between the two generators, such as wave interference on the cable and mutual influences of the rf generator control systems. Moreover, the generators have to run at exactly the same frequency to avoid effects of the difference signal. In either of the above reactors, the rf power is split between the two electrodes, and the ion bombardment can be adjusted from a cathode-coupled mode (maximum ion bombardment) to an anode-coupled mode (minimum ion bombardment). In principle, all systems have the possibility to adjust the potential distribution between substrate and counter electrode, without affecting the total rf power dissipated in the reactor chamber, and hence the chemical composition of the discharge.

By using two different frequencies it is possible to combine the specific features of the various frequency ranges introduced in chapter 3. High-frequency excitation (13.56 MHz) on the counter electrode, allows efficient production of excited species, while high-energy ion bombardment can be brought about by applying a low frequency (100 kHz) to the substrate electrode. Such a reactor is shown schematically on the right hand side of figure 5.13.

The multi-frequency reactor can be useful for etch processes where controlled high-energy ion bombardment is very important. An example is deep-groove etching in monocrystalline silicon [151,152]. A special variant of the multi-frequency reactor appears if the dissociation is done with microwaves (2.45 GHz), and the ion bombardment is controlled by an auxiliary 13.56 MHz source [153]. This situation will be covered in the next section.

5.10 Magnetron and ECR reactors

It has already been mentioned that increasing etch rates is highly desirable for economic wafer processing. The only readily available way to achieve this in conventional parallel plate RIE systems is by increasing the rf power. High power density means a high ion bombardment energy and, hence, significant substrate damage. In the previous chapters, it was shown that the ion sheath plays a major role in sustaining the discharge, as it reflects and accelerates high-energy electrons back into the volume of the plasma. Only these electrons can dissociate and ionize the gas molecules.

Application of a magnetic field gives an additional possibility of reducing the loss of energetic electrons. This results in an increased efficiency of rf power transfer. In other words, more excited species without having to increase surface damage by ion bombardment. The two main categories of systems based on the magnetic confinement principle are the magnetron and the electron-cyclotron resonance (ECR) reactors. In either case, use is made of the fact that electrons moving in crossed magnetic and electric fields are forced into a helical path. To understand this, consider an electron moving with a velocity $\vec{v}_e$ which is affected by both an electric field $\vec{E}$ and a magnetic field $\vec{B}$. The electron will then be subject to a force

$$\vec{F} = e\,(\vec{E} + \vec{v}_e \times \vec{B}) \tag{5.23}$$

The direction of the force is perpendicular to both the magnetic field and the direction of the velocity, meaning that the electron will spiral around the magnetic field lines. If the angle between the momentary electron velocity vector and the magnetic field vector is ϕ, the radius r of the electron path is

$$r = \frac{m_e\, v_e}{e\, B\, \sin\phi} \tag{5.24}$$

At an electron temperature of 2 eV, the mean electron velocity is 9×10^5 m s^{-1}.

For an electron moving with this velocity perpendicular to a moderate magnetic field of, say 5×10^{-3} T, equation 5.24 shows that they will be forced into a circular path with a radius of about 1 mm. This is smaller than the smallest dimension of the average dry etch reactor, and even of the order of the mean free path. Thus, the magnetic field increases the length of the electron path considerably, thereby effectively preventing them from reaching the wall. Any velocity of the electron parallel to the magnetic field will change the circular path into a helix. The presence of an electric field having components perpendicular to the magnetic field and parallel to the electron velocity vector will increase the electron velocity, and give rise to a spiraling path with an increasing radius. This is shown in figure 5.14. The paths of the electrons in the system will change continuously, as a result of collisions

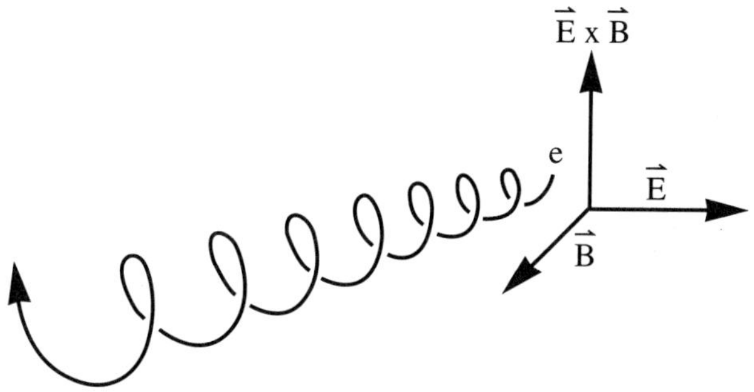

Figure 5.14. Electron path in crossed electric and magnetic fields.

with other particles, and the continuously changing electric rf field. In principle, ions will also move in helical paths. Ions are, however, at least 4 orders of magnitude heavier than electrons. The radius of the helix is, therefore, so large, that the effect can safely be neglected under practical circumstances (see equation 5.24).

From the above considerations, it appears that a magnetic field can lower wall recombination by confining the electrons. Hence, high plasma densities can be obtained in rf reactors for a low energy input. At constant rf power, the electron diffusion in the direction of the electric field decreases. In the presence of the magnetic field, the voltage at which the electron current density reaches equilibrium with the time independent ion current density in the

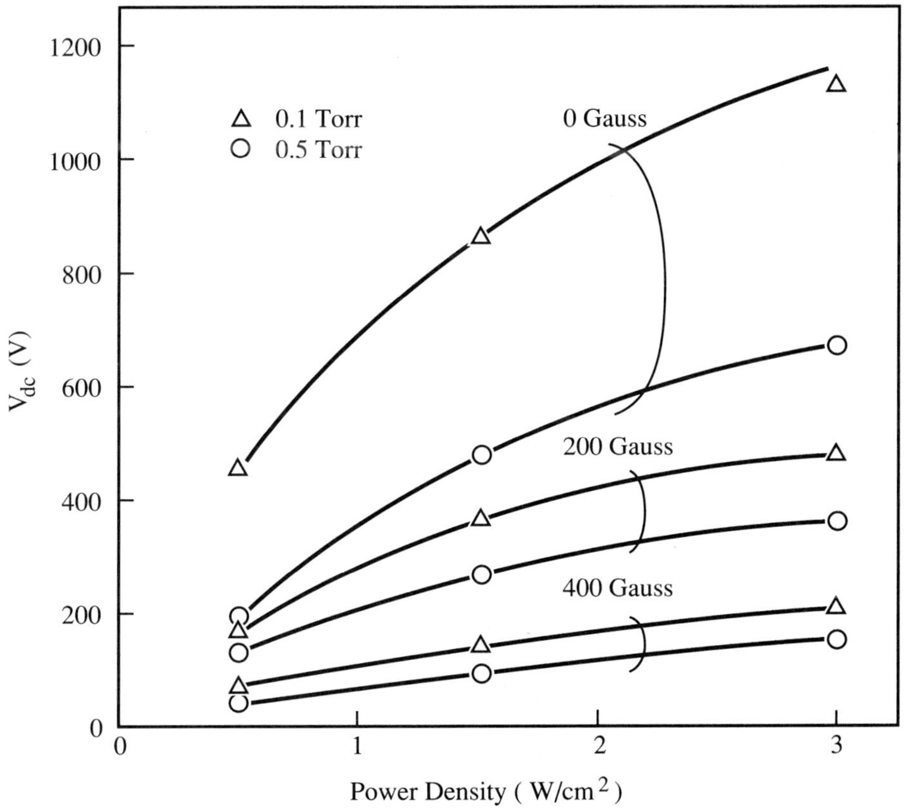

Figure 5.15. Measured dc self-bias voltage for a magnetron reactor.

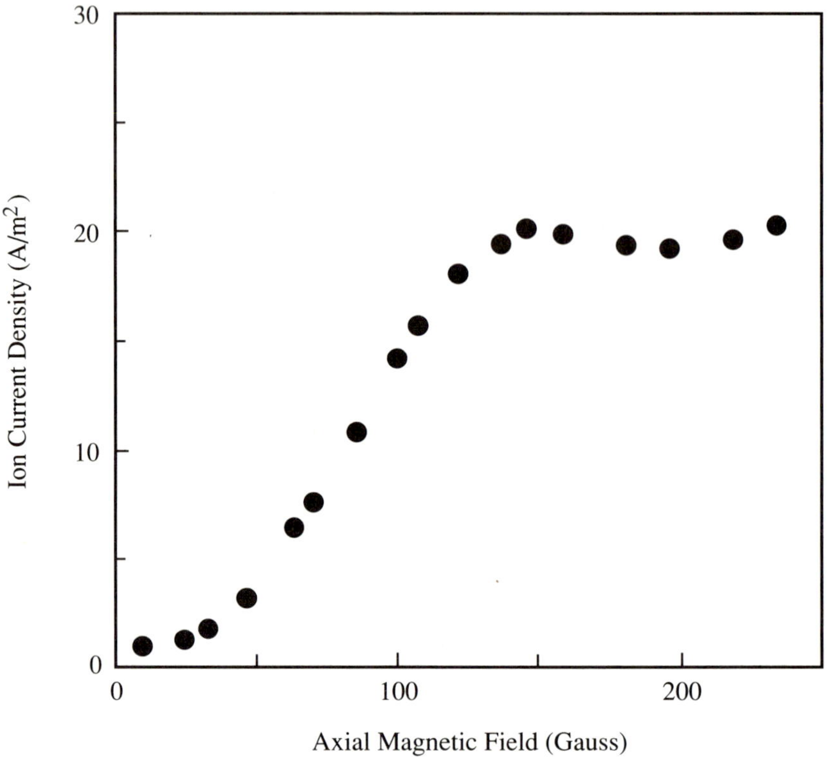

Figure 5.16. Ion sheath current in a magnetron reactor as a function of magnetic field strength.

sheath is lower. Therefore, the dc self-bias, which controls the ion bombardment energy, drops with rising magnetic field strength, see figure 5.15 after Hirobe and Azume [154]. At the same time, the plasma density increases, and hence the ion bombardment flux density, see figure 5.16 after Kuypers and Hopman [155]. An additional feature of rf discharges with magnetic confinement is, that they can operate gas pressures much lower than usual; 0.1–1 Pa is typical. This is also a result of the reduced loss of electrons to the wall.

Dry etching in crossed magnetic and 13.56-MHz electric fields is known as magnetron etching or, by analogy to RIE, magnetron ion etching (MIE). Through combination of a low pressure and a high magnetic field strength, it is possible to design systems where the dominant etching species are ions rather than reactive neutrals. This can have advantages for processes where a very high degree of etch directionality is required. Obviously, the threshold energy of the surface reaction should be not too high; the accelerating potential in the sheath will be low [156,157,158].

A number of options exist for the application of magnetic fields to a parallel plate rf reactor. One possibility is to mount permanent magnets under either the substrate or the counter electrode, lined up in alternating rows of north and south poles. In this way, a situation is achieved similar to conventional magnetron sputtering for metal deposition. From the direction of the magnetic field lines in such a set-up, it is clear that plasma enhancement occurs only in the zones between the north and south poles of the consecutive magnets, as these are the places where the magnetic and electric field cross at right angles. The result is large etch rate non-uniformities. To compensate for this effect, either the magnets or

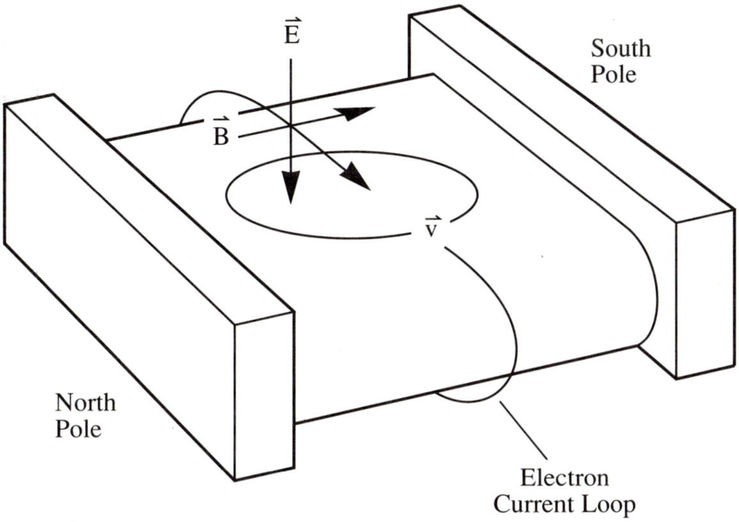

Figure 5.17. Planar magnetron reactor with parallel permanent magnets. Typical conditions: p = 0.5 Pa; P = 5 kW m^{-2}; B = 0.01 T.

the substrates must be moved continuously during etching [159]. An alternative geometry is depicted in figure 5.17. Here, the magnetic north-south direction is parallel, rather than perpendicular to the electrode surface, and a very uniform magnetic cross field between the electrodes is achieved [160]. A slightly different solution is to mount electromagnetic coils around the reactor side wall, as depicted in figure 5.18 after reference [161]. A uniform field is obtained in this case through careful design of the coils and modulation of the solenoid current to induce a rotating magnetic field.

In any magnetically enhanced system, the electron confinement can be improved by adding magnetic mirrors. As can be seen from equation 5.24, the radius of the electron orbit decreases with increasing magnetic field strength. This means that when an electron approaches an area where the magnetic field strength has a positive gradient, it will spiral with decreasing radius in the direction of increasing field. Finally, it will reflected back, spiraling out again with increasing radius; hence, the term mirroring. The coils in the system from figure 5.18 are magnetic mirrors; the magnetic field strength increases from the center to the circumference of the reactor.

The circular motion of electrons in the presence of a magnetic field can be described in terms of an angular frequency, called the electron cyclotron frequency ω_{ec}. The magnitude of this parameter can be deduced from the electron velocity v_e and the radius r of the electron helix as described in equation 5.24,

$$\omega_{ec} = \frac{v_e}{r} \tag{5.25}$$

For the parameter values assumed above, it is found that the cyclotron frequency is about 10^8 s^{-1}, which is in the microwave (MW) range. If the rf excitation frequency is in the same range, resonance effects can occur, causing an additional increase in the rf energy transfer efficiency. From equations 5.25 and 5.24, the system is found to be resonant for

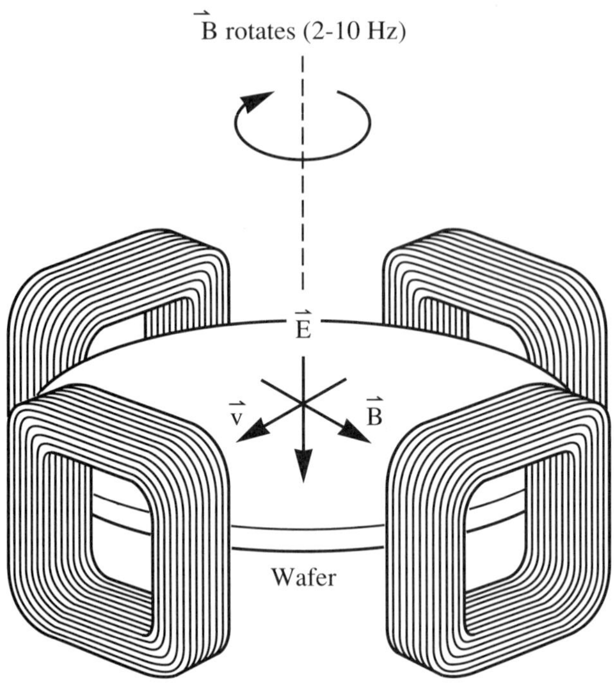

Figure 5.18. Magnetron system with electromagnetic confinement.

$$\omega = \omega_{ec} = \frac{e\,B}{m_e} \tag{5.26}$$

where ω is the angular frequency of the rf excitation. Most microwave systems operate at the industrial frequency of 2.45 GHz; it appears, therefore, that the magnetic field will have to be at 0.0875 T in order to achieve resonance. Reactors based on this principle are called electron cyclotron resonance (ECR) etchers [162].

A conventional ECR reactor is depicted in figure 5.19. The heart of the reactor is a cavity designed to match the exciting wave precisely. Microwave energy is transmitted from the generator to the system through circular or rectangular waveguides. The energy is then coupled into the discharge through a material which is both high-vacuum compatible and transparent for microwaves, such as quartz glass and fused alumina. The transmitter interface can be either a flat window or a pipe with a diameter sufficiently large to match the cavity. Since no powered electrode is present, pollution by electrode material cannot occur. Solenoid coils surround the cavity. The resulting magnetic field is oriented axially; the field strength is adjusted such, that the magnitude is greatest at the microwave input end, and decreases toward the open end of the cavity. Through this divergent magnetic field, the electrons generated in the cavity are pushed out of the discharge zone and towards the substrate. In steady state, ions are attracted by this net flux of negative charge, and will be accelerated out together with the electrons. The substrate is thus subjected to a mild neutralized ion beam with a typical energy of several tens of eV. Note, that the energy will always be larger than the floating potential in the absence of magnetic fields. The exact beam energy can be varied by adjusting the current through the solenoids [163,164].

The divergent-field ion energy will not always be high enough to achieve etch directionality, and an auxiliary rf bias voltage on the substrate holder is needed [165]. This situation

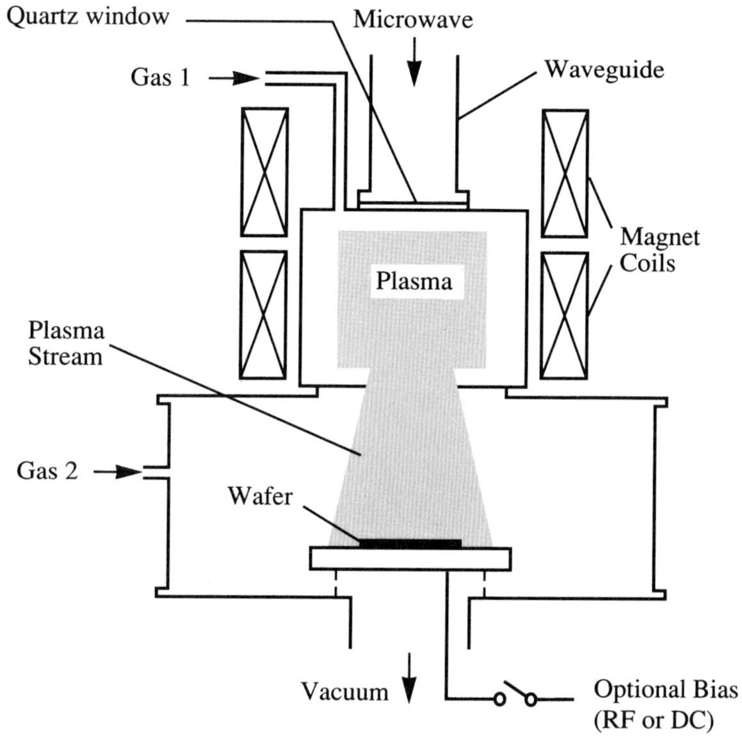

Figure 5.19. Divergent beam ECR reactor with optional substrate bias.

is depicted as a switch in figure 5.19. It is interesting to notice, that the set-up with a dc bias potential is quite similar to a regular ion-beam etcher; only the grid between source (cavity) and chamber is missing. If the bias voltage is sufficiently high, an auxiliary discharge can start at the substrate electrode; in that case, the difference between ECR and regular RIE starts to vanish. An obvious limitation of the ECR reactor described above is the fact that use is made of a beam originating from a source of which the dimension is limited by the wavelength of the excitation frequency, in this case 10 cm. This restricts the effective substrate area to be processed to about 15 cm, although some compensation is possible by mounting accelerating grids at the exit of the discharge zone.

A less common system is the multipole reactor, of which an example is shown in figure 5.20. The main difference between this reactor and the one shown in figure 5.19 is the orientation of the magnetic field, being axial rather than radial. The energy dissipation is now concentrated in lobes connecting the alternating confinement magnets. In this specific set-up, the microwave energy is coupled into the discharge by means of conductive antennas positioned a few millimeters from the wall of the reactor chamber, and directly in front of the magnets [166]. A multitude of magnets, usually of the permanent type, is mounted around the circumference of the discharge zone in order to obtain optimum magnetic mirroring. This implies that the dimension of the discharge chamber is not limited to the wavelength of excitation; very large volumes can be covered. The substrate to be etched is placed in or directly below the center of the system, where a secondary plasma is formed by diffusion of charged particles generated in the lobes. To reduce pollution of the substrate, the antennas must be made of materials with a low sputter yield, as they will be heavily attacked. There is no divergent magnetic field in the direction of the substrate. The energy of ions impinging

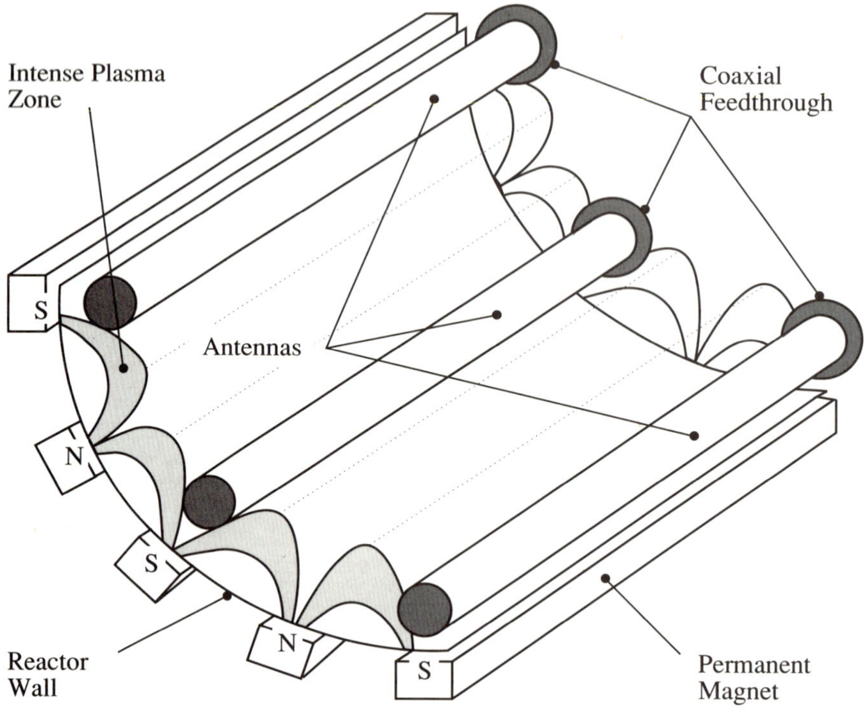

Figure 5.20. Section of a multipole microwave reactor with internal antennas.

on the substrate is, therefore, determined by the true floating potential only, and some rf or dc biasing is usually required to obtain directional etching.

With the above ECR reactors, charge densities as high as 10^{17} m^{-3} have been reported [167]. These reactors usually operate at pressures of about 0.1 Pa, implying that the degree of ionization can be higher than 1%. Consequently, many ECR plasmas are truly "hot" plasmas, and can differ considerably from the family of cold glow discharge plasmas to which all other dry etching systems belong (see chapter 2). This is a factor which should not be neglected when interpreting ECR process data.

Chapter 6

ETCH PROCESSES

6.1 Introduction

Dry etching was defined in chapter 1 as the removal of material by interaction with glow discharges. Of all the species produced in the discharge, two classes have special importance for etching.

1. Reactive neutrals, hitting the surface at random, without any impact directionality. These neutrals can serve two purposes,
   - chemical reaction to form volatile compounds, resulting in transportation of surface material to the system exhaust;
   - produce surface coatings by polymerization, either directly, from the source gas, or indirectly, from the product formed in the reaction with the substrate.
2. Positive ions, hitting the wafers perpendicularly as a result of the dark space electric field. Accelerated ions can:
   - remove material by physical sputtering;
   - locally enhance the surface reaction with reactive neutrals;
   - suppress surface polymerization.

This simple classification is valid only as a first approximation. Complications arise in high-pressure systems, especially those of the single-wafer type. At pressures of 50 Pa and up, charge transfer ionization in the dark space becomes a dominant process. Because this implies that electrons from non-directional neutrals are being transferred to directional ions, a flux of highly directional neutrals is created. If this happens close to the wafer surface, the distance to be covered after transfer can be short enough to prevent randomization, and a directional bombardment of both ions and neutrals takes place. By the same effect, non-perpendicular ion impact is possible [168]. Other complicating mechanisms are present in low-pressure systems. Examples are reflection of ions against pattern sidewalls, and deflection of incoming ions by surface charging. In the latter case, electrical potential differences are developed between masking layer and substrate. Finally, there is a change in ion etch yield when going from a horizontal to a vertical surface structure. If the flux density is counted only, a cosine etch behavior is expected. In practice, the yield function is often more complicated, see figure 6.1 after reference [169].

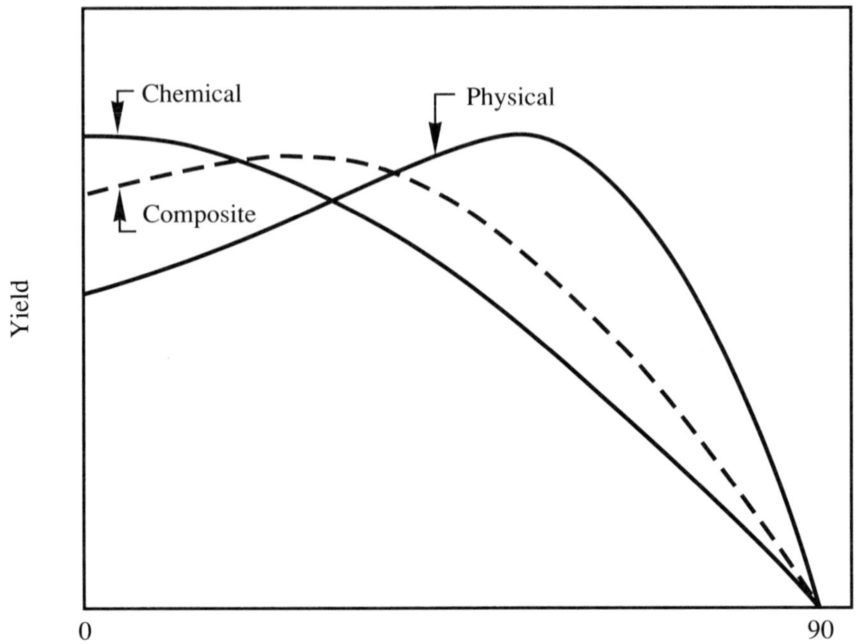

Figure 6.1. Typical curves for angle-dependent yield.

6.2 Requirements

The three most important parameters in dry etching processes are etch rate, selectivity and anisotropy. For production processes, non-uniformity must be added as a fourth parameter. The terms selectivity and anisotropy need some explanation. Selectivity is defined as the ratio of the etch rate perpendicular to the surface for materials not to be removed (such as a masking or underlying layer), relative to the material which is to be etched. The term anisotropy is also not self-evident. It is common practice to use a definition in terms of etch directionality, that is, the ratio of lateral (horizontal) and perpendicular (vertical) etch rate. Thus, a perfectly isotropic process has an anisotropy ratio of 1, while a perfectly anisotropic process has a ratio of 0. This description adequately covers undercutting, which is the effect of removing material from underneath the masking layer.

An anisotropy ratio of 0 does, however, not mean that the resulting etch pattern has vertical sidewalls, contrary to what is often implied in the literature on etch processes. This is exemplified in figure 6.2. The top diagram in this figure is a cross section drawn for a system having an erodible etch mask with a non-vertical side profile. The latter could be a photoresist exposed with low-contrast lithography or subjected to extensive baking. In such a case, even if etching occurs at right angles to the surface only, the foot of the mask moves sidewards during processing, since it is being eroded. The result is a pattern having a bottom width equal to the original mask spacing, but with a wider top width. The second diagram shows the effect of having a fully anisotropic process for the material to be etched combined with isotropic mask erosion. Although the mechanism is different, the result is the same as in the first example. The last diagram illustrates a situation that can occur when surface polymerization is dominant and mask erosion in negligible. This condition, with significant isotropic deposition and anisotropic etching at the same time, leads to patterns where the top

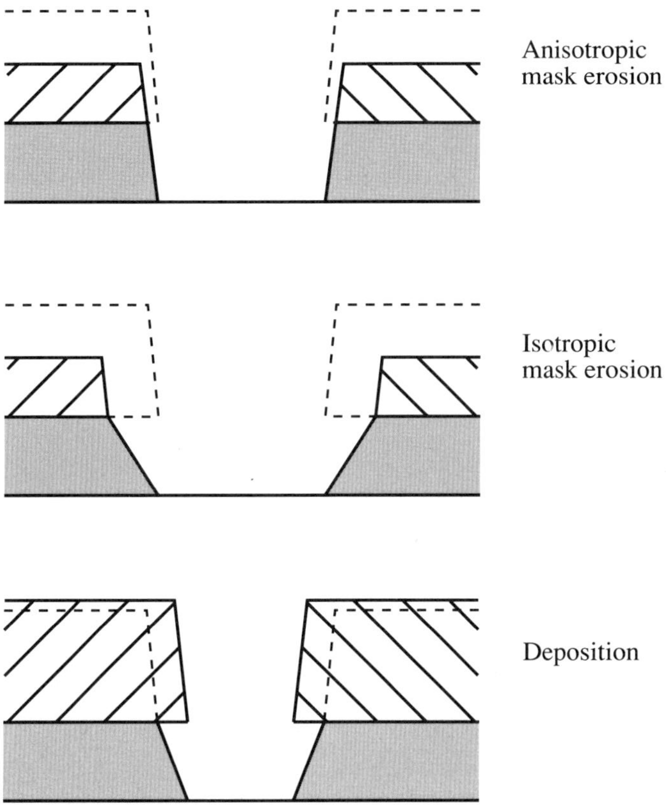

Figure 6.2. Possible mechanisms for non-vertical pattern formation with anisotropic etching. See text for explanation.

width is equal to the mask spacing and the bottom width is less. In extreme cases this will result in V-shaped grooves after which the pattern evolution stops.

Developing a dry etching process basically means optimizing the key parameters introduced above. Improving in one direction means more often than not deteriorating in another field, as Murphy's law is certainly applicable to etching plasmas. In extreme cases, all internal reactor surfaces, including the wafers, are being covered with a heavy polymer deposit, or the mask is etched many times faster than the intended layer. Suitable etch processes should, of course, be sufficiently far away from such situations. Factors influencing the selectivity in a given combination of materials are the following:

- Preferential reaction;
- Preferential polymerization;
- Preferential sputtering.

All of these factors require a dedicated choice of the reaction gas mixture. It is, for example, not always possible to find a gas composition which reacts preferentially with the material to be removed. In such a case, it may instead be necessary to protect the other materials by promoting local polymerization, or by forming intermediate surface products which can be selectively sputtered. Low pressure and a low excitation frequency usually emphasize the importance of ion sputtering, which is not very selective. At high pressure and high frequency, large concentrations of reactive neutrals are formed. Chemical action, then, becomes more important, and higher selectivities are achievable. Other generalizations are not easily made for this aspect of dry etching.

In anisotropy, the same factors as introduced above play a role, but the impact of external settings and plasma parameters is more straightforward. A high energy and a high flux density of ions hitting the surface will help in reaching high anisotropy; the same holds for a low concentration of reactive neutrals. This means low pressure and low frequency, and also a high rf power density. It will be clear that this observation hints at a conflict embedded in all dry etch processes, namely that improving anisotropy is in contradiction with the requirement of high selectivity, at least in this simple model. That good processes can be created at all, has to do with the fact that anisotropy can also be enhanced by preferential pattern sidewall passivation – growth of protective polymer films on surfaces which are not hit by ions. Film formation on exposed areas is inhibited, as the polymer is continuously removed by ion impact. This is the dominant mechanism behind almost all practical etching systems.

Creating a process with high etch rates relies really on applying brute force, that is, high rf power and high pressure. Etch rates are limited in practice by the requirements for selectivity and anisotropy, but also by heat dissipation, as the photoresist will degrade, and certain devices can be damaged by high temperature. The fourth process parameter, the non-uniformity, is the least understood of all. Influences have been ascribed to gas distribution (for instance the location of gas inlets and pump exhausts), rf voltage and current flows inside the reactor, materials used for walls and electrodes, and even the shape and density of the pattern on the wafers.

Processes based mostly on unassisted reaction between wafer and excited neutrals, like the isotropic etching in "barrel" and other afterglow systems, have a major drawback in the "loading" effect. With this term, a change in etch rate with increasing exposed wafer area is meant, usually a decrease. The explanation for this effect is the reduction in reagent concentration associated with the higher degree of source gas conversion. At the low pressures in cold glow discharges, the afterglow will become rapidly exhausted. This leads in turn to etch rate non-uniformities across individual wafers, as well as deteriorated uniformity over the complete batch. In principle, it is possible to explain most loading effects by using principles from conventional chemical engineering, such as bulk diffusion-limited reagent supply.

In the next sections, the most important dry etch processes for VLSI are listed by material. Each group has its specific chemistries and mechanisms; the discussion starts with what is known best, and ascends from there to more recent developments. Unless specified otherwise, all processes involve direct contact between wafer and discharge. The chapter is finalized with a section on damage and residue formation and removal.

6.3 Silicon oxide and silicon nitride

The major application of dry etching of silicon oxide is the anisotropic patterning of contact holes to silicon, and of via holes between various levels of interconnect in multilevel metallization. Isotropic patterning of silicon nitride layers on silicon oxide is part of the well-known LOCOS procedure for lateral isolation in integrated circuit processes. It is one of the oldest industrialized dry etching processes; the first process is, by the way, photoresist stripping, see the section on polymer processing. Less frequently, combinations of anisotropic processing of oxide and nitride are used to create well-defined blocks of dielectric material for lateral device isolation. These are used as a follow-up of LOCOS, and for sidewall spacer formation in small MOS and Bipolar transistors; for example, to achieve graded implantation profiles and other self-aligned features. A slightly different, but equally important area of oxide etching is topography reduction in high-density interconnect structures by etch-back. This involves a sacrificial layer, usually an organic polymer. This, too, is discussed in a later section.

The vast majority of dry etching processes for silicon oxide are based on fluorocarbon (Freon) discharges; the chemical endproducts of the reaction are silicon tetrafluoride and carbon monoxide or dioxide. Patterning is done mostly with hard-baked photoresist. Crucial factor in all processes, apart from the evident anisotropy requirements, is selectivity to silicon. The reaction of silicon oxide and silicon with Freon discharges is probably the best documented system in the short history of dry etching. Much of what will be explained here in terms of fundamentals, will also turn out to be useful in other etch processes [170,171,172].

Fluorocarbon molecules in the ground state are inert towards both silicon and silicon oxide surfaces. Sticking coefficients are in the 10^{-6}–10^{-7} range; even on ion-bombarded surfaces no reaction occurs. Etching only starts after the reaction gas has been fragmented and ionized, for example by striking a glow discharge. Particles relevant to the reaction are charged and uncharged C_nF_m fragments (n, m=0,1,...). The simplest fluorocarbon is CF_4 (Freon-14). The fragmentation and subsequent etching of SiO_2 and Si in a CF_4 plasma can be described according to the following scheme,

$$CF_4 \rightarrow 2F + CF_2 \qquad (6.1)$$

$$SiO_2 + 4F \rightarrow SiF_4 + 2O \qquad (6.2)$$

$$Si + 4F \rightarrow SiF_4 \qquad (6.3)$$

$$SiO_2 + 2CF_2 \rightarrow SiF_4 + 2CO \qquad (6.4)$$

$$Si + 2CF_2 \rightarrow SiF_4 + 2C \qquad (6.5)$$

For simplicity, atomic fluorine and difluorocarbene have been assumed to be the only intermediates, with carbon monoxide and silicon tetrafluoride as the volatile endproducts. The relative importance of the four idealized etch reactions will be discussed first. Next, it will be shown how equilibria can be shifted to obtain the required behavior for a given system.

Atomic fluorine, the reagent in equations 6.2 and 6.3, is a long-lived species in the pressure regime of glow discharges. Even in the absence of ion bombardment, it reacts rapidly with silicon; the reaction rate with SiO_2 is 1-2 orders of magnitude lower. Ion bombardment enhances the reaction rate between F and Si only slightly. This is the reason why silicon etches isotropically in most plasma systems based on fluorine. There are some exceptions to this rule; these will be described in the next section. Adding an oxidizing agent such as O_2 to a CF_4 glow discharge increases the amount of free F through the formation of oxyfluorides from fluorocarbons, e.g.,

$$CF_4 + O \rightarrow COF_2 + 2F \qquad (6.6)$$

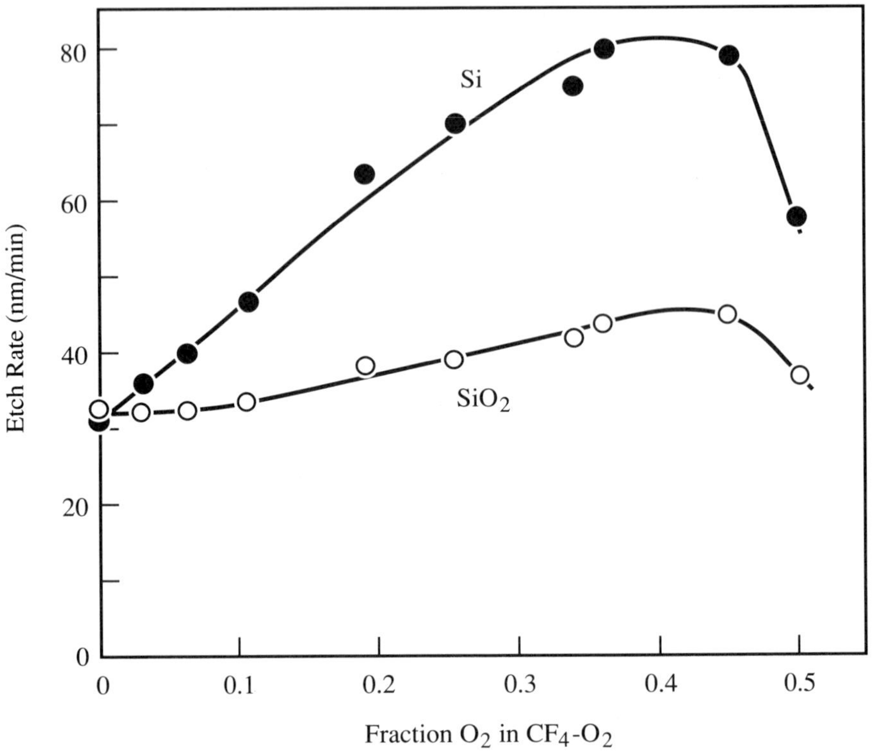

Figure 6.3. Etch rate of Si and SiO_2 in a CF_4–O_2 planar discharge.

Figure 6.3 after van Roosmalen [170], shows that the silicon etch rate in a CF_4 discharge is increased by such an addition, in agreement with equation 6.3. At higher oxygen concentrations, the etch rate drops again. This can be ascribed in part to a lowering in free F because of gas-phase recombination,

$$O_2 + F \rightarrow FO_2 \quad (6.7)$$

$$FO_2 + F \rightarrow F_2 + O_2 \quad (6.8)$$

However, the largest F atom density in CF_4–O_2 occurs at a different – and higher – oxygen concentration than the maximum Si etch rate. The explanation for this behavior lies in surface oxidation competing with the fluorine-based etch reaction.

Free fluorine radicals in a glow discharge can be scavenged by adding hydrogen,

$$H_2 \rightarrow 2H \quad (6.9)$$

$$H + F \rightarrow HF \quad (6.10)$$

The resulting hydrogen fluoride can be considered as an effective thermodynamic sink, owing to its high bond strength; the HF molecule dissociates at a much lower rate than its parents

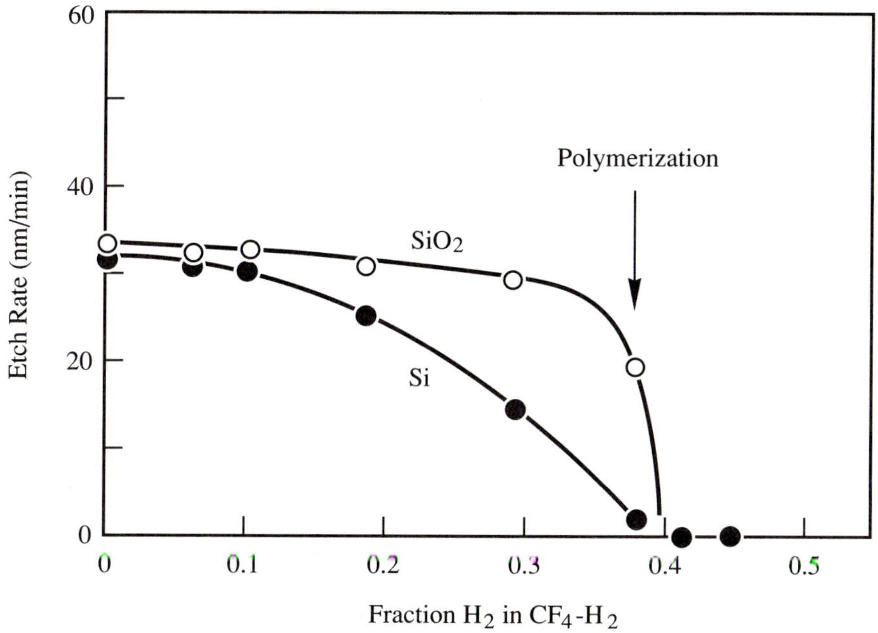

Figure 6.4. Etch rate of Si and SiO_2 in CF_4–H_2. Same conditions as in the previous figure.

H_2 and CF_4. Figure 6.4 after van Roosmalen [170], demonstrates that adding H_2 to a CF_4 discharge lowers the Si etch rate. It appears, therefore, that etching of silicon in fluorocarbon plasmas can be described conveniently by equation 6.3. That is, through interaction with free F [173,174].

Meanwhile, the etch rate of silicon oxide is virtually unaffected by moderate additions of either O_2 or H_2. From the same argument as given above for silicon, this means that free F does not contribute significantly to the etching of SiO_2 in the presence of ion bombardment. Instead, carbon-containing compounds have to be taken into account, see equation 6.4 and 6.5. Fluorocarbon fragments are strongly chemisorbed on both bare silicon and oxide surfaces, but react only slowly without ion bombardment. In the presence of enough free fluorine, the adsorbate is steadily removed by back-formation of CF_4. Hence, with a CF_4–O_2 plasma it is possible to etch silicon and, at a low rate, silicon oxide, even if ions are not allowed to reach the surface. The latter is the case in the "barrel" reactor with internal shielding, see the previous chapter. Making the discharge fluorine deficient by, for example, adding hydrogen, suppresses this recombination, and a polymer film is formed that will eventually stop the etch reactions,

$$n\ CF_2 \rightarrow [CF_2]_n \qquad (6.11)$$

Whether surface inhibition occurs for a given gas phase or not depends on two factors: the chemical composition of the surface and the intensity of the local ion bombardment. Pattern sidewalls are not hit by ions, and lateral etching of both silicon and silicon oxide stops at relatively low hydrogen concentrations. In CF_4, a few percent is typically sufficient to obtain a perfectly anisotropic process. At a sufficiently high flux density and energy level, impinging ions will damage the surface layers, and the fluorocarbon adsorbate can react with

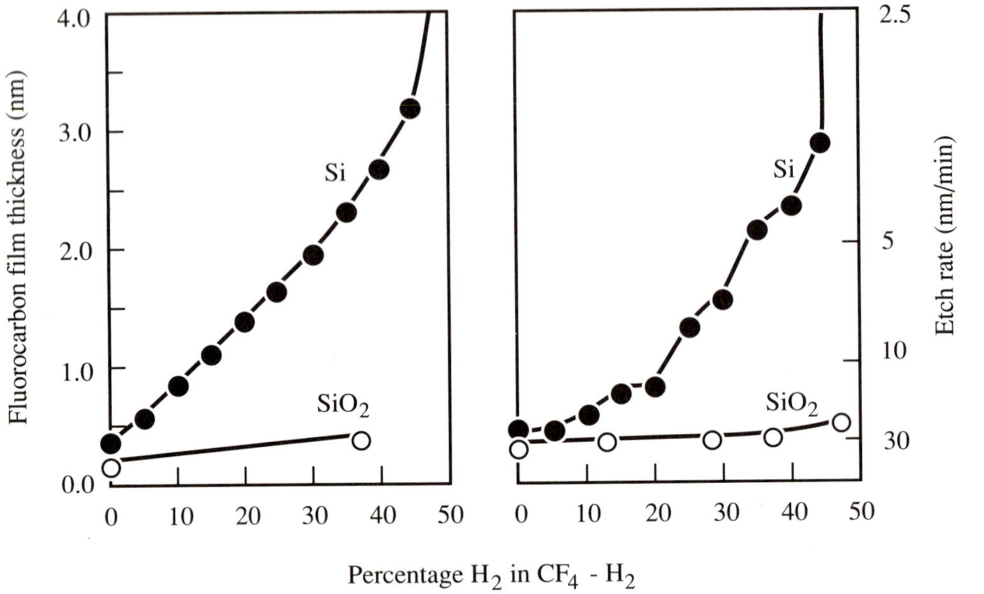

Figure 6.5. Fluorocarbon film thickness (left) and inverse etch rate (right) for Si and SiO_2 as a function of hydrogen concentration in a CF_4–H_2 discharge.

the substrate. If the substrate is silicon oxide, oxygen from the lattice is available to form volatile carbon oxides with carbon-containing species, see equation 6.4. At the same time, F is released to remove the remaining Si atoms. This explanation is supported by the fact that there exists a positive relation between silicon oxide etch rate and gas-phase difluorocarbene concentration [175]. A similar mechanism is not available for silicon, and a carbon-rich film will be formed there, see equation 6.5. Carbon is a material with a low sputter yield, which suppresses further Si removal. This is illustrated in figure 6.5 after Oehrlein et al. [172].

The silicon etch mechanism explained above is also applicable to organic layers, such as photoresist. Silicon nitride takes an intermediate position between silicon and silicon oxide. Nitride reacts faster with atomic F than the oxide because of the intrinsically weaker Si-N bond, although not as fast as silicon. It also appears to react with CF_2 species, as it is etched in CF_4–H_2 at roughly the same rates as silicon oxide. By analogy with equation 6.2, the surface carbon may be removed as CN.

The ion-induced reactions in CF_4–H_2 glow discharges require a minimum bombardment energy of 100–200 eV. The exact nature of the ions appears to be of lesser importance. Both the dependency on the angle of incidence and the yield per ion indicate that the surface reaction is of a chemical nature, with the ions just serving as reaction initiators [145,169].

Figure 6.4 shows that above a certain H_2 content, 40% in this example (arrow), all etch rates suddenly drop to zero and extensive polymerization occurs on all surfaces exposed to the discharge. Below this concentration, a delicate balance exists between desorption of the CF_2 monomers, recombination (with F), and reaction (with Si and O) on one side, and polymerization (to $[CF_2]_n$) on the other side. The exact location of this so-called polymerization point depends critically on pressure, residence time, dc bias, substrate composition, and reactor conditioning. Unfortunately, it can also be seen that the most favorable setting for etching silicon oxide with a high selectivity relative to silicon is situated directly before this point [176,177].

Coburn and Kay [128,178] recognized for the first time, that fluorocarbon systems for the selective etching of silicon oxide on silicon can be described completely with just one parameter, namely the F/C ratio. This ratio is calculated from the theoretical number of F and C atoms in the gas mixture after making corrections for the formation of CO (with O) and HF (with H), see also equations 6.6 and 6.10. The onset of polymerization can be approximated by the point where F/C = 2, in other words, when the dominant reagent in the discharge is CF_2. A large variety of etch gas mixtures has been compared and found to behave quite similarly for equal F/C ratio. Examples of mixtures with F/C = 2 are CF_4–50%H_2, CF_4–33%CH_4, C_2F_4, and CHF_3. Not all mixtures contain hydrogen; apparently, it is not an essential constituent. Apart from the direct reaction gases, also the materials from which the reactor chamber and especially the rf-powered electrode are made can affect the working point. A carbon electrode consumes F and lowers F/C. Quartz is etched, which releases oxygen-containing compounds in the discharge and increases F/C. Teflon releases gaseous unsaturated fluorocarbons, leading to a lowering of F/C.

Practical processes in use nowadays for the selective patterning of silicon oxide are based almost exclusively on CHF_3 (Freon-23), usually in combination with a noble gas, such as Ar and He, as diluent and a more fluorine-rich compound, such as C_2F_6 and CF_4, as moderator. Less frequently, O_2, N_2, and NF_3 are added to reduce polymer formation in the reactor or to control the shape of etched patterns by lateral erosion. See the section on organic processing for details on the latter. The silicon oxide etch rate with cathode-coupled RIE is about 50 nm/min; for conventional single-wafer etching 500 nm/min is usual. In either case, a selectivity to silicon between 10 and 20 is readily achievable.

Once the range for reasonable results has been determined, finding the optimum still requires a considerable effort, as best etching relies on working close to complete inhibition by polymerization. Optimization involves a large amount of parameters; statistical methods such as response-surface methodology are often required to achieve the best possible settings. This is important especially for single-wafer reactors with their severe constraints on high and highly uniform etch rate. Figure 6.6 after Riley et al. [179], illustrates the complexity involved in optimizing these processes. Shown is the effect of pressure, a parameter known to have a large impact on etch rates and selectivities in the high-pressure single-wafer systems commonly used for oxide etching. The same holds for electrode spacing. Single-wafer reactors for oxide etching operate mostly with the top electrode at less than 1 cm from the substrate plane, typically 4-7 mm. This is very close to the limit below which the discharge extinguishes, but also the point where the etch rates are at their maximum. To understand the effect of pressure and electrode spacing, the product of the two has to be considered, see figure 6.7 after Mathad [180]. The limiting value of pressure times spacing, about 30 Pa cm in this example, roughly indicates the point where the discharge is compressed into two dark spaces and no plasma (see the section in chapter 3 on the dc tube discharge). Under these conditions the electrons to sustain the discharge will be generated mostly by secondary emission from the electrode surfaces; the wall potential will be high. These electrons bounce back and forth across the two dark spaces, undergoing large accelerating voltages on the way. Hence, the electron distribution contains an unusually large number of highly energetic electrons, and the degree of gas dissociation, ionization and the etch rate will also be high [181].

In the range of conditions suitable for selective silicon oxide etching, highly selective etching of silicon nitride relative to silicon oxide and silicon cannot be achieved. This is

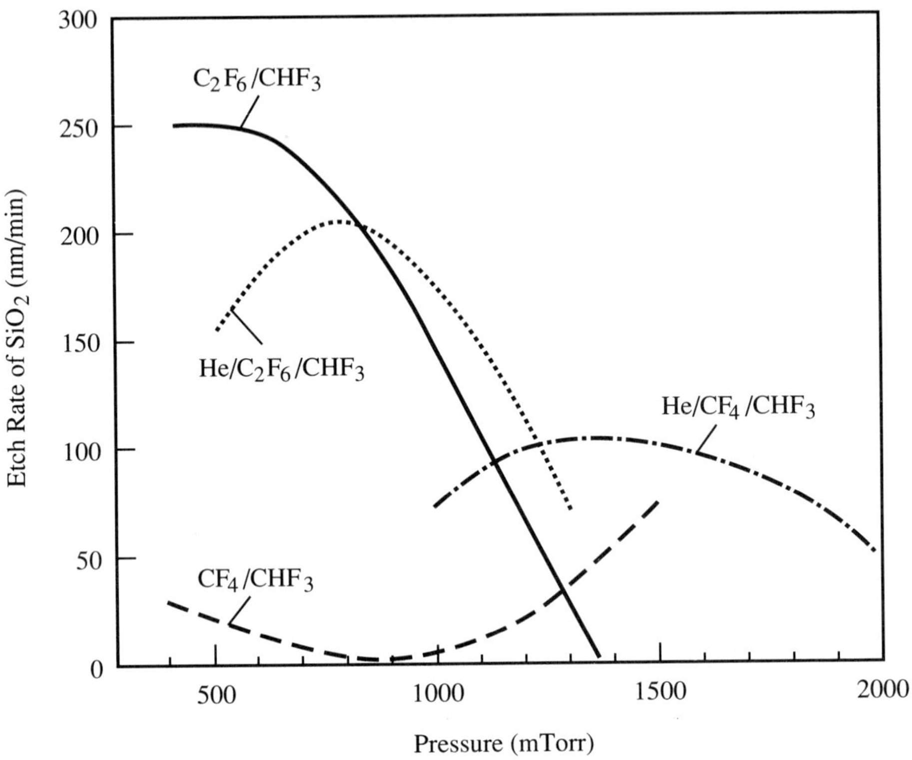

Figure 6.6. Silicon oxide etch rate trends for various fluorocarbon systems in a single-wafer reactor.

caused by the intermediate position of the nitride in terms of silicon bond strength and electronegativity between the oxide and the pure silicon. If the object is just to etch with high selectivity to oxide, and anisotropy is not a concern, for example when etching the thin nitride prior to the high-temperature surface oxidation in the LOCOS procedure, the reagent can be limited to free fluorine. The latter is known to have a much higher affinity to nitride than to oxide. The original process to achieve this is by using CF_4–O_2 in a barrel reactor. Unfortunately, the reaction between the fluorocarbon fragments and oxygen as per equation 6.6 is incomplete, and atomic and excited molecular oxygen will still be present in the discharge, even at low O_2 concentrations. This causes partial oxidation of the silicon nitride surface during etching, which deteriorates both the etch rate and the selectivity. Adding heavy halogen compounds to the discharge, such as CF_2Cl_2 and especially CF_3Br, removes the residual oxygen by formation of stable halogen oxides,

$$CF_3Br \rightarrow CF_3 + Br \tag{6.12}$$

$$O + Br \rightarrow BrO \tag{6.13}$$

A less desirable effect of these additives is the reduction of the overall etch rate; as the above reaction occurs with free fluorine too. The exact amount of additive has to be balanced against the acceptable rate loss. In adjusting the process, it must be taken into account that oxygen comes not only from the gas supply, but also from quartz reactor walls, photoresist, and water adsorbed on wafers and wafer carriers. A very high intrinsic selectivity can be reached by

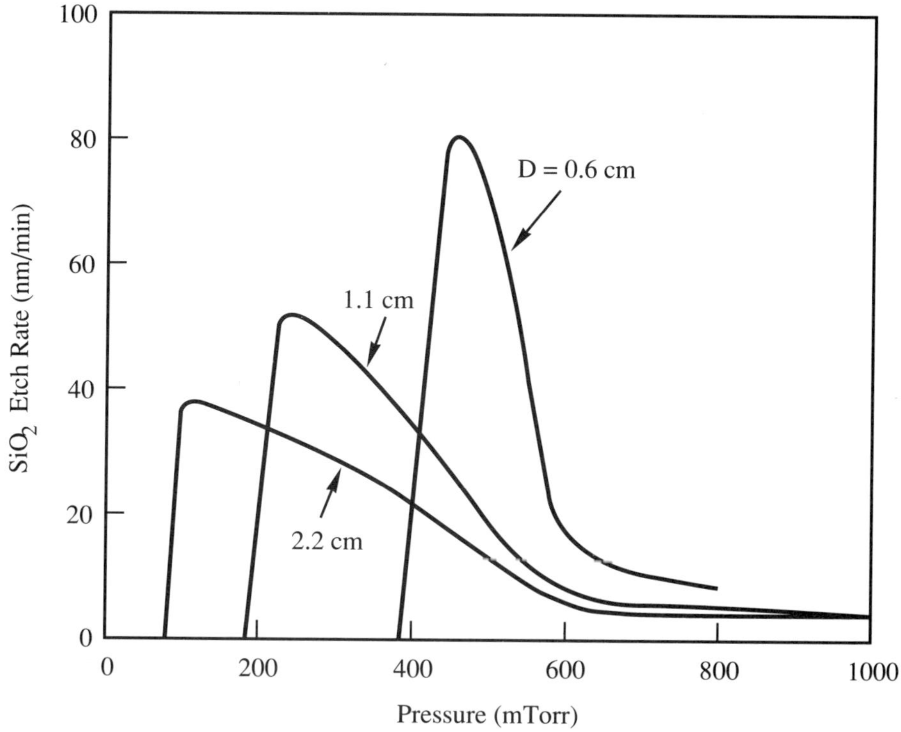

Figure 6.7. Effect of electrode spacing on silicon oxide etch rate in CF_4.

using gases which do not contain oxygen and still provide a sufficient supply of free fluorine, such as SF_6 and NF_3. Gas price, the fact that these gases are hazardous, and the ever-present sensitivity to external conditions, such as humidity, have prevented them from replacing the traditional processes based on CF_4–O_2 [182].

Selective and anisotropic etching of silicon nitride relative to oxide and silicon at the same time is seldom required, but not impossible as such. If the hydrogen concentration in a CF_4–H_2 discharge is increased to a value well above the polymerization point, etching of all relevant materials starts again. For the parameter settings in the example of figure 6.4, the second polymerization point lies around 70% H_2. Near that point, conditions can be found where attack of both silicon and silicon oxide is suppressed, while silicon nitride continues etching. The same behavior is found for hydrogen-rich fluorocarbon gases such as CH_2F_2 and CH_3F. As etching can be continued up to almost pure hydrogen, the reaction product in this range might be SiH_4 rather than SiF_4, but so far little is known in detail [183,184].

Although this was not explicitly mentioned, the above discussion implied 13.56 MHz excitation (HF), in either a planar or barrel reactor. In principle, the conclusions are also valid for the 100–400 kHz range (LF), although use must be made of the rule of thumb, that changing frequency and pressure has the same effect. Thus, a low-frequency, high-pressure process is roughly equivalent to high-frequency, low pressure conditions. The extrapolation fails for microwave or otherwise enhanced dissociation plasmas, such as magnetron (MIE) and hollow cathode systems. The extremely high degree of fragmentation in the latter systems is the source of this deviating behavior. With CF_4, the etch rate of Si is already high in the pure gas because of the high free fluorine concentration. Other than in the 13.56-MHz case, addition of oxygen decreases the Si etch rate. No fluorocarbon surface adsorbates are formed, and carbon is deposited on the oxide. Therefore, the etch rate of silicon oxide is relatively low, and addition of hydrogen does not improve the selectivity [156,185].

6.4 Mono and polysilicon

Next in importance to silicon oxide dry etching comes the patterning of silicon itself. Two fields of application can be distinguished. The first is the etching of grooves ("trenches") with a high depth to width ratio in monocrystalline silicon. These grooves are used for circuit isolation as an alternative or addition to LOCOS, and, more and more frequently, as vertical capacitors in high-density Dynamic Random Access Memory IC's. The literature on monosilicon dry etching is almost as abundant as that on oxide patterning. There are relations, but also large dissimilarities between these two classes of plasma systems. One of the more evident differences is the fact that conventional silicon plasma processes are based on chlorine rather than on fluorine. The second, much wider field, is the controlled etching of polycrystalline silicon films for gates and high-temperature interconnect in MOS technology. In recent years, poly-Si tends to become replaced by refractory metals and metal silicides for the same applications. A specific section on these materials can be found later in this chapter.

Grooves in silicon for isolation and vertical capacitors need to be at least a few μm deep, with steep sidewalls and clean surfaces. To facilitate subsequent filling and planarization steps, there are often additional process requirements, such as smooth bottom corners (U-grooves). During the etch process, the top surface should remain clean and free of damage, as transistors are often made after this stage. For these reasons, the etch mask is usually made by first dry etching a thermally grown top oxide and stripping the photoresist prior to starting the actual groove etch process. An additional advantage of the oxide mask is, that it enables the use of the high rf power density required to keep the process time within reasonable limits [186,187,188,189].

In the discussion from the previous section, the purely fluorine-based plasmas appeared to be less suitable for the anisotropic patterning of silicon. Instead, many mono-silicon processes involve mixtures containing heavier halogen compounds, usually chlorine. The silicon tetrachloride formed in the surface reaction is volatile, but less so than the fluoride; at room temperature, it is a liquid, whereas silicon tetrafluoride is a gas. Apart from the volatility aspect, an important difference between chlorine and fluorine plasmas is the much shorter lifetime of chlorine radicals under usual plasma conditions. Chlorine species must, therefore, be created near the wafer surface, which excludes the use of barrel or other afterglow reactors.

The etching of mono-silicon in a chlorine discharge can, to a first approximation, be described with the following reaction schemes,

$$Cl_2 \rightarrow 2Cl \tag{6.14}$$

$$Si + 2Cl \rightarrow SiCl_2 \tag{6.15}$$

$$SiCl_2 + 2Cl \rightarrow SiCl_4 \tag{6.16}$$

Atomic chlorine can be considered to be the dominant reagent in this case. Other etching mechanisms are probably less important, because of the lower abundance of the relevant reactive species. Conceivable examples are the reaction with chlorine ions and molecular

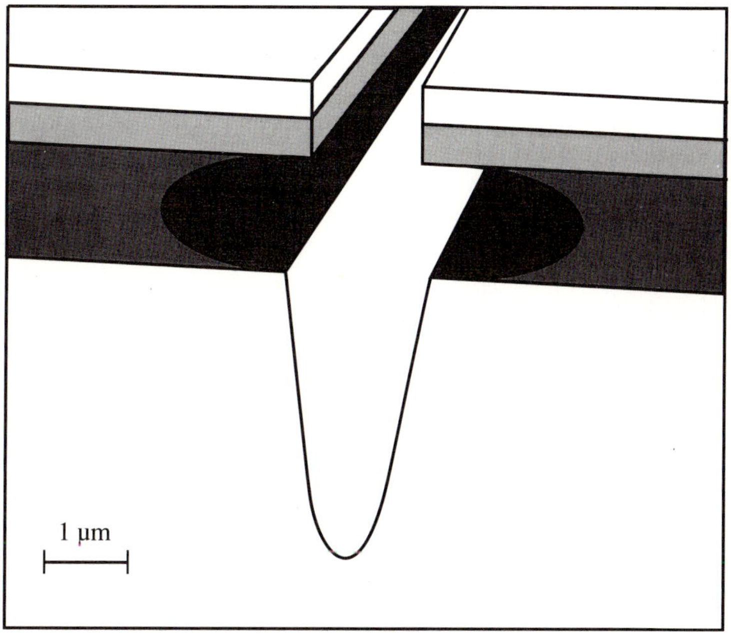

Figure 6.8. Undercut of an n^+ buried layer in $SiCl_4$–Cl_2.

chlorine in the presence of ion bombardment. Silicon dichloride appears to be the primary species formed upon reaction.

Contrary to the fluorine systems, etch anisotropy in trench etching is not normally created by formation of organic deposits. Instead, pattern sidewall passivation, and hence profile control, is achieved by plasma-enhanced chemical vapor deposition of chlorinated amorphous silicon from silicon subchlorides,

$$n\ SiCl_2 \rightarrow [SiCl_2]_n \tag{6.17}$$

The silicon chloride is always present as reaction product, but can also be supplied externally by adding silicon tetrachloride, or by exposing bare silicon wafers to the discharge during etching. The latter two actions can indeed affect the anisotropy in the etching of silicon with chlorine. In any practical process, the isotropic deposition needs to be balanced against the etching by atomic chlorine, the latter having both a non-directional (the radicals) and a directional component (ion bombardment). Evidently, obtaining reproducible profiles requires careful process tuning. Too low a deposition rate will cause lateral etching; too high deposition rates lead to ingrowth of the pattern sidewalls and limit the achievable groove depth, which is the same as the "overcutting" effect from figure 6.2, bottom.

Additional complications can come from the presence of heavily doped buried layers, as is the case in many practical Bipolar devices. In extreme cases, the etch process can be completely isotropic in the buried layer, and anisotropic elsewhere, as shown in figure 6.8 after Powell and Chambers [190]. The main reason for this is that the chemical reaction between silicon and halogen atoms (Cl and F) is a function of the dopant level. The highly doped surface is probably more susceptible to attack because of the charge dislocation induced by donor or acceptor sites. It should be noted that the vertical etch rate in chlorine plasmas

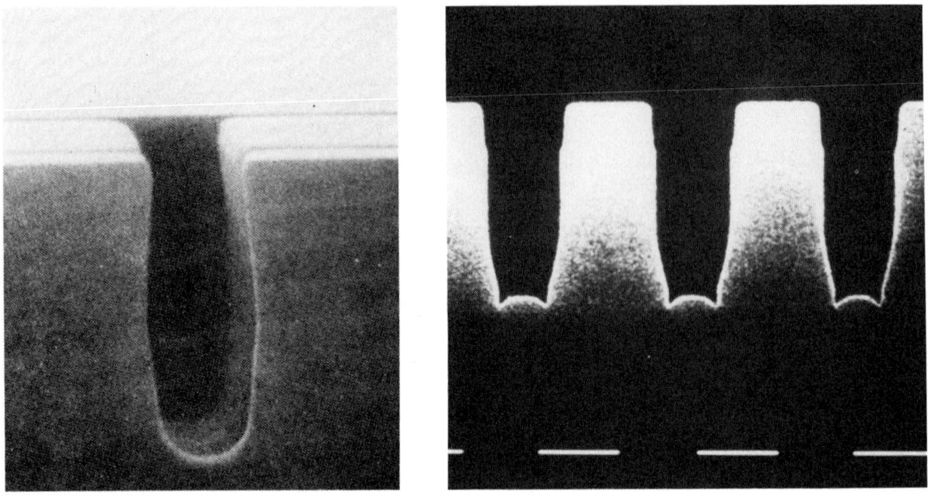

Figure 6.9. Cross section of a silicon trench etched in CCl_4–Cl_2 showing bowing (left) and crevice formation (right).

hardly depends on the presence of dopant, as it is determined largely by ion bombardment. With fluorine being chemically much more reactive, the ratio between ion-induced etch rates of doped and undoped material can be as large as 5 [191,192].

Other phenomena found in silicon trench etching are bottle-shaped profiles, also known as "bowing", and the formation of "crevices", narrow vertical slits in the bottom corners of the trench. This is illustrated in figure 6.9. The extent to which these effects are important, depends not only on the reactor conditions, but also on the pattern width and depth, as well as the composition of the masking layer. The profile artifacts cannot be described solely by a balance between a vertical and a random etch rate component. The most plausible model, which is supported by Monte Carlo simulations of ion trajectories, explains both effects from ion scattering. First, the ions can hit the surface off-normal because of collisions in the dark space region. As molecular chlorine has a large collision cross-section for electron attachment, large amounts of negative ions will be present. Hence, the discharge potential is relatively low, see chapter 3 and the electric field near the electrode surface will be insufficient to compensate for randomizing sheath processes, especially at higher pressures. Although silicon is not attacked by molecular chlorine as such, a beam of hot molecular chlorine etches silicon at high rates [11]. Fast neutrals from ion charge transfer processes in the dark space can, therefore, add to the lateral etch enhancement by scattered ions. The second, and probably most important effect, is direct reflection of ions against the sidewalls of the trench profile. The result is an increased bombardment intensity in sharp bottom corners (crevice formation), as well as a moderate flux of ions reflected from the top edge of the trench mask toward a region somewhat lower on the opposing wall. The latter induces a local disruption of the passivating layer and, hence, lateral etching (bowing) [168,193,194,195].

Chemically, the silicon etch anisotropy in chlorine plasmas can be influenced by the addition of hydrogen to the discharge. Hydrogen will lower the concentration of the reactive atomic chlorine by reaction to the relatively stable HCl, see equation 6.18. It also enhances the deposition reaction by producing more silicon subchlorides, see equation 6.19.

$$H + Cl \rightarrow HCl \tag{6.18}$$
$$SiCl_4 + 2H \rightarrow SiCl_2 + 2HCl \tag{6.19}$$

Carbon containing species, such as carbon tetrachloride, chloroform, and methane also reduce the lateral attack through additional surface film formation, as can be seen in equations 6.20 and 6.21. The latter two reactions are analogous with the CF_4 polymerization reactions in the previous section on oxide etching (see equations 6.1 and 6.11).

$$CCl_4 \rightarrow 2Cl + CCl_2 \tag{6.20}$$
$$n\,CCl_2 \rightarrow [CCl_2]_n \tag{6.21}$$

If the amount of carbon in the discharge is chosen too high, dark deposits are formed on all internal reactor surfaces. This material, a chlorinated carbon, is sometimes referred to as "glow polymer". Contrary to the situation in fluorocarbon discharges, this polymerization does not inhibit vertical etching at the wafer surface [70].

For a given gas mixture, the trench profile is a function of about any parameter that can shift the balance between deposition and etching. High pressure enhances deposition and will lead to V-shaped grooves. High rf power increases the electrical field in the dark space, resulting in better ion alignment and therefore steeper profiles. Low-frequency excitation has the advantage of combining high ion bombardment energy with a low dissociation rate, hence a low concentration of atomic chlorine. In the LF range, reproducible etching of high aspect ratio U-grooves is possible even in pure chlorine. Since about 1985, reactors have become commercially available which combine the advantages of low-frequency (high anisotropy) and high-frequency (high etch rate) for trench etching by using two independently powered electrodes and a common ground [190,152,196].

Finally, the wafer temperature is known to play an important role. Figure 6.10 after Carlile et al. [187], shows the transition from V-shape to U-shape trench etching in $CHCl_3$–N_2 by increasing the temperature. The precise trend appears to depend on the chemistry of the process; for Cl_2–CCl_4 low temperatures are needed to obtain vertical profiles [197].

Systems commonly used for trench etching are Cl_2–$SiCl_4$–H_2 and Cl_2–CCl_4–N_2. For environmental reasons, the carbon tetrachloride in the second mixture is gradually being replaced by chloroform, $CHCl_3$. Etch rates obtained are around 50 nm/min for conventional cathode-coupled RIE and up to 500 nm/min for single-wafer reactors. Aspect ratios of 1:10, corresponding with a sidewall angle of 85 degrees can be obtained routinely.

Apart from the chlorine discharges discussed so far, a number of trench etch systems have been developed that are based on some fluorine component, such as SF_6, NF_3, and CF_4. Because of the risk of undercutting built-in with the presence of the highly reactive atomic fluorine, these systems are more difficult to control. One big advantage is that fluorine prevents the attack of metal parts in a plasma reactor, parts which are often severely corroded by chlorine. Silicon hexafluoride produces $[SF_2]_n$ surface polymers similar to those formed by Freon discharges, but the deposition rates are seldom sufficient to prevent undercutting; for nitrogen fluoride, no deposition process is available at all. Therefore, in either case the addition of substantial amounts of a polymerizing agent is required for anisotropy control, usually CF_3Br or CF_5Cl_2. The carbon in these components provides polymerization while the

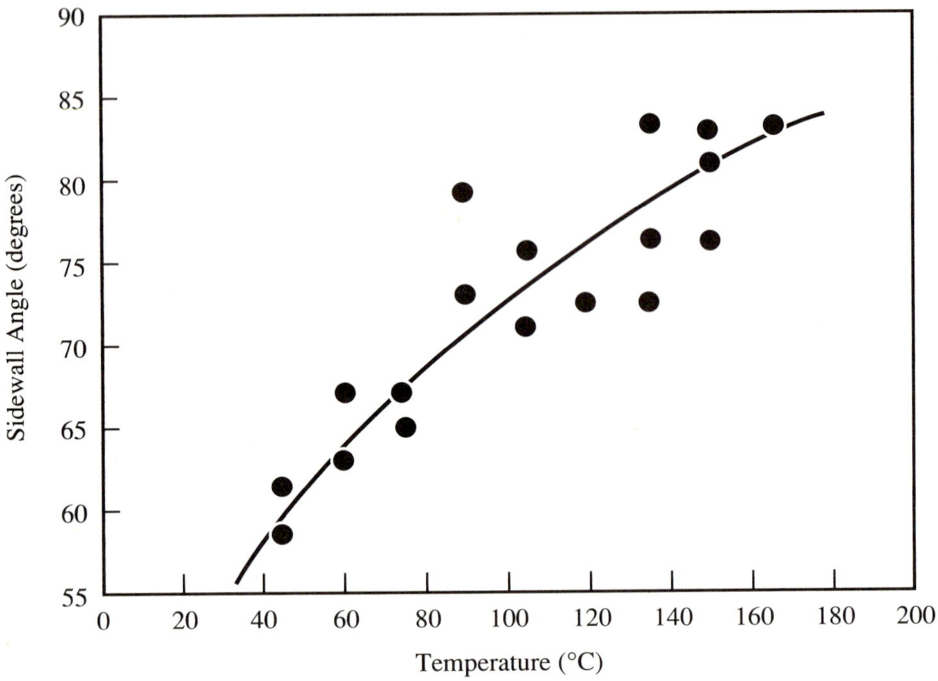

Figure 6.10. Sidewall angle for silicon trench etching in $CHCl_3$–N_2. Angles are measured relative to the surface.

heavy halogen scavenges atomic fluorine by reacting it to the relatively stable BrF and ClF. CF_3Br can also be used on its own, without using one of the fluorine compounds mentioned earlier; it is especially popular in single-wafer type reactors [198,199,200,201,202].

Non-undercutting silicon etching in plasmas containing fluorine compounds only is possible, but requires rather extreme conditions. At very low pressure (1 Pa and less) and sufficiently high ion flux the lateral reaction stops because the concentration of gas-phase fluorine radicals drops below a threshold limit [203,204]. Very low temperatures can also be used to suppress the non-directional reaction, see figure 6.11 after Tachi [202].

What has been said so far about monocrystalline silicon is to a large extent also valid for the patterning of polysilicon: the same mechanisms apply and similar gas mixtures are used. The differences lie mostly in the application requirements. Trench patterning asks for a very steep etch over several μm; in MOS gate patterning, the etch selectivity to the underlying oxide is a dominant factor. As the polysilicon is seldom thicker than a few hundred nm, the shape of the profile is less important, provided that the lateral loss is less than, say, 50 nm. Masking is for practical reasons done almost exclusively with photoresist.

Crucial here is the behavior of the process during overetching. As soon as the silicon clears from the underlying oxide, two effects occur. Firstly, the amount of silicon in the direct vicinity of the wafers drops dramatically, which reduces the deposition rate of the sidewall protecting film. Secondly, the concentration of atomic chlorine increases because it is not consumed anymore by the etch reaction, thereby increasing the lateral etch rate. Obviously both reactions affect the anisotropy in a negative sense. Indeed, a seemingly anisotropic process can become completely isotropic during overetching.

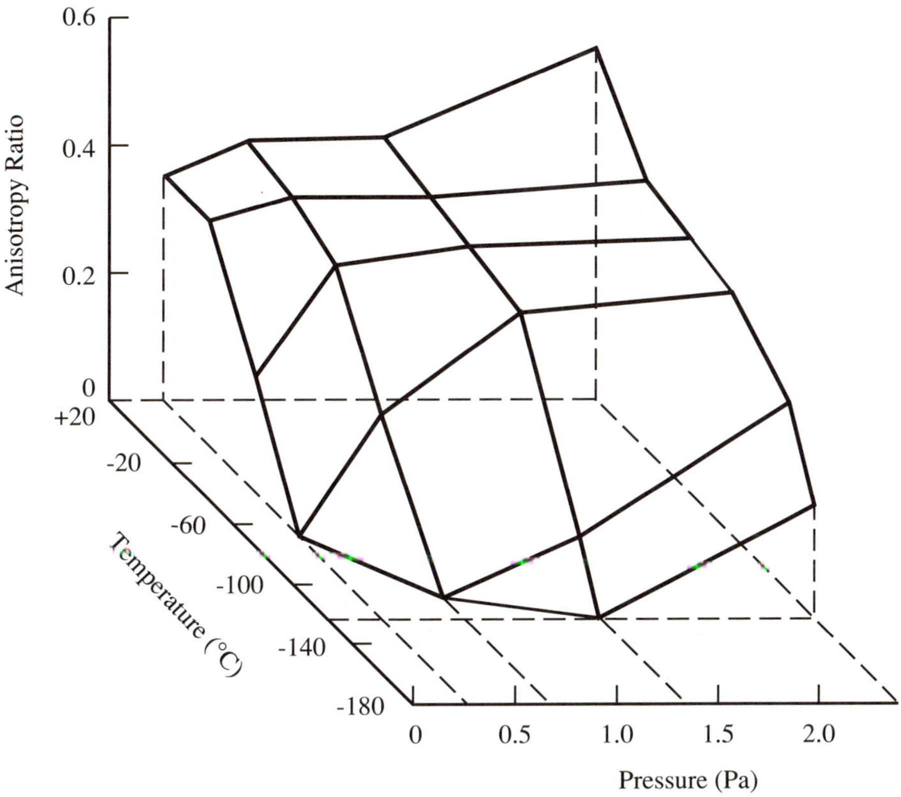

Figure 6.11. Anisotropy ratio for silicon etching in SF_6 as a function of pressure and wafer temperature.

On the other hand, the required oxide selectivity limits the allowable concentration of carbon-containing anisotropy controllers. Silicon oxide hardly etches in pure chlorine, but it is readily attacked by chlorocarbons in much the same fashion as in the Freon-based oxide etch processes,

$$SiO_2 + 2CCl_2 \rightarrow SiCl_4 + 2CO \quad (6.22)$$

The chlorocarbene intermediate in equation 6.22 can come from gas-phase additives, but also from erosion of the photoresist. If an oxide mask and pure chlorine are used, the oxide selectivity can be more than 200, but anisotropy control becomes doubtful. With regular resist masking, on the other hand, it is very difficult to achieve better than 15. In this respect, the presence of resist has been found to be equivalent to the addition of a few % by volume of methane or carbon tetrachloride. Chemical effects of resist masking in silicon etching are not limited to the material directly next to the mask pattern; resist is "seen" over distances of centimeters. For more information about the interactions with photoresist in silicon etching with chlorine plasmas, see the section on residues and damage [186,201].

Although all chemistries for trench etching have been applied to polysilicon too, some of the more popular mixtures for gate definition have not been mentioned yet, namely SF_6–CCl_4 and SF_6–$CHCl_3$. Increasing the SF_6 concentration raises the silicon etch rate and, thereby, the oxide selectivity; it does reduce the anisotropy, however. Increasing the CCl_4 or $CHCl_3$ content enhances the anisotropy, but reduces the oxide selectivity and promotes residue formation on wafers and reactor walls. The reason for using chloroform rather than carbon tetrachloride is its less pronounced carcinogenic properties. The $CHCl_3$ mixtures being used mostly in single-wafer reactors; the etch rate can be as high as 1 μm/min, with selectivities to gate oxide ranging from 10 to 20. Like in any other gate patterning process, finding the optimum conditions requires careful balancing between anisotropy and selectivity [205].

6.5 Aluminum and aluminum alloys

Aluminum, either pure or as an alloy with a few percent of silicon and/or copper, is used commonly as the metal in the upper interconnect layers in integrated circuits. It has a very low electrical resistance and an excellent adhesion to oxide layers; another argument is, that it can be made at low costs in very high purity. The requirements for dry etching are roughly the same as those in silicon gate patterning, but the dimensions are more relaxed and the underlying dielectric usually thicker than the gate oxide below polysilicon [206].

Pure, clean aluminum reacts rapidly with many gases and liquids. The practical observation that aluminum is in fact quite stable under normal atmospheric conditions results from the immediate formation of a very inert native oxide when it is exposed to air or water, for example,

$$4Al + 3O_2 \rightarrow 2Al_2O_3 \tag{6.23}$$

Only few aluminum compounds exist which are both stable and volatile enough to be removed by chemical reaction in a dry etching discharge. The fluorides are almost as stable as the oxide in equation 6.23; the same holds for the oxyfluorides. The chloride is a solid with a vapor pressure of around 100 Pa at room temperature. This is sufficient to get it out of the reactor chamber, but too low for uncomplicated transport to the vacuum pump exhaust; aluminum dry etchers always require rather special, corrosion and deposit resistant pumping units. Aluminum hexachloride is formed readily on exposure of aluminum with molecular chlorine,

$$2Al + 3Cl_2 \rightarrow Al_2Cl_6 \tag{6.24}$$

The etch rate can in this case be as high as 10 μm/min, and appears to be limited by the rate of evaporation only, see figure 6.12 after Efremov et al. [207]. This reaction will only start after the native oxide has been removed; this is also known as the incubation period or break-through step. In practical etch systems this is brought about by reagents which are strong reductors in a glow discharge. Probably the most effective component in this respect is the Lewis acid BCl_3, for which the surface reaction can be written as

$$O + BCl_3 \rightarrow BOCl + 2Cl \tag{6.25}$$

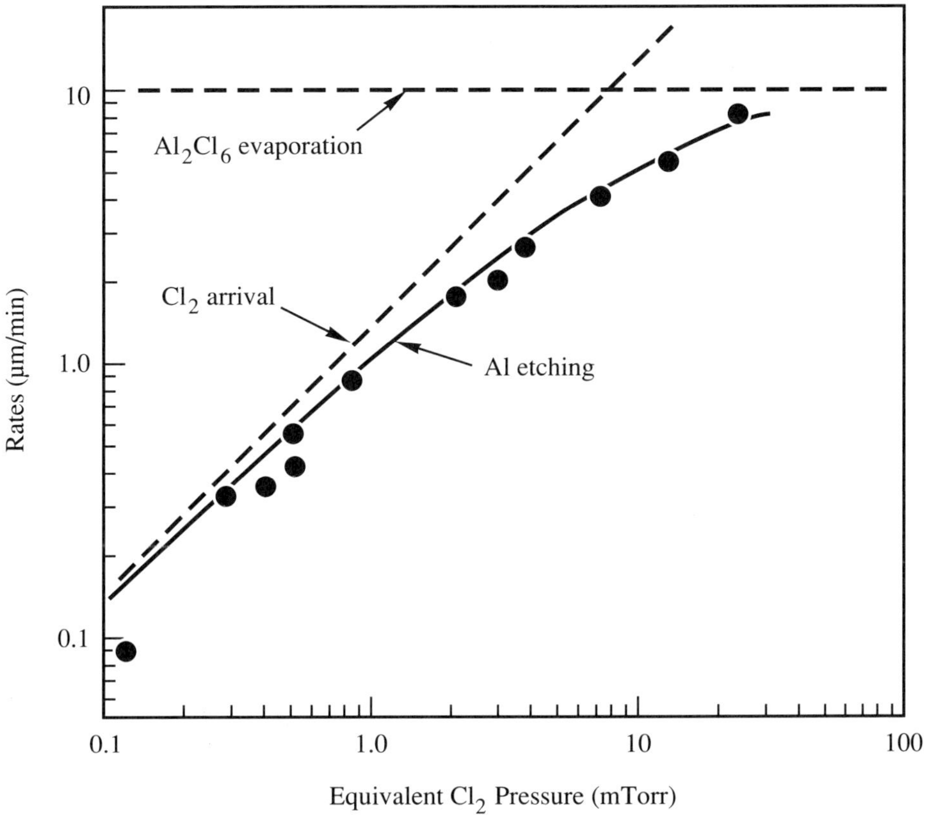

Figure 6.12. Aluminum etch rate at 25 °C as a function of the equivalent pressure of a molecular chlorine beam.

Other suitable gases are $SiCl_4$ and CCl_4 (or $CHCl_3$). In either case, bombardment with ions is necessary for surface activation, but the ion energy can be low. Oxide reduction is not significant in pure Cl_2, nor in compounds such as H_2 or HCl. Because the etch rate of the aluminum oxide is very low, all dry etch processes for aluminum start with an incubation time of about 30 seconds. Once the reaction has started, the presence of ions, atoms, or otherwise activated chlorine particles are not necessary. Apart from removing the native oxide, the boron, silicon, and carbon chlorides serve as scavengers for water and other oxygenated species in the glow discharge. The latter compounds, which can originate from reactor desorption or resist erosion, would otherwise react preferentially with the clean aluminum surface and inhibit further etching [208,209].

To achieve anisotropy in aluminum dry etching, some additional sidewall-forming chemistry must be used. In analogy with silicon, $SiCl_4$ and CCl_4 (or $CHCl_3$) can be used to this purpose. The above-mentioned boron trichloride does not form surface films and, hence, cannot provide anisotropic etching. As the reaction between photoresist and chlorine leads to the formation of chlorocarbons, the masking layer influences the degree of anisotropy too. This effect is much more pronounced here than it is with silicon. Even with pure carbon tetrachloride, the presence of a photoresist pattern is often necessary to provide fully anisotropic etching, see figure 6.13 after Wolters [210]. The distinct chemical nature of the dry etching of aluminum makes it sensitive to loading. With a large wafer load (in a batch reactor), or a

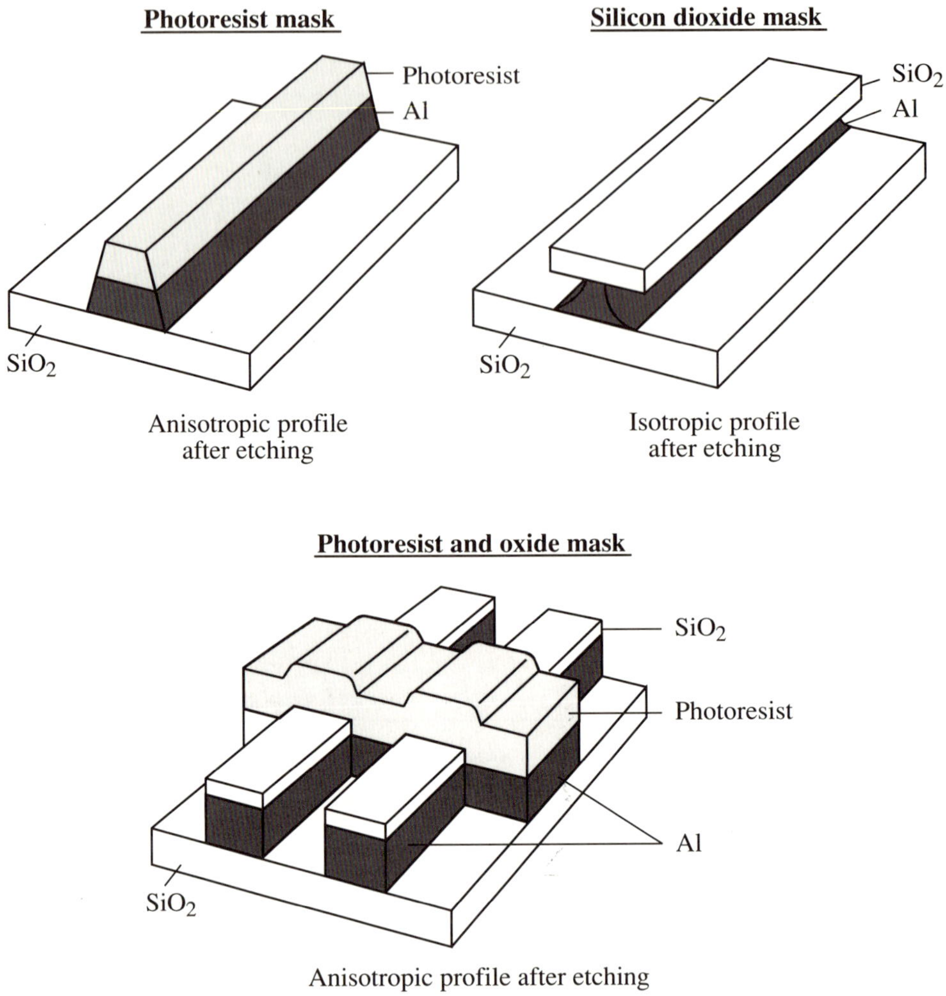

Figure 6.13. Schematic drawing of mask effects on undercutting in aluminum etching with a 380-kHz carbon tetrachloride plasma.

large exposed aluminum area (in a single-wafer reactor) the relative chlorine concentration is low because of depletion, and etch rates will be low. With few wafers, or a small exposed area, the etch rate will be high, as will be the degree of undercutting [211,212,213].

In many practical aluminum etch systems molecular chlorine is added to the gas mixture to improve the oxide selectivity. The reason is the severe topography of the underlying dielectric in conventional metallization schemes, which causes residues on steps if the aluminum is removed anisotropically, see figure 6.14. A reasonable selectivity is needed in order not to remove too much of the underlying oxide. To achieve this, some isotropy can be accepted. A practical solution to this problem if linewidths are $> 2\ \mu m$, can be found in a short dip in a wet etchant for aluminum after patterning the layer in a plasma without overetch. Apart from chlorocarbons, fluorocarbons have also been used as additives to improve anisotropy. Such components have the intrinsic risk of causing complete inhibition; relatively high ion bombardment energies are necessary to keep the etching surface clean. This negative aspect can be turned into an advantage, as it will serve to protect all areas within the reactor which are not

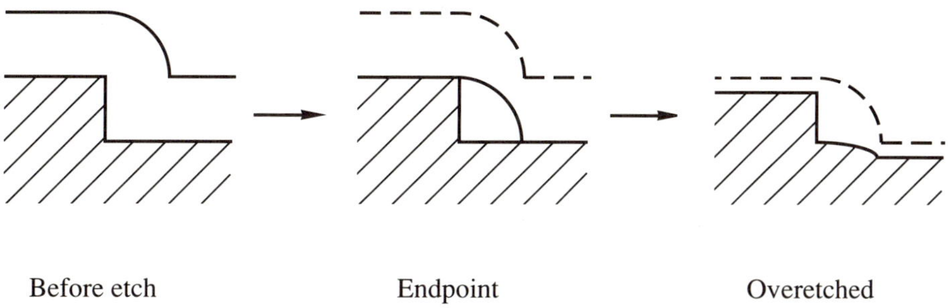

Figure 6.14. Attack of underlying material in anisotropic etching of layers on surfaces with a high topography.

exposed to high energies, such as walls and other grounded surfaces. Another frequently used additive in aluminum dry etching is nitrogen. Here, the nitrogen molecule is not a reagent as such, but serves as a second body for electron detachment processes. It thereby reduces the concentration of the negative ions which are the dominant species in strongly electronegative plasmas such as CCl_4, and especially BCl_3. This in turn raises the bombardment energy on the wafers and, hence, improves anisotropy and reduces the incubation period. A relatively new chemistry for etching aluminum and its alloys is BBr_3. As opposed to the corresponding chloride, the bromide is reported to be able to provide anisotropy without further additives, apparently by a surface condensation mechanism. The selectivity relative to silicon oxide is also significantly improved [214,215,216].

The aluminum chloride reaction product is very reactive toward many organic compounds. With aromatic polymers such as positive photoresist, the following reaction is possible,

$$\langle\bigcirc\rangle + AlCl_3 \rightarrow \left[\langle\bigcirc\rangle \cdots AlCl_2\right]^+ Cl^- \qquad (6.26)$$

Baked positive resist can absorb up to 10% of $AlCl_3$ by weight. Complexes as depicted in equation 6.26 hydrolyze readily in water vapor to give hydrochloric acid, which attacks the underlying aluminum. This can be traced back on wafers in the form of post-etch corrosion, a process which can destroy aluminum patterns in minutes. One of the more effective ways to suppress post-etch corrosion is rinsing with a large excess of deionized water or by passivation with a protective film produced in a polymerizing Freon plasma. Another possibility is a short dip in a wet etchant for aluminum. Heating on a hot plate in an inert atmosphere or exposure to a N_2 plasma also helps to reduce corrosion. In all cases, the photoresist must be removed as soon as possible, preferably by wet stripping [217,218].

The effectiveness of all dry etching systems for aluminum can be seriously influenced by the conditions in the deposition process. Usually, aluminum films are deposited by dc sputtering from a high-purity aluminum target. If the partial vapor pressure of oxygen is too high during sputtering, aluminum oxide will be built in. This need not affect film properties such as resistivity or reflectivity, but it will leave massive residues when the film is dry etched. Another problem occurs when the film is sputtered under conditions where the surface mobility of the aluminum atoms is low. When this is the case, the film growth will be determined by the rate at which aluminum particles arrive from the gas phase. At sharp bottom corners in the substrate topography, this arrival rate is close to zero. Consequently, the deposited film will

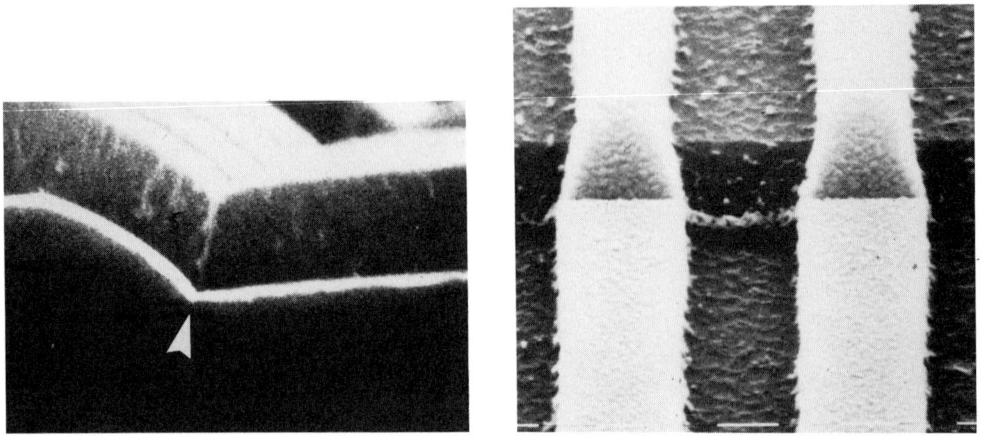

Figure 6.15. Crack formation (arrow) in sputtered aluminum layers (left), resulting in wings causing shorts between aluminum lines (right).

have a "crack" at these spots, as shown in figure 6.15. On exposure to air, the cracks rapidly oxidize, leading to built-in oxide wings. These wings can be the source of persistent line shorting between dry-etched aluminum lines crossing over steps in the underlying dielectric, even after excessive overetching. Crack formation can be reduced by sputtering under "hot" surface conditions, such as substrate heating. Alternatively, the substrates can be exposed to energetic ion bombardment by biasing the substrate holder with an auxiliary rf or dc voltage source. It will be clear that negative angles in the underlying topography should be prevented anyway [219,220,221].

Small additions of silicon to the aluminum, have no influence on the etching behavior. The silicon is generally added to prevent diffusion of silicon into the aluminum at places where contact is made to the silicon substrate, resulting in "spiked" junctions. Using chemistries as described above, it is also possible to etch aluminum with up to 4% copper, a component which is added to reduce electromigration-induced circuit failure. To achieve this, the power density should be high and the pressure moderate (20-40 Pa). This is rather surprising, since copper has no volatile chlorides, and the pressure used is much too high for physical sputtering to be of any importance for the removal of material. Etching of pure copper in a reactive glow discharge is, indeed, very difficult [222]. Possibly, complex reaction products such as $AlCu_xCl_y$ are formed, since some of the mixed chlorides do have substantial vapor pressures. The process is critical; metallic etch residues are found if $AlCu_x$ intermetal compounds are present in the film as a result of inappropriate sputtering or annealing. Post-etch corrosion on etched Al(Cu) layers is much more severe than in the case of Al or Al(Si), due to the formation of local electrochemical cells. Even more problems are encountered when etch and diffusion barriers such as TiW are used, as this introduces yet another cell [212,223,224].

6.6 Refractory metals and silicides

For many high-density MOS integrated circuits, the conductivity of conventionally doped polysilicon is not sufficient to meet the required switching speeds at the gate interconnect level. This problem can be solved by shunting the silicon with another conductor,

Table 6.1. Boiling or sublimation (*s*) points of some refractory metal halides in °C at 1 atm. Values for silicon and aluminum are given for comparison.

Fluoride	Temperature	Chloride	Temperature
WF_6	20	WCl_6	345
WOF_4	190	$WOCl_4$	230
MoF_6	35	$MoCl_5$	270
$MoOF_4$	180	$MoOCl_4$	180
TiF_4	285 *s*	$TiCl_4$	135
TaF_5	230	$TaCl_5$	240
SiF_4	-85	$SiCl_4$	60
AlF_3	1290 *s*	$AlCl_3$	180 *s*

such as a refractory metal silicide. The silicides are real compounds rather than alloys, and combine a good conductance with good resistance to the high temperatures required in this stage of the process. In order not to disturb the transistor characteristics and stability, which depend strongly on the conditions at the gate oxide, a polysilicon/silicide sandwich is almost always used; this is known as the polycide structure. Conventional silicides are made either by cosputtering from two targets or by direct sputtering from a composite target. The deposited film is a mixture rather than a compound; to create the silicide an annealing step is required. Popular are the silicides of tungsten and molybdenum, WSi_2 and $MoSi_2$, but $TaSi_2$ and $TiSi_2$ are also used [225,226].

More recently, the pure refractory metals have found application. In part, this is caused by the fact that the resistivity of the metals is often significantly lower than that of the corresponding silicide. An example is the use of a Mo–TiW bilayer, which has the additional advantage that TiW is impermeable to dopant material from underlying junctions, whilst rapid diffusion can occur in the silicide. Another factor is, that a material like tungsten can be conveniently made into thin films by chemical vapor deposition from the gas tungsten hexafluoride. The intrinsic good step coverage of CVD processes – much better than sputtering – makes deposited tungsten very suitable for filling deep and narrow contact windows. In the latter process, contact holes are being filled with either a silicon-selective or a blanket deposition, after which the excess material is removed by what basically is an unmasked dry etch-back process [227].

In order to gain understanding of the dry-etch processes used for refractory metals and their silicides, it is very useful to have a look at the volatility of the potential reaction products. These are presented in table 6.1. This table lists the boiling points of a number of fluorides and chlorides of the above-mentioned metals, in conjunction with the corresponding silicon and aluminum compounds. It should be noted that some of the halides do not melt, but sublime directly. A first observation is that tungsten and molybdenum form highly volatile fluorides, just like silicon. The mechanism of tungsten (and molybdenum) etching in fluorine-based plasmas is, indeed, quite similar to that of silicon. Still, some differences do exist. An example is the oxygen concentration in CF_4–O_2 needed to reach maximum etchrate, which increases in the order silicon-silicide-metal. Interestingly enough, the optimum for the pure metal lies far above the point where a maximum free fluorine concentration is reached; the silicide takes an intermediate position between silicon and metal, see figure 6.16 after Oehrlein et al. [172]. Ion bombardment increases the vertical etch rate in fluorine plasmas considerably.

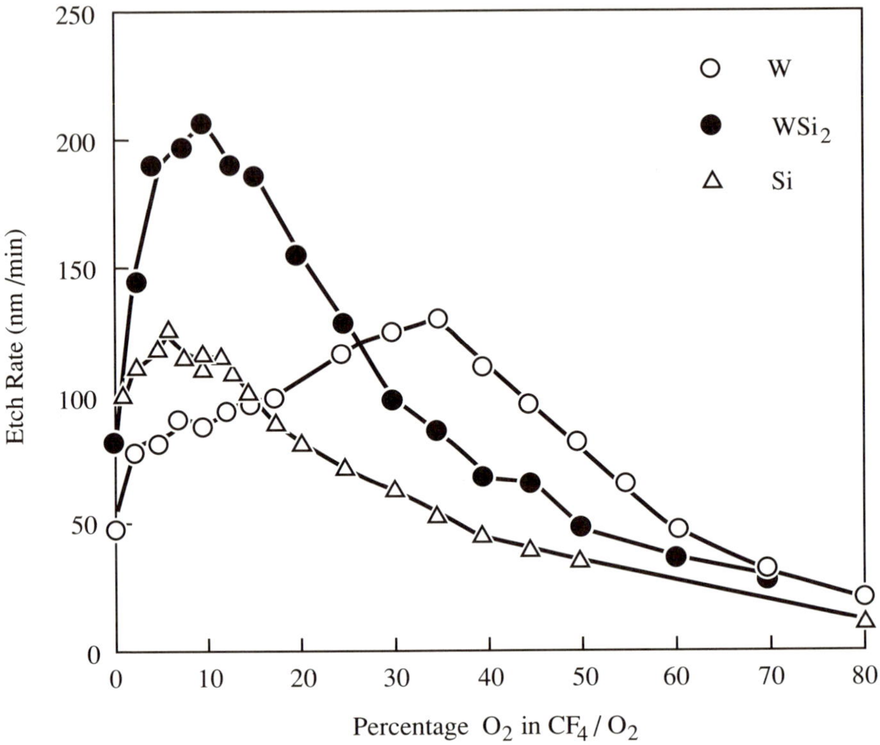

Figure 6.16. Etch rate of Si, W, and WSi_2 in CF_4–O_2 dry etching.

The required energies are very low, 5 eV appears to be sufficient. Obviously, some highly reactive metal oxyfluoride surface compound is involved [172,228,229,230].

An unusual situation occurs with titanium, of which the fluoride has a vapor pressure at room temperature of only 2×10^{-3} Pa. The result is a very high temperature dependence of the etch rate in fluorine plasmas, see figure 6.17. From the figure, the activation energy for chemical etching can be shown to be about 0.45 eV. This value is close to the heat of evaporation of TiF_4, which is 0.6 eV, indicating that the rate limiting step in this case is evaporation of the reaction product rather than a surface reaction. In the etch processes discussed so far, such a mechanism has not been encountered yet; it is a well known process in GaAs and InP dry etching, though [231,232].

For reasons similar to those mentioned earlier for silicon etching, the pure fluorine plasmas are not suitable for non-undercutting pattern definition of refractory metals and silicides. They can, however, often be used conveniently for processes like tungsten etch-back in the contact-hole filling process. Table 6.1 shows that all refractory metal chlorides have a low volatility, just like aluminum. This effect can now be used to achieve anisotropy. Most systems for etching refractory metal and polycide interconnects use indeed a mixed halogen chemistry with a high fluorine concentration to reach high etch rates, and some chlorine to keep the process anisotropic. In pure chlorine the rates are often very low. Because the volatility of both the chloride and the fluoride of tantalum is relatively low, it would be expected that etching of this material would be problematic with any reagent. Surprisingly enough, mixing chlorine and fluorine also raises the etch rate in this case to acceptable values while maintaining anisotropy [233,234].

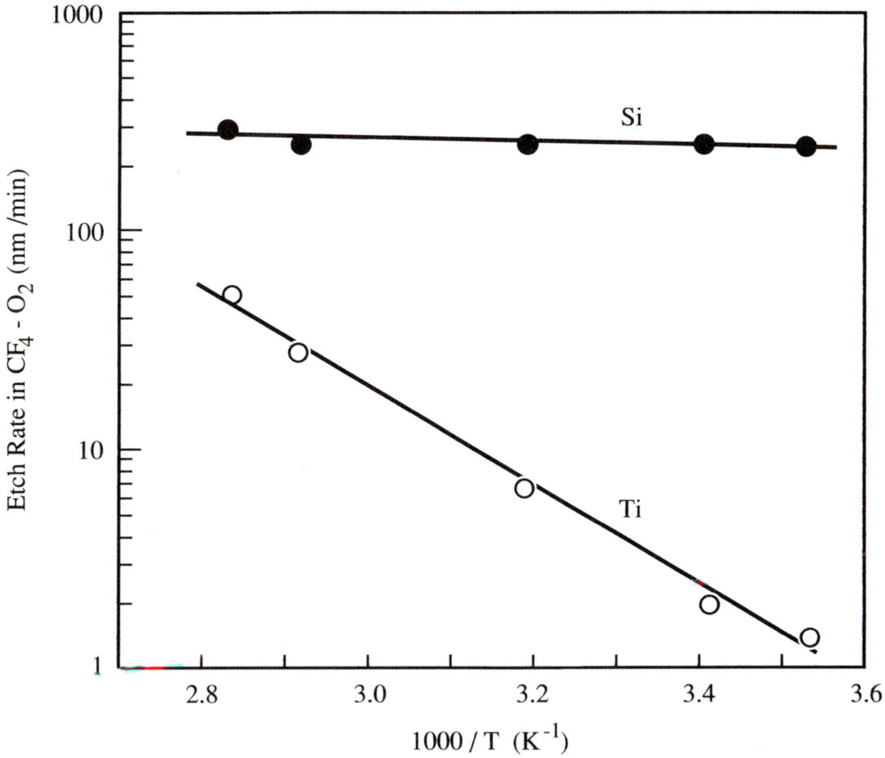

Figure 6.17. Etch rate of poly-Si and sputtered Ti in a CF_4–O_2 plasma at 40 Pa.

The fact that the boiling point of the refractory metal halides is higher than that of the corresponding silicon compounds, implies that the metals are less sensitive to undercutting than polysilicon. This can be seen readily from the temperature dependence of the lateral etch rate, see figure 6.18 after Tachi [202]. The difference in reactivity between silicon and refractory metal silicides is the cause of an intrinsic problem in polycide patterning, namely selective undercutting, see figure 6.19. In principle, both profiles drawn in the figure are possible, but in most cases the polysilicon undercuts in the chemistry necessary to etch the top silicide. The result will be a marginal step coverage of subsequent layers, and often insufficient edge definition for self-aligned implantation steps to define source and drain areas. Many practical processes use a multistep sequence, with endpoint detection to determine the switching points [235,236,237]. A process typical for polycides, in this case tantalum polycide, is

(a) $TaSi_2$ etch in SF_6–CCl_2F_2–He

(b) surface cleanup in Cl_2 to remove adsorbed fluorine

(c) poly-Si etch in Cl_2–O_2

Apart from the mixed halogen systems, pure chlorine-based chemistries are also being used. In this case it is important to realize that the refractory metal chlorides are not very volatile. Combined with an only moderate reactivity of the metals, this implies that removal of the reaction products can be a limiting factor. Apart from the obvious redeposition in etch reactor and exhaust systems, as is the case with aluminum, redeposition on the wafers

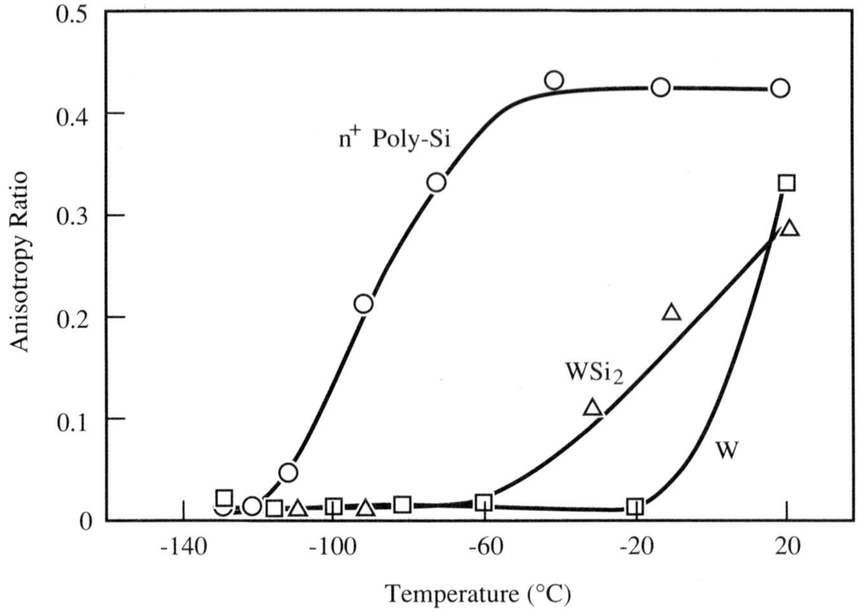

Figure 6.18. Temperature dependence of the anisotropy ratio for Si, W, and WSi_2 in an SF_6 plasma.

is possible, and substrate heating is normally required to obtain reproducible results. In principle, the reaction between chlorine discharge and refractory metal is similar to that between chlorine and silicon. That is, a mechanism where atomic chlorine is the most important reactive species, a sidewall passivating reaction product (metal chloride), and a

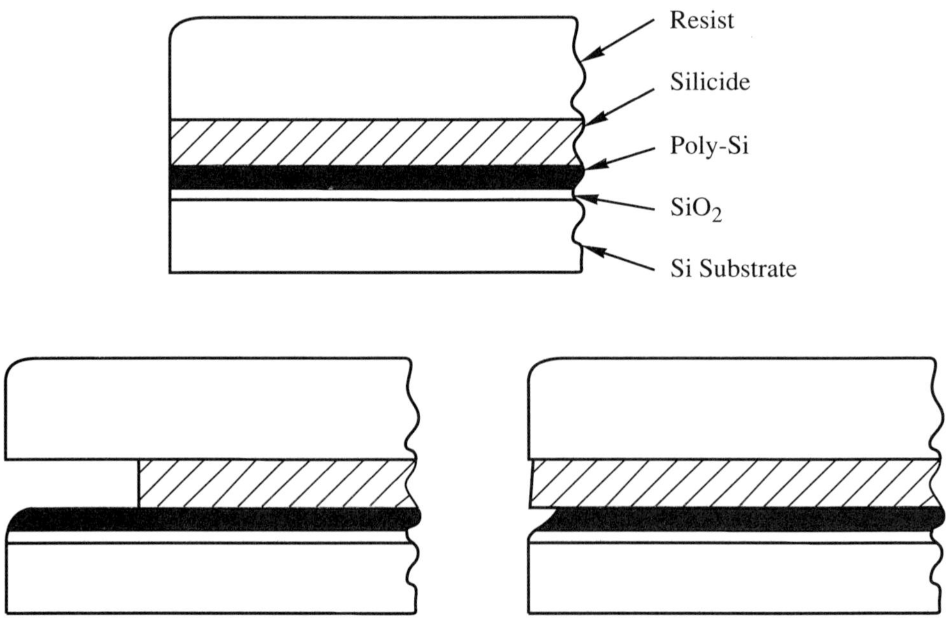

Figure 6.19. Possible cross-sectional edge profiles for etched polycide structures.

large influence of ion bombardment. Tungsten and molybdenum can even be etched in BCl_3–Cl_2 mixtures without undercutting, in contrast to aluminum. Titanium chloride is much more volatile than the other three refractory metals; it is therefore more sensitive to isotropic etching [238,239,240,241,242].

The absolute etch rate of the refractory metal silicides can be as high as that of silicon under comparable conditions, of the order of 0.5 μm min^{-1}. The metals etch slower by a factor of 2 to 5. Because silicides are compound layers, the etch rates and the degree of residue formation are very sensitive to the procedure by which the films were made. Built-in oxygen can have a negative effect on rate, as is the case with titanium. It can, however, also lead to enhanced etching. This happens with molybdenum and tungsten, where the oxychloride is more volatile than the corresponding chloride [243,244].

6.7 Polymer processing

The original application of plasmas for organic material processing was photoresist stripping in oxygen with a barrel reactor, which is actually the first dry etch process introduced in silicon integrated circuit fabrication. After the introduction of the parallel plate reactors, it was extended with contact hole etching in polyimide dielectric layers, and more complicated procedures involving organic polymers, such as etch-back planarization and dry photoresist development. The latter processes will be discussed later in this section [245].

At elevated temperatures, most organic materials react readily with excited oxygen. Ion bombardment enhances the reaction rate, but is not a necessity. Therefore, in order to prevent ion-induced surface damage, stripping of photoresist is done almost exclusively in afterglow plasmas. The thermal activation energy of the reaction in an oxygen plasma is approximately 0.5 eV, indicating that an evaporation-limited mechanism is involved. Adding a few percent of some fluorine-containing component, such as CF_4 or SF_6, dramatically increases the achievable strip rates. At the same time, the activation energy drops to about 0.1 eV, which is close to the energy involved in the following equation,

$$R_nH + F \rightarrow R_n + HF \qquad (6.27)$$

where R_n denotes an alkyl group. The surface radical created in equation 6.27 will react rapidly with atomic oxygen, which leads to material removal and more radicals,

$$R_n + O \rightarrow R_{n-m} + R_mO \qquad (6.28)$$

The etching of an organic layer in an O_2–CF_4 plasma with a low fluorine content can, therefore, be depicted as a chain reaction with equation 6.27 as initiation and equation 6.28 as propagation reaction. For fluorocarbon concentrations above about 20%, the etch rate drops again because of recombination of surface radicals with atomic fluorine,

$$R_n + F \rightarrow R_nF \qquad (6.29)$$

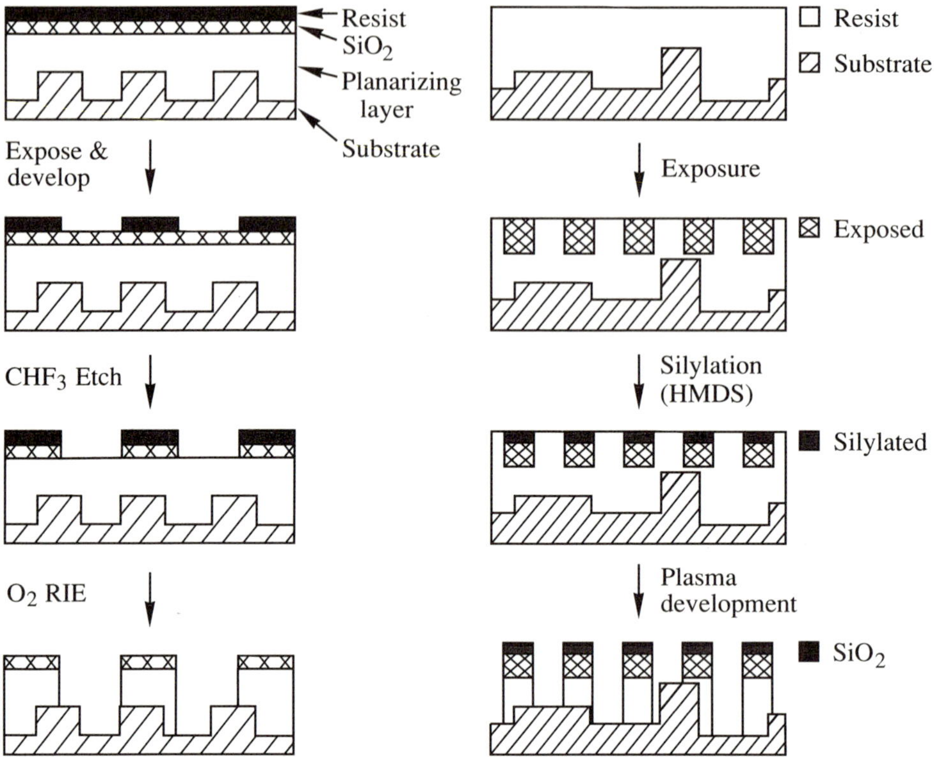

Figure 6.20. Process steps in trilevel (left) and DESIRE (right) dry photoresist development.

Strip rates with oxygen plasmas to which fluorine is added are about 1 μm min^{-1}; without fluorine a factor 10 lower [133,246,247]. The highest strip rates, 8 μm min^{-1}, are reported for the combination of fluorine addition, elevated temperature (200 °C), and microwave excitation [248]. Because of the positive effect of fluorine on the etch rate of organics, fluorine-containing polymers will etch rapidly and can have a large effect on the etching of other organic layers in the same reactor [249]. Under bombardment with energetic ions, all organic polymers are etched quite fast in oxygen, chlorine, and all non-polymerizing fluorine compounds. Also nitrogen and hydrogen discharges etch organics [250].

Dry photoresist development originates from a procedure which is known as trilevel (or trilayer) lithography. In trilevel lithography, the surface to be etched is coated first with a thick planarizing organic layer. Secondly, a thin silicon oxide layer is applied, for example by rf sputtering or from a silicate solution. The third layer is the actual photosensitive material. This sandwich structure is then patterned as depicted in the left-hand side of figure 6.20 after Hartney et al. [245]. Because the photosensitive top layer is applied on a flat surface, and needs to withstand only the moderate etch required to remove the silicon oxide locally, it can be made very thin. Thus, the dimension of patterns defined in that layer can also be very small, and almost any material can be used that suits the lithographic process. If the final oxygen dry etch is completely anisotropic, the resulting etch mask will be vertical and a very precise transfer of the initial pattern to the underlying layer is possible.

Although trilevel lithography is straightforward, it is also complicated. In dry development, it is attempted to combine all processing into a single sequence. An example is the DESIRE (Diffusion Enhanced Silylating Resist) procedure, shown schematically in the right-hand side of figure 6.20. Here, the top part of a positive photoresist is exposed. Subsequent treatment with hexamethyl disilazane leads to selective silylation of the exposed areas. When the samples are brought into a low-energy oxygen plasma, the silylated parts are oxidized to form a silicon oxide etch barrier, provided that enough silicon has been build in (at least 4%). In the second step of the oxygen plasma treatment the energy is increased, and the final pattern created by etching of those areas not protected by silicon oxide [251]. This final step is highly anisotropic, since the etch behavior is governed by ion induced chemical reaction [252].

Anisotropic etching of polymers in an oxygen plasma involves the formation of a protective film against the pattern sidewalls. The sidewall film consists mostly of redeposited silicon oxide originating from sputtering of the primary mask at the top surface. During etching, the polymer immediately below the protective layer will become extensively crosslinked because of the continuous oblique implantation of ions to which it is exposed. After clearing of the bottom surface, the sidewall film will be mixed with backsputtered oxidized substrate material, such as silicon or aluminum oxide. The major difficulty in trilevel lithography and dry development is hidden in the requirement that all layers are to be removable from the surface once the substrate has been etched. Protection of sidewalls against isotropic etching by (re)deposition of oxygen-resistent layers will leave residues that can, of course, not be removed in conventional oxygen plasma stripping. On the other hand, preventing formation of these layers will cause undercutting. In practice, a combination of limited sidewall formation and extensive cleaning must be used [253].

So far, a number of procedures have been mentioned in this chapter that cope with defining patterns over high substrate topographies. Probably the most widely spread process in this category is etch-back planarization of silicon oxide. Here, the dielectric is coated with an organic layer, usually straight positive photoresist, baked, and dry etched in a chemistry providing equal etch rates for resist and silicon oxide; for example a carefully adjusted mixture of CF_4–O_2. By applying multiple photoresist layers with varying viscosity, and using combinations of UV exposure and temperature treatment, the surface topography of the resist prior to etching can be reduced to less than 0.1 μm over a range of 10 μm [254,255].

Planarization as described above has two important drawbacks. First, although the flatness of the photoresist coating can be made perfect on the short range, the absolute thickness will vary considerably with the feature sizes in the underlying topography, see figure 6.21 after Schiltz and Pons [256]. The largest difference is found between large "high" areas, such as bonding pads, and small "high" areas, such as isolated lines; it amounts to the original step height in the substrate. After etch-back, the surface of the oxide may be smooth, but the thickness differences remain since these have simply been copied from the photoresist into the oxide (see figure 6.21), thus causing problems in subsequent contact or via hole etching. The second problem is the fact that the dry etch reactions of photoresist and silicon oxide in CF_4–O_2 and similar chemistries are influenced by each other. The presence of photoresist increases the amount of carbon in the plasma, and thereby reduces the silicon oxide etch rate; silicon oxide raises the oxygen concentration and increases the photoresist etch rate. Thus, adjusting the composition of the discharge to create equal etch rates is not that simple. Moreover, as more and more silicon oxide is exposed when the photoresist is etched, the balance shifts during the process. The impact of these drawbacks can be reduced by adding more process steps, such as applying an inverted dummy resist pattern with the correct thickness in all "low" areas. Unfortunately, this reduces the attractive simplicity of

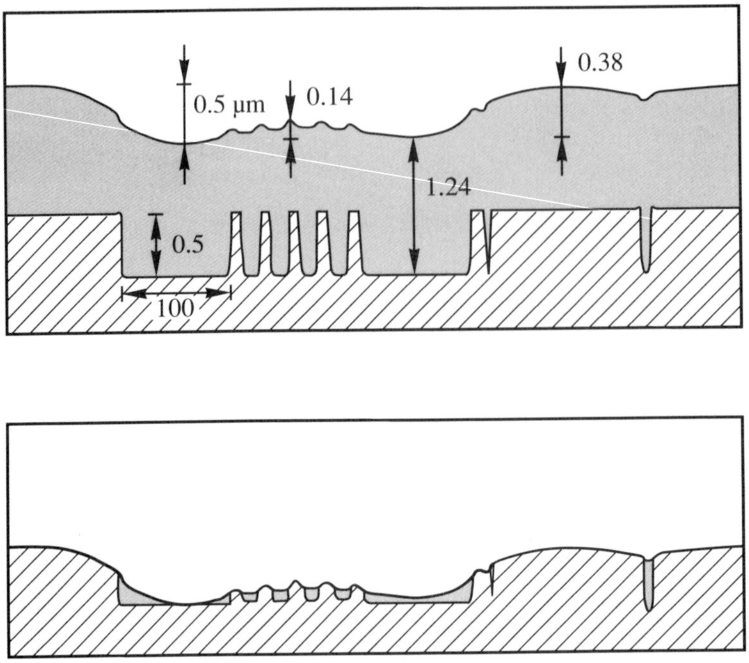

Figure 6.21. Typical surface profile before and after resist spin-on planarization and subsequent etch-back.

etch-back planarization, and brings other procedures, such as deposition and etching of silicon oxide at the same time, within competitive distance [257].

Step coverage-induced continuity problems in contact and via holes can be reduced by tapering. Instead of the usual straight etch profile, a slope is created in the dielectric by extending the silicon oxide etch process with controlled resist erosion. This is exemplified in figure 6.22. In the procedure at the left-hand side of this figure, oxygen is added continuously to the etch plasma during the first half of the process to create a lateral etch rate component in the photoresist. In the right-hand procedure the silicon oxide etch process is stopped halfway through. The photoresist is then etched isotropically in pure oxygen, after which the silicon oxide is etched to completion. In either case, care must be taken to have sufficient photoresist thickness to start with, as the vertical resist erosion will also be significant. Apart from these procedures, which focus on resist erosion, it is possible to create tapered silicon oxide holes by using NF_3 discharges. This compound promotes under certain conditions isotropic etching of silicon oxide, especially if the oxide is heavily doped with phosphorus or boron [169,258,259,260,261].

6.8 Residues and damage

As the etch anisotropy in reactive glow discharges is brought about by the combination of disruption of surface bonds by ion bombardment, and the deposition of passivation layers, damage and residues are inevitable side effects of dry etching. Also, the energetic photons emerging from the plasma zone can cause deep-lying lattice traps. Fortunately, many apparently damaging processes have been found not to affect the final integrated circuit char-

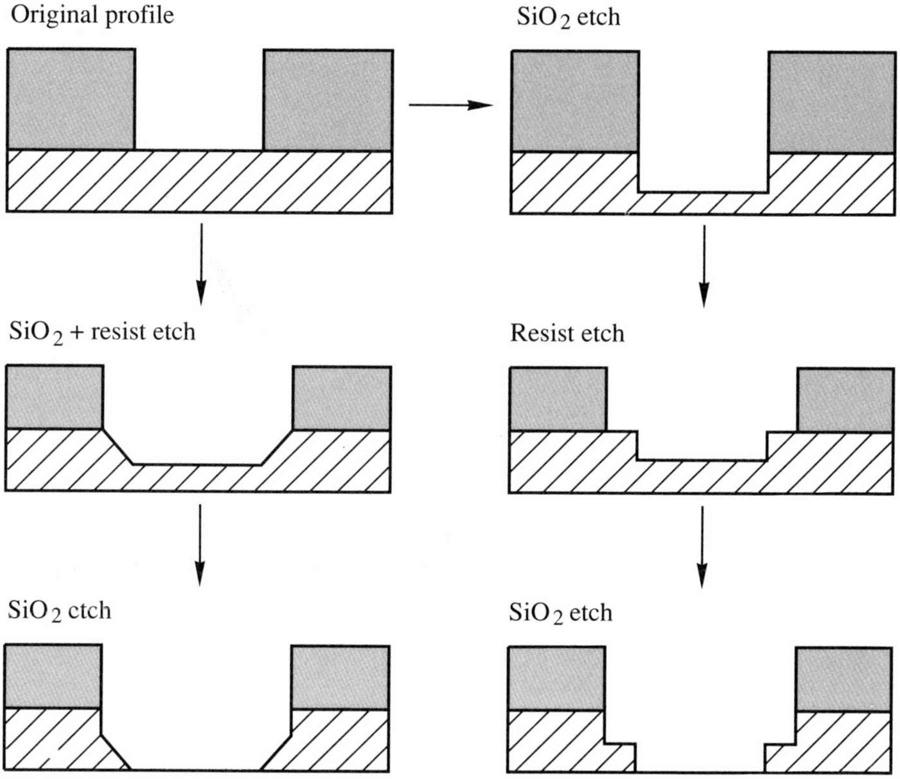

Figure 6.22. Process steps in two-step taper (left) and step (right) contact hole etching.

acteristics; once the damage has been done, the original properties can often be restored by a proper post-treatment. A number of mechanisms and procedures in the creation and removal of residues and damage in dry etching will be discussed below. The examples are for silicon oxide and silicon only. Nevertheless, they also apply to other similar systems, such as the etching of silicon nitride and refractory metal silicides [161].

A characteristic example of the surface composition of material which has been subjected to dry etching is given in figure 6.23 after Oehrlein et al. [262]. This schematic drawing depicts the surface of monocrystalline silicon after exposure to a plasma used for silicon oxide patterning. Starting top down, the surface consists of

(a) a deposited teflon-like film;

(b) a heavily distorted and roughened silicon layer containing entrapped carbon and fluorine;

(c) a lightly distorted layer with deep-lying traps created by implanted hydrogen.

The thickness of the distorted layers (b) and (c) can be correlated directly with the penetration depth of the ions involved. Figure 6.24 after Oehrlein et al. [262], shows the projected range of carbon, fluorine, and hydrogen, as a function of ion energy. With bombardment energies used in dry etching, it appears that a depth of several tens of nm can easily be reached.

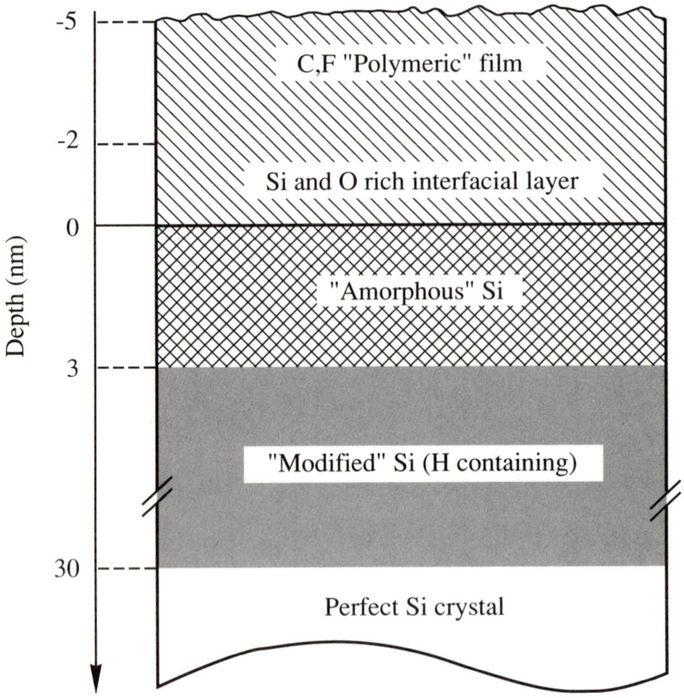

Figure 6.23. Schematic of silicon near-surface region after dry etching with CF_4–H_2.

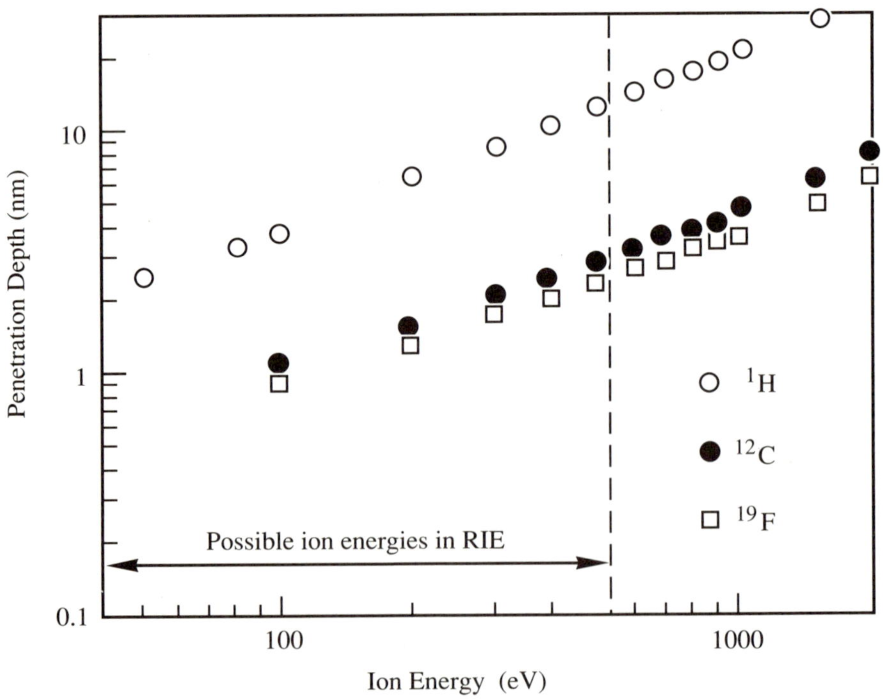

Figure 6.24. Projected range of ion implantation in amorphous silicon from a Monte Carlo simulation. Ion energies in dry etching are given for comparison.

The organic top film (a) and most of the carbon implanted in the amorphized layer (b), is readily removed in an oxygen afterglow plasma. Subsequent heating of the wafers at 400 to 800 °C is then sufficient to anneal out the lattice damage. Removal of carbon before the high-temperature treatment is necessary to prevent formation of silicon carbide at the silicon surface. If the silicon surface is to be used to make contact with a metal interconnect layer, annealing is seldom necessary. Generally, the teflon layer is both necessary and sufficient to reach an acceptable contact resistance. The interfacial oxide produced in the oxygen plasma can be removed with an HF solution, if required. It should be noted that the sputter processes used for metal deposition involve interaction with a plasma, and are, therefore, producing surface damage themselves [263,264,265,266].

Very deep-lying damage can be found in boron-containing silicon after exposure to ion bombardment. This is related to the exceptionally high diffusion coefficient of hydrogen in silicon doped with boron. The hydrogen can be found as far as 10 μm from the surface, thereby deactivating the boron dope. The hydrogen involved in the deactivation can come from the chemistry used, but it can also result from photoresist erosion, as similar effects have been found in CF_4–O_2 plasmas. Heating at temperatures of the order of 200 °C reactivates the boron [267,268,269].

A completely different and much more severe source of device damage is the photoresist stripping process. All photoresists contain traces of heavy metals and alkali, which are left on the substrate surface after stripping in an oxygen plasma. If the wafers are exposed to ion bombardment at this stage, the trace metals will enter the substrate by knock-on implantation. Other than the carbon, fluorine, and hydrogen mentioned above, metal contamination creates permanent traps which cannot be annealed out. This is one of the main reasons why oxygen plasma stripping is done in reactors with cages or screens to prevent ions from reaching the wafers. It also puts an upper limit to the allowable temperature level in stripping, as 300 °C can already sufficiently enhance the diffusion rate of these metals [270].

Another "killing" process is accumulation of charge in the gate area of MOS transistors. This occurs when a net current of positive or negative charge is drawn through a wafer containing floating gates; the charge stored during etching can result in field-induced breakdown of the gate oxide. A mechanism as described here is important especially in low-frequency-operated reactors. Some LF systems do not have a blocking capacitor in series with the discharge, which allows dc currents to flow. But also reactors with external wafer biasing, like many ECR systems, are in principle suspect [271,272].

Residues are defined here as microscopically visible artifacts left on the etched surface, resulting directly from the plasma process. Leftovers from inclusions and contaminations, such as oxide and copper in aluminum, are excluded. Residues under this definition are caused almost exclusively by locally enhanced polymerization under otherwise normal process conditions. Relevant factors are the substrate electrode material, the photoresist procedure, the masking material, and the underlying layer. The right hand side of figure 6.25 shows a photograph of the voluminous cauliflower structures left after etching a molybdenum-masked silicon oxide layer on an aluminum electrode. The same process with a quartz electrode produces no residues, see figure 6.25, left. The residues here consist of unetched substrate material in a thin organic envelope. It seems that a catalytically active surface contaminant serves as a chain initiator for polymer growth. The locally produced polymer then serves as a mask, leading to the formation of the observed structures. Likely candidates for chain initiation are non-volatile metal fluorides, in this example AlF_3 from the electrode; MoF_6 is

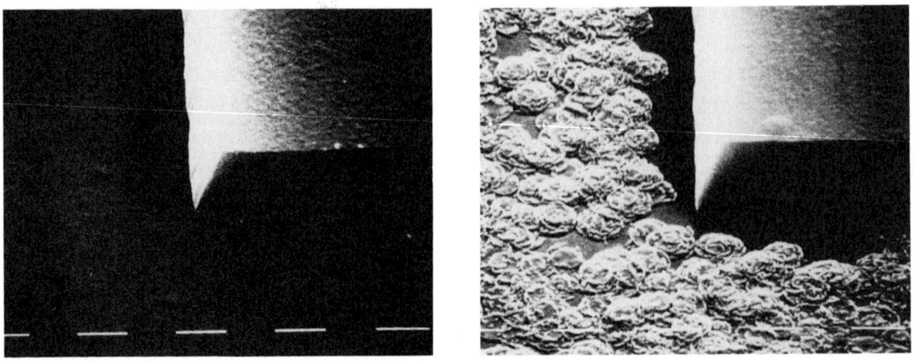

Figure 6.25. Residues after etching silicon oxide in CHF_3–Ar in a cathode coupled reactor. Electrode material quartz (left) and aluminum (right). White bar is 1 μm.

sufficiently volatile. Similar problems can occur if silicon oxide is etched with an aluminum mask, or if the underlying layer is aluminum, as is the case in via hole etching [170,273,274].

Where the polymer creating the residues in silicon oxide etching is teflon, it is mostly chlorinated amorphous silicon in the anisotropic patterning of mono and polycrystalline silicon. Traces of contamination or native oxide at the top surface of monosilicon can cause what is known as "black silicon" after trench etching. This is a structure resembling a grassland in cross section, see figure 6.26. Black silicon is, in fact, unetched substrate material, protected from attack by chlorine radicals by the high redeposition rate for the process. Because inci-

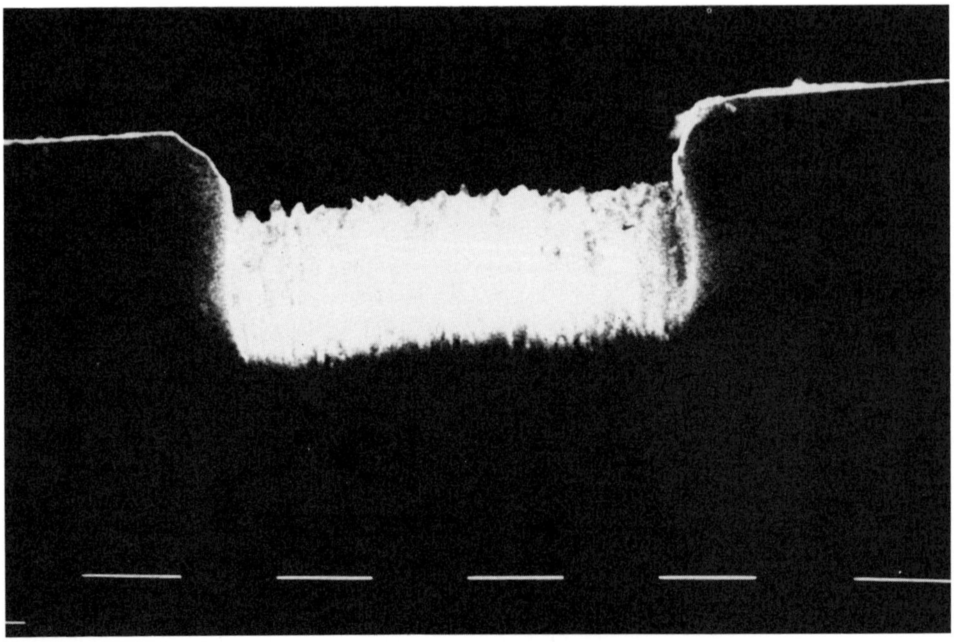

Figure 6.26. Cross section of black silicon formed on etching monosilicon in a chlorine plasma. White bar is 1 μm.

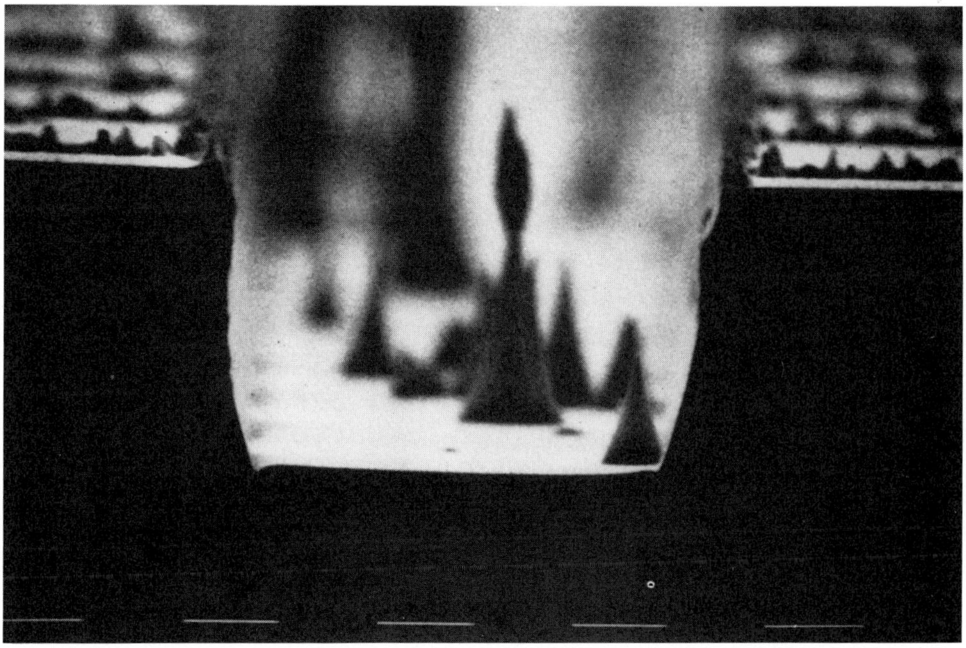

Figure 6.27. Cones left after etching <100> silicon in CCl_4. White bar is 1 μm.

dent light is complete scattered on such a surface, it has a very dark appearance when looked at with the unaided eye. If the redeposition is less severe, or the top surface cleaned, "cones" can be formed locally (see figure 6.27). As would be expected, this happens preferentially in areas where the relative silicon concentration is high, such as large open areas. Even if the conditions and the wafer pretreatment have been adjusted such that grass or cones are prevented during trench etching, extensive residue formation will normally still take place against the trench sidewalls. These layers can be up to a few hundred nm thick. As the chlorinated silicon is readily hydrolyzed, it can be removed easily with an HF solution [190,275].

The chlorosilane redeposition described above occurs not only on the etched silicon itself, but also on the sidewalls of all other sufficiently steep steps on the wafer surface. The most likely candidate is the photoresist pattern used in polysilicon etching. Whether or not a net formation of silicon residues occurs, depends on the pattern sidewall angle, see figure 6.28 after van Roosmalen et al. [186]. In this figure, r_d and r_e are the deposition and etch rates, respectively. The deposition is assumed to be isotropic, r_d is constant. Etching is anisotropic, hence r_e will decrease when the sidewall angle θ increases. This means that a critical angle θ_c exists above which net deposition of material will occur. The figure depicts the situation where the angle of the photoresist foot θ_f is larger than the critical angle, implying that a silicon film will be formed from the photoresist foot up to the point where $\theta = \theta_c$. After etching, and especially after photoresist stripping in an oxygen plasma, these films become visible as "wings" on top of the etched pattern with a thickness of about 50 nm, see figure 6.29. The critical angle in this and similar processes is near 70 degrees, just like the yield cut-off point in sputtering.

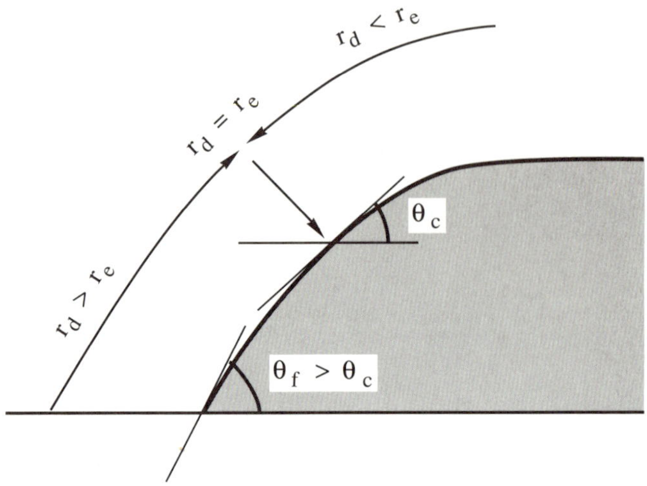

Figure 6.28. Schematic drawing of sidewall redeposition process. See text for explanation.

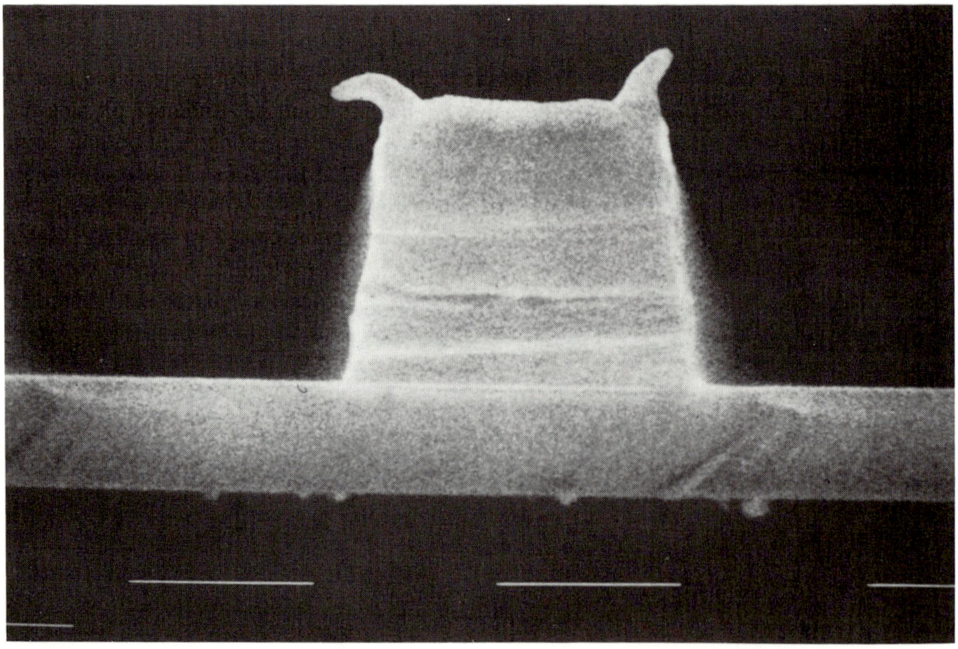

Figure 6.29. Wings along a polysilicon line etched in Cl_2–CCl_4. The photoresist has not been stripped; wing height corresponds with the original photoresist thickness. White bar is 1 μm.

A side-effect of the resist passivation is, that the resist need not be perfectly vertical to provide perfect linewidth control. Even if the (vertical) resist etch selectivity is low, the wing will protect the resist completely against lateral attack. For certain photoresist profiles and etch selectivities, the wing can be patterned into the underlying polysilicon, making the effect a permanent, and usually device-killing one. Otherwise, the wings can be removed rapidly in most conventional wet cleaning procedures [276].

Chapter 7

DIAGNOSTICS AND ENDPOINT DETECTION

7.1 Introduction

An enormous amount of data and information on plasmas has been gathered up till now. How has all this information been obtained? This point is not only of academic interest, but measurement of plasma properties instead of reading machine settings reduces the amount of "black-magic" involved in etching. In fact, certain diagnostic methods are a prerequisite for precise control of etching processes. Over the years a large number of techniques have been developed; these usually require some in situ "sensor" to measure the etch rates or to detect the endpoint when a film has been etched through to expose the underlying material.

Optimization of macroscopic phenomena such as the gas composition, pressure, and rf power, has produced many successful "recipes" for etching. With the advent of increasingly sophisticated techniques and computer modelling, however, it is now possible to obtain detailed information on the chemistry and physics of these plasmas and their interaction with surfaces. A vast number of diagnostic techniques have emerged. These have been used in the study of microscopic plasma properties and measurements on the surface of the material being etched. The different fields distinguish themselves for very different reasons. Optical methods are probably the most widely used. They are particularly suited for in situ plasma diagnostic studies because they are non-intrusive and can yield information which is location specific and time-resolved. Optical diagnostics can globally be split into two groups. Optical spectroscopy, which analyzes the light emitted or absorbed in plasmas, and laser probes, which are specifically suited for the analysis of surface modifications during the etching. Mass spectrometry provides a method for analyzing the particles emanating from plasmas. This technique is highly specific, but also suffers from some serious drawbacks. Electrical probes, on the other hand, are probably the simplest of diagnostic techniques, since it involves the insertion of a solid object into the plasma and the measurement of the particle and energy fluxes to that object; that is, a probe. The interpretation of the probe signals are however extremely difficult, even for the simplest of plasmas.

Not only the study of fundamental plasma properties has been greatly enhanced by these developments, but also the practical implementation of plasmas through the provision of real-time signals which are representative of the progress of etching. The most important of these is undoubtedly endpoint detection; a signal indicating the completion of the etch process. Any practical reactor used in semiconductor processing today cannot be termed practical if it is not equipped with some sort of endpoint detection system. It is therefore important to consider what actually makes a good signal for endpoint detection:

- Foremost: reproducibility. Wafer to wafer, batch to batch, day to day, irrespective of machine conditions.
- It is important to use a signal with a well defined origin. If the source of the signal (emission, pressure variation, and so on) cannot be determined unambiguously, the risk that the signal changes or disappears for unknown reasons is great. Also, the effect of transfer to another etching system, or the effect of small process adaptations, is unpredictable.
- Signals directly related to a chemical species responsible for etching, or to an etching product, are good candidates. However, product signals are often very weak; while the signals from the reactants may be large, for process reasons (depletion) large variations are undesired.
- In all cases the response time of the signal must be considered. If a product is monitored, the residence time in the plasma is of importance. This time must be short compared to the etching time.
- The signal must have a clear relation to the end point on the wafer (or batch) as a whole. Often this has to be determined experimentally.
- Signal recognition must be operator independent, thus making it possible to automate.

In this chapter an overview will be given of the various methods used and, to some extent, the theoretical background of the many techniques. The emphasis will lie on those diagnostics used in endpoint detection. For more in-depth literature on the subject matter, especially the diagnostics used for the study of fundamental plasma properties, the reader is referred to a recent series of books edited by Auciello and Flamm [277,278].

7.2 Macroscopic phenomena

Electrical measurements

Several methods have been developed to measure the electrical properties of plasmas. The plasma potential, DC-bias, electron and ion densities and temperatures are important parameters determining the chemical processes in the plasma, as well as interactions with the walls of the chamber and, last but certainly not least, the wafer. Some electrical parameters related to the plasma can be deduced from machine settings; others are relatively easy to measure; power dissipation, DC-bias, and network settings. The theory of plasma impedance and DC-bias in relation to chemical composition is complex and precludes direct predictions. It is, nevertheless, possible to use variations in power, bias and impedance for endpoint detection of etching processes.

Reflected power measurements for endpoint detection has been reported for polysilicon and silicon nitride etching [279]. This technique is simple to apply. Its main limitation is that it requires a discontinuity of the plasma impedance at the end of the etching process to give the required signal in the reflected power. This will depend strongly on the layers being etched. Plasma impedance monitoring has been applied for endpoint detection during an aluminum etch process in a low frequency commercial etcher. The power variation at constant current due to changes in plasma impedance towards endpoint are monitored. In practice, the method appears to be very sensitive to process variations (of unknown origin!). When applied to load-locked systems, impedance monitoring appears to give more stable results.

A number of authors [280,281,282,283] have made the assumption that the electrical series resistance of the rf glow discharge is determined to a large extent by electron drift in the bulk of the plasma. Under this condition the electron density is readily calculated from a knowledge of the electron mobility in the gas under observation and the measured resistance. Unfortunately this assumption is incorrect; most of the energy is lost in the reactor and the matching network and the dominant loss in the discharge is acceleration in the dark spaces. Consequently, the references mentioned state unrealistically low electron densities.

In the case of very asymmetric systems, the DC-bias is more or less equal to the potential difference between substrate and "ground". In that case, it is relatively easy to measure. In other cases, a probe technique needs to be used (see further on). It has been shown [59] that in commercial etching equipment there can be high power losses, rendering comparison of power dissipation in the plasma difficult. Comparison of DC-bias, on the other hand, is far more reliable.

Reasonable estimates for the electron density in systems not containing significant amounts of negative ions can be calculated directly from the measured DC-bias and dark space thicknesses at both the grounded and the powered electrode. This relies on the validity of the Child-Langmuir equation, as well as the Bohm criterion for describing the ion flux density in the dark spaces (see chapter 3). In low-pressure systems the agreement between these simple measurements and other more elaborate techniques is encouraging [39].

A non-intrusive method to determine electron density is, for example, microwave resonance. With this method the (average) electron density can be determined from the resonance frequency of a plasma filled volume. The method is based upon the fact that electromagnetic waves can propagate freely in a plasma if their frequency exceeds the plasma frequency. For these waves, the plasma acts as a dielectric medium with a permittivity dependent on the electron plasma frequency ω_e, and hence on the electron density n_e. The propagation of waves through the plasma can be measured; the frequency at which attenuation is observed is then the electron plasma frequency [54]. This method has been applied in commercial systems [66]. Also, the shift in resonance frequency $\Delta\omega$ of a cavity when a plasma is present, as shown in figure 7.1, can be related to the electron density [284,285]. An advantage of the resonance method is that the resonator need not be in contact with the plasma, unlike other electrical methods. A limitation of the resonance technique is that it is sensitive to the total number of free electrons inside the resonator, and the method gives no information on spatial distribution.

Temperature measurement

One of the most obvious effects of ion bombardment of the surface from the plasma, is the heating of the film being etched. Even if the wafers come out cold from the reactor, in the plasma they may have been heated to substantial temperatures (as is sometimes seen with "exploding" photoresist). Clamping to some form of heat-sink is often insufficient. Dependent on the heat capacity and conductivity of the substrate and its heat contact to the surroundings

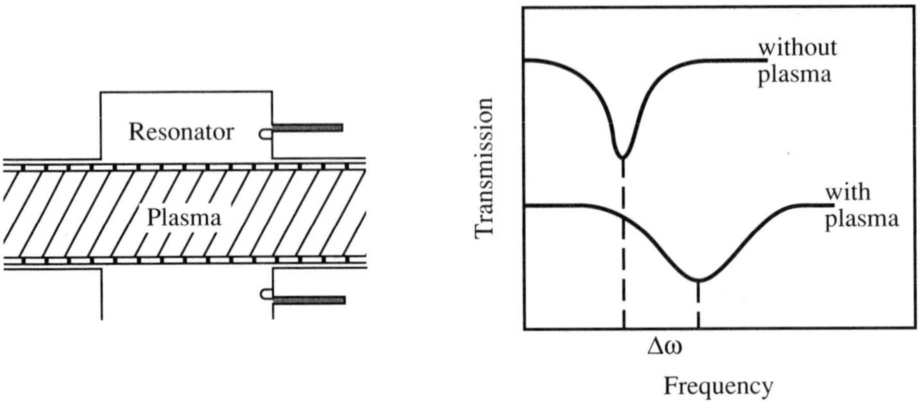

Figure 7.1. Schematic representation of the microwave resonance technique (left) and the resulting signals (right).

(substrate table, backflow, etch gas) an equilibrium temperature will be reached after some time, see figure 7.2 after Visser [286]. This temperature will, of course, also depend on the amount of power being put into the plasma and the equipment set-up. Measurement of the surface temperature renders comparisons between etch systems and transfer of processes easier, since a direct parameter is being measured in contrast to, for example, the power at the input of the system [287,288].

Until recently, measurement of the surface temperature was very difficult, mainly due to electromagnetic interference from the plasma. In most cases, only the substrate holder temperature could be measured. For surface measurements special thin film thermocouples

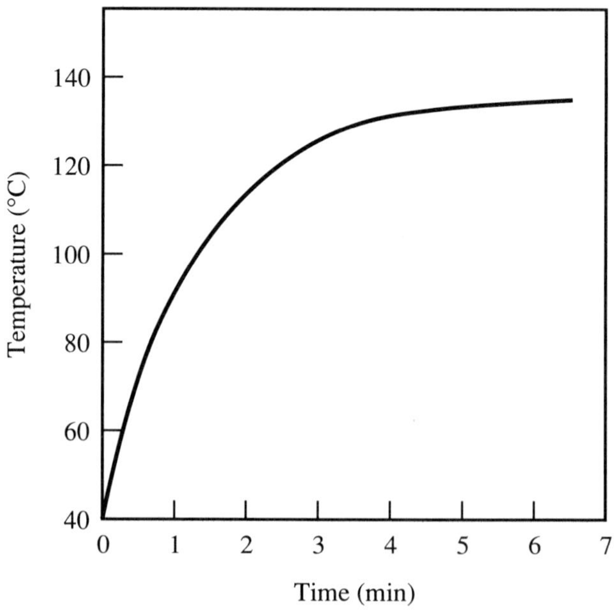

Figure 7.2. Temperature rise of a wafer during exposure to an argon plasma (p = 6.5 Pa, P = 3.5 kW/m^2).

have to be made and these can only be used with the plasma off. Optical methods, such as pyrometry, suffer from the very limited accessibility usually found in plasma etching systems.

Much more convenient is the fluoroptic thermometer which has been used to measure surface temperatures in studies of the effect of different cooling methods [289]. With this technique Visser [286] analyzed the heat transport to and from a silicon wafer when exposed to an rf plasma. In this study it was shown that such an analysis provides a convenient and accurate method to measure the total energy flux on the electrodes of a low pressure rf plasma reactor.

Pressure measurement

Pressure as an endpoint detector is often very easy to implement. The method is based on the following principle. On ignition of the plasma the gas is dissociated and heated. At constant pumping speed, gas heating will cause a pressure rise which will stabilize after equilibration with wall losses. Dissociation of the gas will bring more molecules in the volume and chemical reactions will alter this number. To be useful as an endpoint detector, the processes during etching have to result in an increase or decrease of the number of molecules towards endpoint. This will strongly depend on the composition of the reactive components and the etch products. A problem is that in general depletion of gases is avoided, implying that pressure variations will be small (less than 5%). Hence, contamination or small process variations in the samples may lead to the total disappearance of the endpoint signal.

A further drawback is that pressure monitoring is very non-specific, since the measured variation in pressure is generally the sum of a large number of different processes, which are generally not known in detail. It is also highly process and reactor dependent, as well as being dependent on the vacuum pumps used and their condition in time. For these reasons, pressure variation is seldom used as a diagnostic tool.

7.3 Optical spectroscopy

Emission spectroscopy

It is well known that many gas plasmas emit light. The bright colors of gas discharges motivated early research on this subject; spectroscopy is one of the oldest fields in plasma physics. The emission of light from the plasma is caused by the large number of radiative paths causing decay of an excited state. The information contained in emitted light is information on the nature of the emitting, i.e. excited, species. In general the bulk of the heavy particles in the plasma will be in their electronic ground state. Thus, in order to obtain absolute densities for the total population, calculation of the ground state population from the measured excited state population is necessary. Reference [290] gives a good overview of the various processes involved.

Examples of set-ups for the measurement of emitted light are shown in figure 7.3 after Degenkolb and Griffith [291]. Measurement of the emitted light is relatively easy to do and is a non-intrusive technique. Derivation of the electron energy distribution function from this, though, is extremely complex and not readily measured or calculated. To overcome this, some authors have employed a technique called actinometry [292,293,294]. This consists of adding

Table 7.1. Emission lines used for end point detection

Material to be etched	Etch chemistry	Species monitored	Wavelength (nm)
SiO_2	CF_4–CHF_3–Ar	CO	488
Si_3N_4	CF_4–CHF_3–Ar	N_2	452–650
Si_3N_4	CF_4–CF_3Br	Broadband visible	
Si	CF_4–O_2	F	704
Al	Cl_2–CCl_4	Al	396
Resist	O_2	CO	450

a small quantity of a noble gas (the actinometer) to a reactive plasma. The actinometer is chosen such, that the threshold energy and the collision cross section energy dependence are similar to that of the species of unknown concentration. The ratio of the emission intensities of the unknown species and the actinometer are then assumed to be equal to the actual concentration ratio. Since the amount of actinometer is known, the density of the unknown species can be readily calculated. For atomic fluorine, good results have been reported [295, 296,297]. With other species, the situation is more complicated; one should be aware of that fact that many excited states do not, in fact, have an allowed radiative decay path. An example of the use of actinometry is given in figure 7.4 after Coburn and Chen [294]. This shows the intensity of the F atom emission at 703.7 nm and the F atom density variation as determined by actinometry (Ar at 750.4 nm was used as reference) as a function of pressure in a CF_4 plasma.

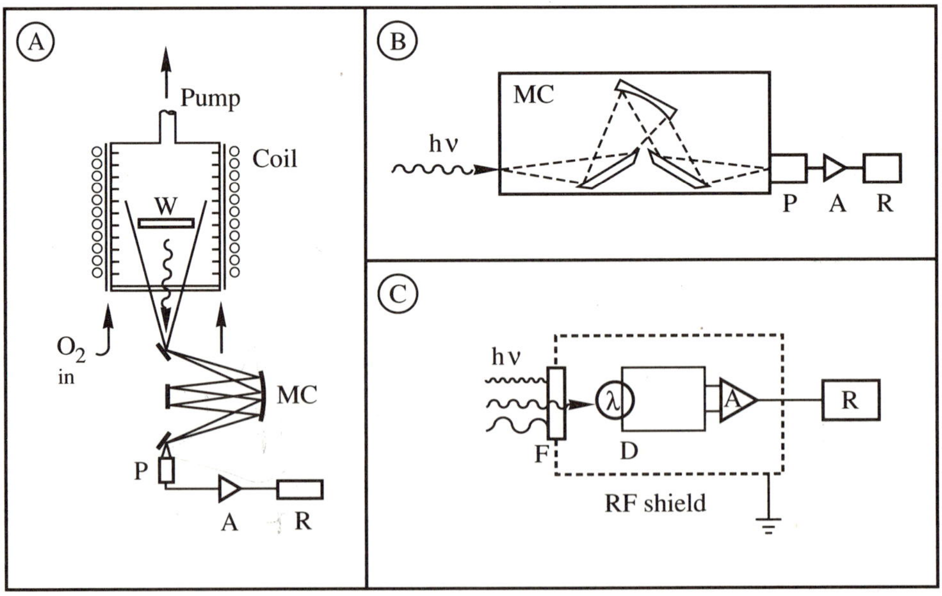

Figure 7.3. Experimental set-ups for the detection of light emitting species in photoresist stripping on a silicon wafer with an RF oxygen plasma (W = wafer; MC = monochromator; P = photomultiplier; D = photovoltaic detector; F = filter; A = amplifier; R = recorder).

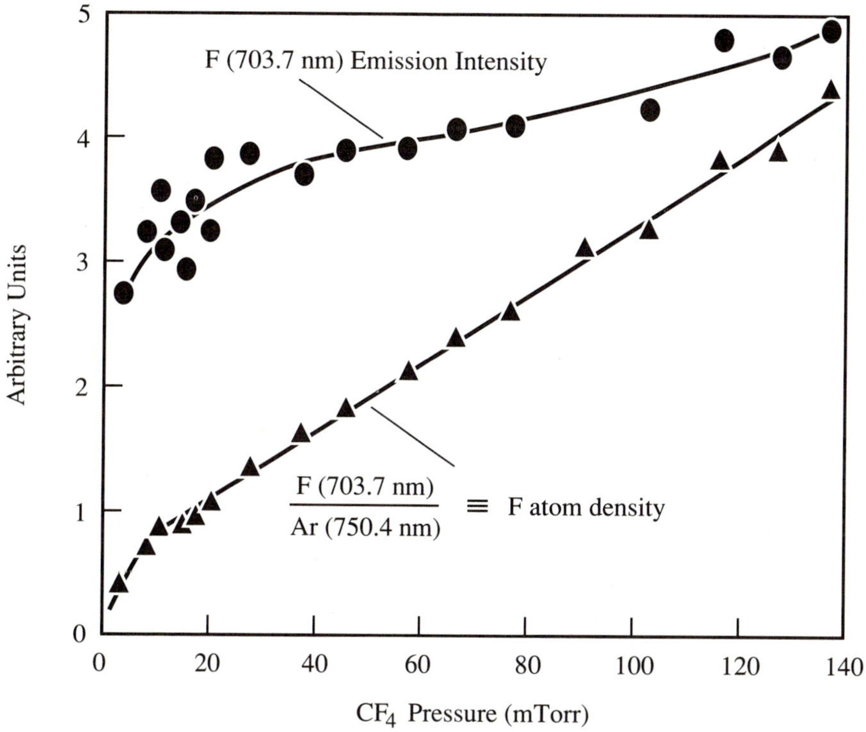

Figure 7.4. The F atomic emission (703.7 nm) and the F atom density as a function of pressure in a CF_4 discharge.

Time-resolved optical spectroscopy has been used to gain insight into the discharge dynamics of rf glow discharges. An example is the determination of the time required to accelerate ions across the sheath towards the electrode [298].

Apart from the more fundamental applications of emission spectroscopy described, the most common application of this method is endpoint detection of etch processes, by sensing changes in the emission from etch products and/or reactants. By comparing with other techniques, it has been shown that emission spectroscopy is a reliable monitor of the processes taking place in the plasma [299]. Many different applications have been found for emission spectroscopy as an endpoint monitor. For example, endpoints for etching aluminum with a photoresist mask have been determined by following the emission lines of AlCl [300] and CO [301]; the erosion of the photoresist can be monitored with the CCl line [302]. Some of the emission lines most commonly used for endpoint detection are summarized in table 7.1. An example of signals obtained, in this case for a resist stripping process, is given in figure 7.5.

Absorption spectroscopy

Although rf discharges are suited for absorption spectroscopy, relatively few diagnostic studies have been performed using this technique. The principle advantage is that the method is quite simple to implement in most cases and yields absolute concentrations, provided that the absorption cross-section is known [290]. Absorption spectroscopy has been widely used

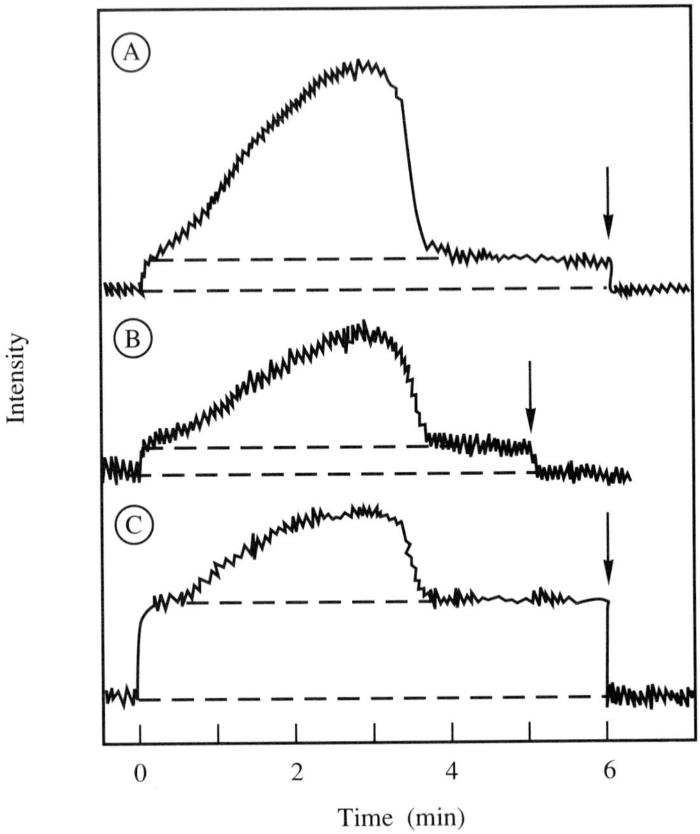

Figure 7.5. Intensity of CO radiation (483.5nm) versus time, during the stripping photoresist on silicon. The letters A, B and C refer to the measurement set-ups described in figure 7.3 (after Degenkolb and Griffith [291]). The arrows indicate the time at which the rf was switched off.

in gas-phase kinetic and photochemistry studies. Examples of the use of this technique in semiconductor processing can be found in the study of the dissociation of Br_2 [303], Cl_2 [304], and the study of fluorocarbon-oxygen plasma used to etch silicon [305,306]. It is important to measure the integrated absorption over the entire band to remove the effects of temperature broadening and shifting [303].

Advanced techniques, such as second derivative tunable laser absorption or Fourier transform-infrared spectroscopy (FTIR), can extend the detection limits by several orders of magnitude, making it possible to distinguish atoms and excited species. The advantages of FTIR are rapid acquisition of the full spectrum and improved signal-to-noise ratios; studies have been reported of N_2O discharges, and Cl_2 etching of aluminum [307,308] and polycrystalline silicon [308].

Optogalvanic spectroscopy

In optogalvanic spectroscopy the change of discharge parameters, such as current, voltage, electron density, induced by the absorption of light is monitored. Spectroscopic information is obtained by tuning the light source through atomic or molecular resonances.

A large number of mechanisms give rise to optogalvanic signals; only some will be considered here. The simplest mechanisms for creating light-induced changes in discharge characteristics are photo-ionization and photo-detachment. Photo-ionization of neutrals results in additional charge carriers and photodetachment of negative ions results in free electrons. These increases in electron density are easily detected by monitoring discharge current, voltage, and impedance, as well as by electrical probes. Spectroscopic information can then be obtained with tunable lasers to probe the frequency region near the threshold of the photo-induced process. Greenberg et al. [71] showed by this technique that the principal negative carriers in NF_3 discharges are negative ions rather than free electrons.

Excitation of atoms or molecules to highly excited states will result in subsequent collisional ionization, which results in easily detectable changes in discharge impedances.

Optogalvanic measurements are normally performed along a column in a manner similar to absorption spectroscopy; the spatial resolution is limited. Full three dimensional spatial resolution can be achieved by using a two-step excitation scheme, where the effect of interest occurs in a region where two laser beams intersect. This has been demonstrated by Doughty et al. [309]. By a similar technique, optogalvanic spectroscopy can provide a sensitive probe of ion motion in the cathode sheath region of the plasma [310].

Species specific optogalvanic spectroscopy has been demonstrated for a wide variety of neutral atoms and molecules, as well as positive and negative atomic and molecular ions. Furthermore, basic discharge parameters can be studied using light as a means of generating charge carriers. For example, ambipolar diffusion rates can be monitored by generating a local excess of electron-ion pairs by photoionization and then monitoring the time-dependent change in plasma density. The role of secondary electrons can be investigated by generating secondary electrons at an electrode surface with a pulsed laser and then monitoring the subsequent change in discharge current [311]. In fact, the same kind of technique has been applied as an in situ method for plasma electrode surface characterization [312].

A pulsed laser may also be used to generate photoelectrons, which increase the ion current in the discharge [313]. This is often referred to as the photoelectric effect, and has been shown to be suitable for endpoint detection in rf and dc plasmas, as it provides a real-time, in situ monitor of the surface during plasma exposure [314]. Unfortunately, for an endpoint tool it is very expensive.

Laser-induced fluorescence spectroscopy

Laser-induced fluorescence spectroscopy (LIFS), accesses the ground state population directly. In LIFS a well-defined gas volume is irradiated with a focussed, pulsed laser beam, to excite all particles of a selected species to a well-defined radiating state. The laser-induced emission is measured at right angles with the laser beam, as shown in figure 7.6. If all the selected species are excited, the emission intensity is then directly proportional to the density. With the aid of tables [315,316] the particle species can be identified by its characteristic lines. It is then in principle possible to calculate the ground state density, provided that the formation paths are known. With careful calculation of the constants involved, it is possible to obtain estimates for the electron density [53]. In plasmas with negligible concentrations of negative ions, absolute values of the electron density can also be obtained by measuring the

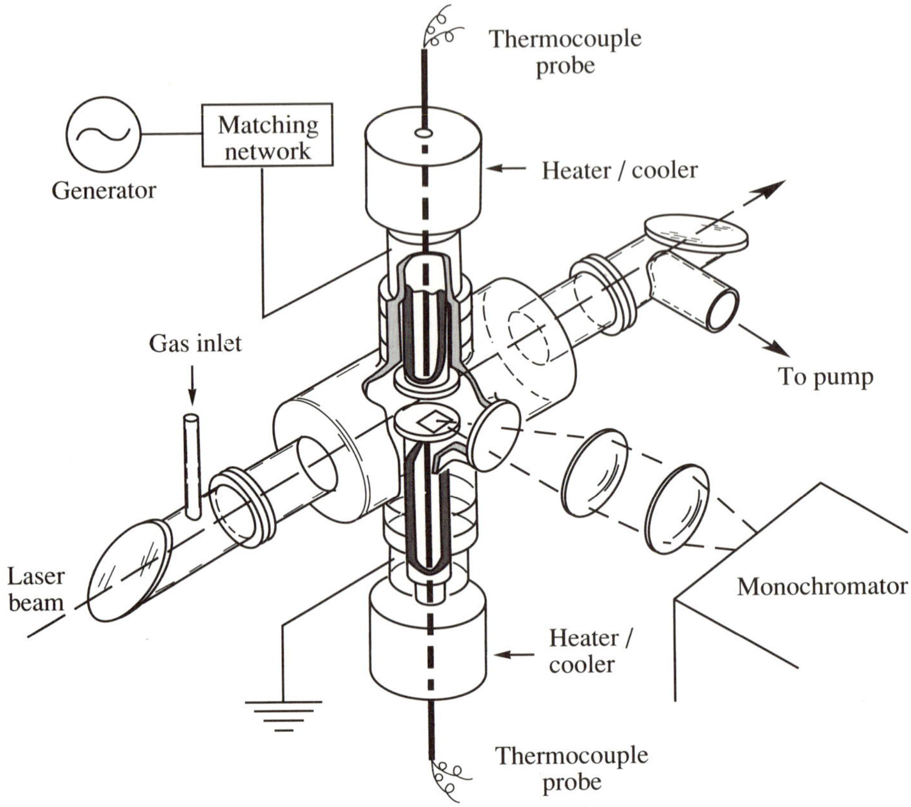

Figure 7.6. Schematic overview of an apparatus for Laser-induced fluorescence spectroscopy.

decay rate of the ion density after switching off the rf power and using known rate coefficients for the different decay mechanisms [40]. See, for example, equation 4.34 and figure 4.15 for N_2^+.

LIFS can also be used to determine plasma ion temperature. This can be found by the Doppler broadening of a moving ion's spectral line caused by the pulsed laser. A problem encountered is that if the intensity of the laser is too high, the spectral line is also broadened by an instrumental effect called power broadening. Ignoring this power broadening may result in measurement errors of several orders of magnitude. This problem is avoided by reducing the intensity, but this will also reduce the fluorescence signal; techniques have been developed to determine the optimum condition [317]. LIFS has also been used to measure velocity distributions of atoms and small molecules in plasmas [318,319,320].

The advantages of LIFS are that it is selective and extremely sensitive, densities below 10^{14} m^{-3} can often be detected. It has, for example, been used to detect the presence of sulphur dimers (S_2) in SF_6 plasmas [321]. A further example of its sensitivity is the detection of dopant As in the discharge when etching silicon [322]. Due to its sensitivity, LIFS is an excellent technique for plasma diagnostics. This sensitivity, combined with the complexity of the equipment set-up render it completely impossible for endpoint detection. Many studies of uncharged species, neutral radicals, as well as positive ions, have been performed with the aid of LIFS in rf discharges, see references [290,323].

7.4 Laser probes

Recently, several optical techniques, extensively used in fields other than plasma, have found their way into the field of plasma diagnostics. These optical "probe" techniques make it possible to analyze real-time in situ changes in the surface exposed to the plasma.

Laser interferometry

The simplest laser probe technique is laser interferometry, whereby changes in the surface film thickness are monitored by changes in the reflectance of the film. No chemical information is obtained, but it does give an accurate measure of the etch rates and can easily be used for endpoint detection [324]. The technique is simple to implement. It is required that the film under observation be at least partially transparent and that it is deposited on a smooth substrate. Furthermore, the film and substrate must have different optical constants; the interface must reflect a sufficient fraction of the incident light. The reflectance changes during etching are caused by interference of the laser beam directed at the surface with reflected light from the interfaces between the various layers (vertical interference) or reflected from different regions (etched/non-etched) of the surface (lateral interference).

For a beam striking the surface at an angle of incidence θ with respect to normal incidence (perpendicular to the surface), the amplitude of the reflected beam will oscillate sinusoidally due to constructive and destructive interference with a period Δ, given by:

$$\Delta = \frac{\lambda}{2\sqrt{n_R^2 - \sin^2\theta}} \tag{7.1}$$

where n_R is the refractive index of the film and λ the wavelength. Maximum depth resolution is obtained at normal incidence, where $\Delta = \lambda/2n_R$, with resolution limited only by the wavelength of the laser used.

This technique can be used for endpoint detection. At endpoint, the interference pattern will be disrupted, which can be easily recognized. When more layers are etched or lateral interference with, for example, an etching mask are present, the situation is more complicated. An example of the signals obtained is shown in figure 7.7 after Heimann [325]. Note that the problem of lateral interference is very real, since the lasers generally used in interferometers have a spot-size of 1 to 2 mm, or larger; the patterns being etched are much smaller than this. From figure 7.7 it is clear that the determination of the actual endpoint can be extremely complex, and signal processing may be required to be able to recognize when a specified depth (endpoint) has been reached. It is also clear from this figure, that interferometry can be used as an in situ monitor of the etch-rate, by counting the interference periods in time.

Ellipsometry

Ellipsometry is a very useful technique for measuring small changes in thickness or thin layers [326]. Ellipsometry uses polarized light; the change in the angle of polarization and the amplitude of the light due to reflection at a surface can be related to the optical constants of the film under observation.

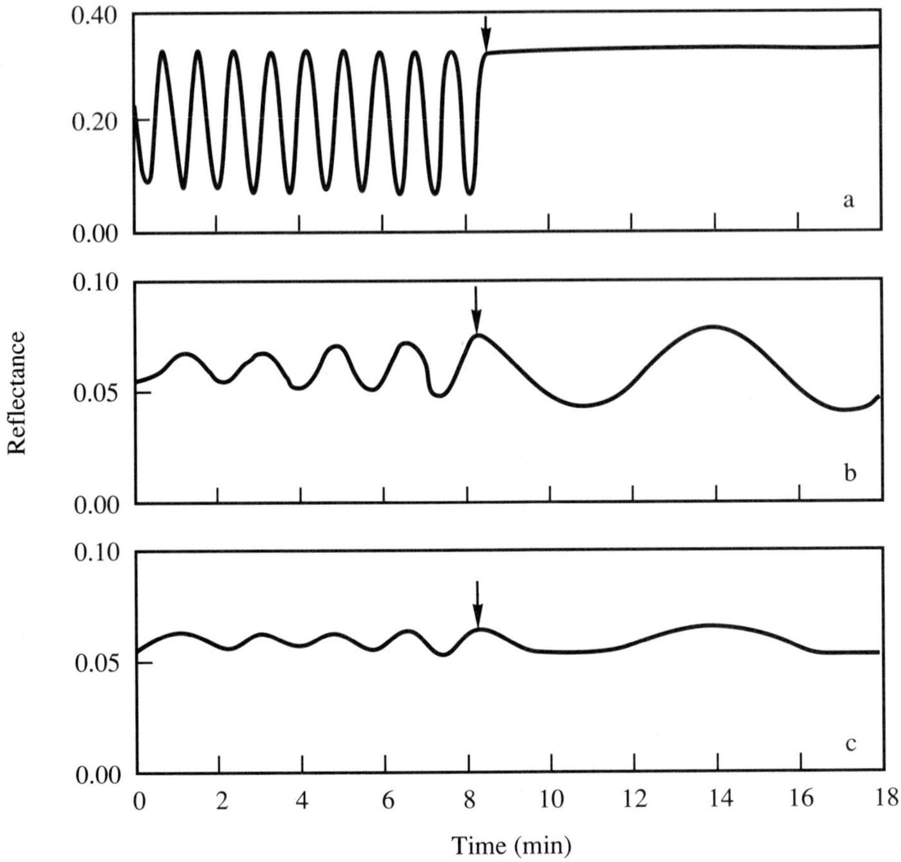

Figure 7.7. Laser interferometric signals at 253.7 nm for endpoint detection in silicon oxide etching: no resist coverage (a), 95% resist coverage (b) and 99% resist coverage (c), with a 5:1 selectivity of oxide to resist. The arrows indicate the point of etching through the oxide.

One of the foremost aspects of ellipsometry is its extreme temporal resolution. The measurement of the angle of polarization, initially performed with mechanically rotating analyzers within 10 ms, was firstly reduced with piezobirefringent elements to approximately 20 μs [327,328], and has now been reduced to the order of 1 ns [329]. This makes ellipsometry suited for real-time monitoring of plasma etching. Real-time ellipsometry has been used to study the plasma enhanced CVD of amorphous Si and Ge [330,331].

Spectroscopic ellipsometry is less suited for real-time analysis, but can be useful for measuring minute changes in the surface under observation. Monatomic coverage of films has been reported, see reference [332]. Reports have appeared on its use during the hydrogen plasma etching of compound semiconductors [333,334], as well as the study of damage caused by plasma processing [335].

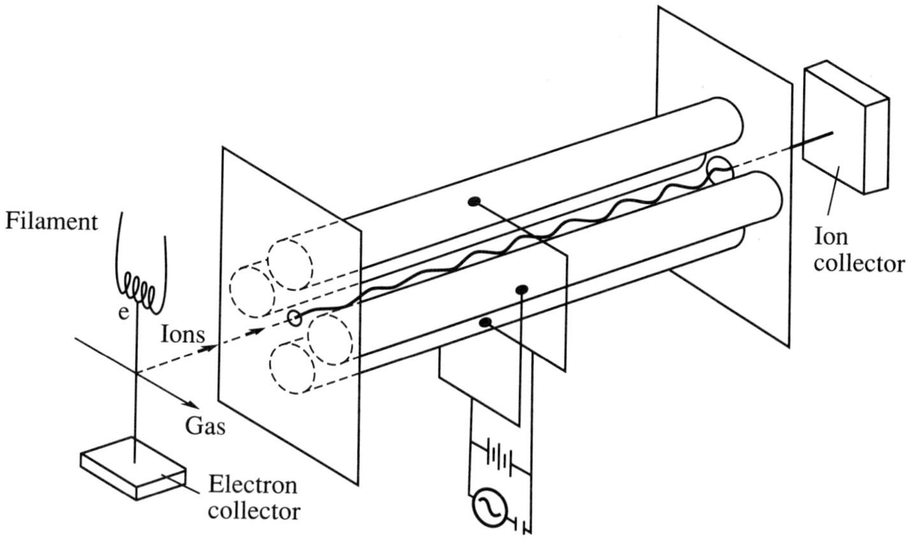

Figure 7.8. Quadrupole mass filter.

7.5 Mass spectrometry

Another means of monitoring reactant and product species is mass spectrometry. Particles from the plasma are sampled through a hole in the chamber wall. After passing this orifice, the neutrals are generally ionized by electrons from a filament, though lasers have also been used. Above the appearance potential, molecules will also be fragmented. An example is the formation of the CF_2^+ ion. This can originate from the CF_2 radical, with an appearance potential of 11.7 eV, but also from the molecules CHF_3 (17.5 eV) and C_2F_4 (15.2 eV).

Fragmentation patterns and appearance potentials can often be found in the literature [336]. The principle of detection is by separation of the various masses in electromagnetic fields. Dependent on their mass, the ionized particles may move in a path leading to the particle counter; by varying the fields, the mass range can be scanned. Figure 7.8 shows a common set-up for mass separation, the quadrupole mass filter.

To ensure that the particle trajectories are influenced by the electromagnetic field only, collisions have to be reduced. This is achieved in medium vacuum conditions; mass spectrometers operate at pressures below 10^{-5} Torr. This means that special techniques, such as differential pumping, need to be implemented to use this method in dry etching systems. The method is very sensitive and highly specific. Detection of some components is difficult, due to the fact that they have the same mass/charge ratio (N_2, CO, Si). A further problem to be aware of in sampling species from a plasma through a dark space, is the possibility that short lived species may have reacted before reaching the orifice.

In high vacuum systems, mass spectrometry combined with time-of-flight spectroscopy has been used to determine primary etch products of several etching [102] and deposition reactions [337]. Mass spectrometry can also be used in the study of etching mechanisms, as shown in the discussion of plasma-surface interactions in chapter 4. A further application is mass spectrometry of ions in combination with energy analysis, which gives information on the processes in the plasma sheath and on the bombardment of the wafer.

Mass spectrometry is seldom used for endpoint detection, due to the sensitivity of parts such as the filaments to the reactive gas species.

7.6 Electrical probes

Electrical probes are probably the simplest of diagnostic tools; they may amount to no more than the insertion of a bare wire into the plasma. They are generally easy to construct and can be used over an extremely wide range of plasma densities. These probes measure electrical currents, which will depend on their bias with respect to the plasma potential. The current "collected" by the probe consists of a combination of incoming ion and electron currents from the plasma, and secondary electron currents produced by (electron, ion, photon) bombardment of the probe. The current-voltage (I-V) characteristics can, in principle, be related to plasma parameters such as electron temperature and density. Care needs to be taken, though. When a probe is immersed in a plasma, it strongly perturbs the potential over a small region by creating a sheath. This means that the data analysis is always ambiguous to some extent, as a full interpretation requires knowledge of the interaction between plasma and surface, the charged carrier flow in the space charge dominated sheaths and the transition between plasma and sheath. Assumptions need to be made with respect to the collection radius of the probe. If there is any form of electrical interference, the probe will not be floating. Since the collection radius is determined by the ion sheath thickness corresponding to the local bias potential, interpretations are tricky.

For example, electron temperatures in dc-excited glow discharges obtained with electrostatic probes are usually of the range of 1-3 eV [338,339]. Several authors have reported electron temperatures of the order of 10 eV for rf-excitation [339,340,341]. The data in the latter references have probably been obtained without proper prevention of electrical interference between excitation and measurement circuits. In that case the I-V characteristics of the probe are determined by the rf ion-sheath formation rather than by direct plasma properties. Hence, the conclusions in these papers are questionable. Electron temperatures obtained by Cantin and Gagne [338] in rf tube discharges and by Mosburg et al. [342] in rf planar discharges are similar to that in the dc case; Okuno et al. [41] came to the same conclusion by momentarily switching the plasma off. Electron density measurements from probes also need to be treated with extreme care.

The example given only serves as an illustration of the difficulties encountered in the interpretation of probe measurements. The following are a number of the most common problems overlooked when dealing with electrical probes.

- Collecting probes do not work well in sheaths.
- Probes do not distinguish between kinetic energy and plasma potential.
- Secondary electron emission can cause electron temperatures to be underestimated.
- Positively biased probes can produce significant ionization in glow discharges.

Extensive literature is available on the ideal characteristics of probes and many reviews have been published [343,344,345]. In this section, probe theory will only be dealt with on a qualitative basis and only a small number of techniques will be dealt with, in order to provide some insight into the methodology of electrical probes.

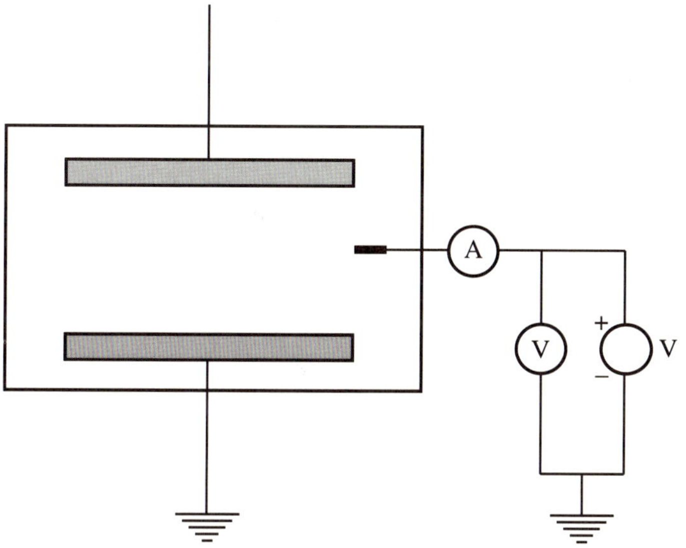

Figure 7.9. Schematic representation of a single probe.

Probe theory and the single probe

Consider a small electrode (probe) introduced into a dry etch reactor. For a complete electrical circuit, the wall or an electrode must be included in the circuit to pick up the return current from the probe. This is shown schematically in figure 7.9. The current-voltage characteristics of this electrical circuit are shown in figure 7.10. If the electrode is large compared to the probe, the characteristics will be influenced mainly by current collection at the probe. This is called the single probe method and is the simplest to understand. In the literature it is often referred to as a "Langmuir probe", though that is a special case of the single probe as applied to low pressure regimes, where the mean free path between collisions is large relative to probe and ion-sheath dimensions.

Assume that the current being drawn by the probe from the plasma is positive; let V_p be the plasma potential with respect to the probe ground. When the probe bias V is very negative with respect to V_p, the electric field around the probe will prevent all but the most energetic electrons from reaching the probe, effectively reducing the electron current to zero. The current collected by the probe will then be entirely due to positive ions, since these encounter only an attracting electric field; this is called the "ion-saturation current" I_{is}.

As the probe bias is increased, the number of electrons which is able to overcome the repelling electric field and contribute a negative current increases exponentially. Eventually, the electron current collected is equal to $-I_{is}$ (total current 0). At this point the floating potential V_f is reached. This potential is smaller than V_p, since the electron thermal velocity is greater than that of the ions. When the probe is allowed to "float", independent of bias, it quickly develops the potential V_f to repel electrons. Further increase of the probe bias to V_p allows the electron current to totally dominate the probe current. At V_p, electrons are unrestricted from being collected by the probe. Any further increase of the bias will simply add energy to the electrons, not the current drawn. Hence the term "electron-saturation current" I_{es} is used at this limit.

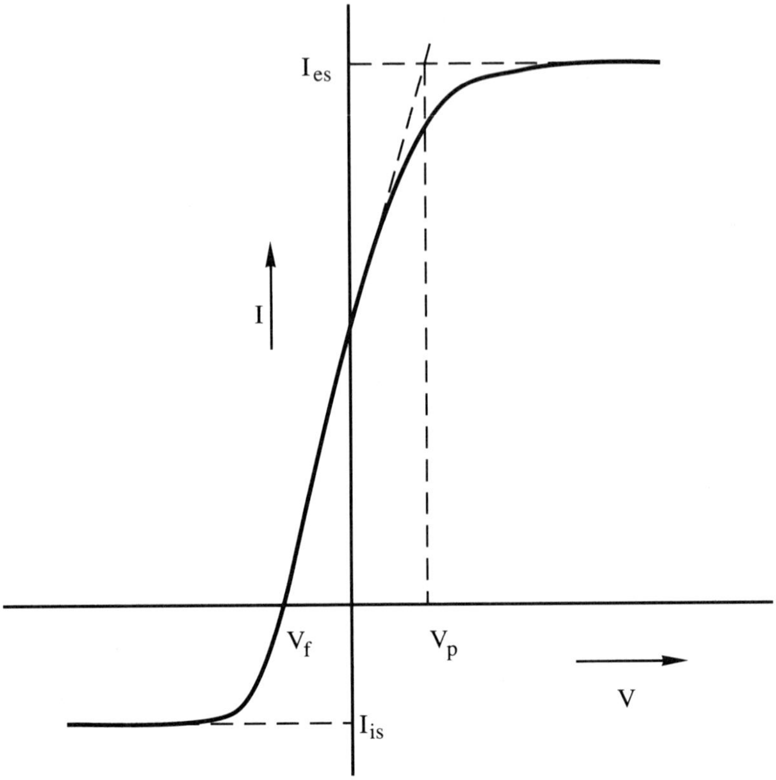

Figure 7.10. Current-voltage characteristics for a single probe.

Note that this is a simplified description of the ideal I-V characteristics of the probe, without taking into account many of the "disturbing" processes such as the bombardment of the probe by energetic electrons and the emission of secondary electrons from the probe. From the floating potential, the electron temperature can be determined; the ion-saturation current then gives the electron density. The relevant equations have been derived in chapter 3 [344].

Multiple probes

In rf and electrodeless discharges there is often no suitable pick-up electrode for the return current. In these cases double or multiple probes are used. These probes are completely floating; the external probe circuit is isolated from ground. Thus, the probes should follow changes in the plasma potential. Due to unavoidable stray capacitances, this only holds true if the changes are not too fast.

The most commonly used multiple probe scheme is that of symmetric double probes, which was first analyzed by Johnson and Malter in 1950 [346]. More recent data has been reviewed by Cherrington [343]; the basic configuration is shown in figure 7.11. Provided all plasma parameters are the same at both probes, and that the spacing of the probes is such, that the regions distorted by the probes do not overlap (that is, the spacing is large compared to the Debye length), and a Maxwellian distribution function can be assumed for the electrons, then the current-voltage characteristic can be deduced to originate from two opposite single probes, and the plasma parameters can be calculated. One drawback is, that the symmetric double

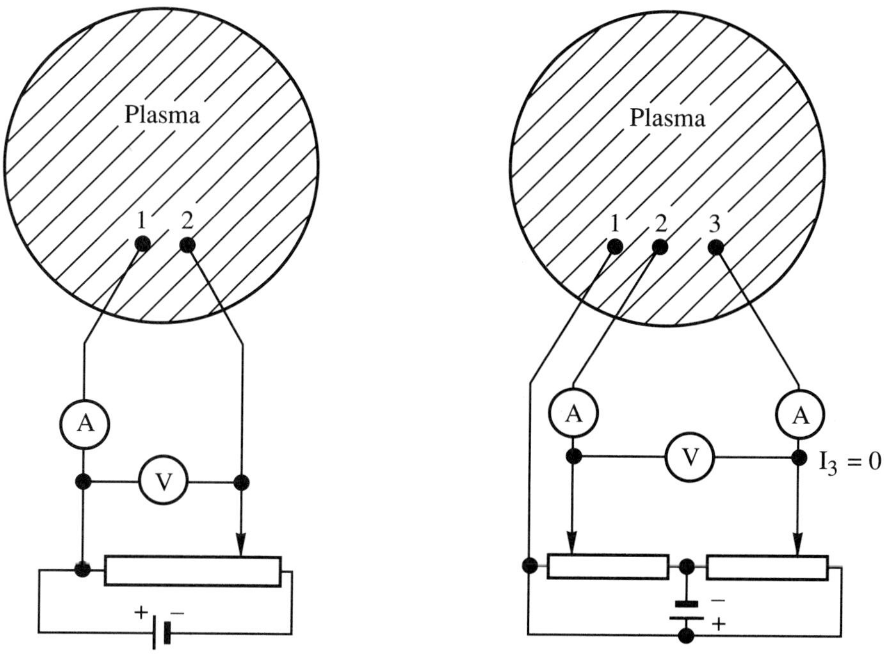

Figure 7.11. Basic circuits for double (left) and triple probe (right) measurements.

probe measures only those electrons which can overcome the floating potential of the probes. In other words, the high energy part of the electron distribution function. Unfortunately, that is where the Maxwellian distributions generally do not hold, see chapter 4.

Making one of the probes much larger would help. If one of the probes were sufficiently large, a single probe characteristic would result. However, it can be shown that the area ratio would need to be at least of the order of one thousand, which is of course not practical.

This difficulty can be overcome if a small third probe is introduced, which is held at the floating potential. This was first introduced by Okuda and Yamamoto [347], and improved later [348]. The third probe acts as a reference electrode and in the symmetric case results in a true single probe characteristic for probe voltages more negative than the floating potential. For the asymmetric case the area of the current pick-up probe needs only be larger than the measurement probe by a factor $\sqrt{m_i/m_e}$ to obtain the single probe characteristic, where m_i and m_e denote the ion and electron mass, respectively. The basic circuit is depicted in figure 7.11. The current to probe 2 is measured as a function of the voltage between probe 2 and 3 while probe 3 is kept floating. In this way it is possible to obtain the electron distribution function.

An interesting modification for fast scanning of plasma parameters is the "instantaneous triple probe method" proposed by Chen and Sekiguchi [349]. In this case the voltage is fixed between probes 1 and 2, with the third probe floating. Assuming a Maxwellian energy distribution, it is possible to obtain temperature and density from the potential across probes 2 and 3 and the current. The big advantage of this method is that it does not require a voltage sweep, thus enabling a direct display of T_e and n_e as a function of time, even in rapidly changing plasmas. The errors and shortcomings of this method are the same as in the symmetric double probe technique.

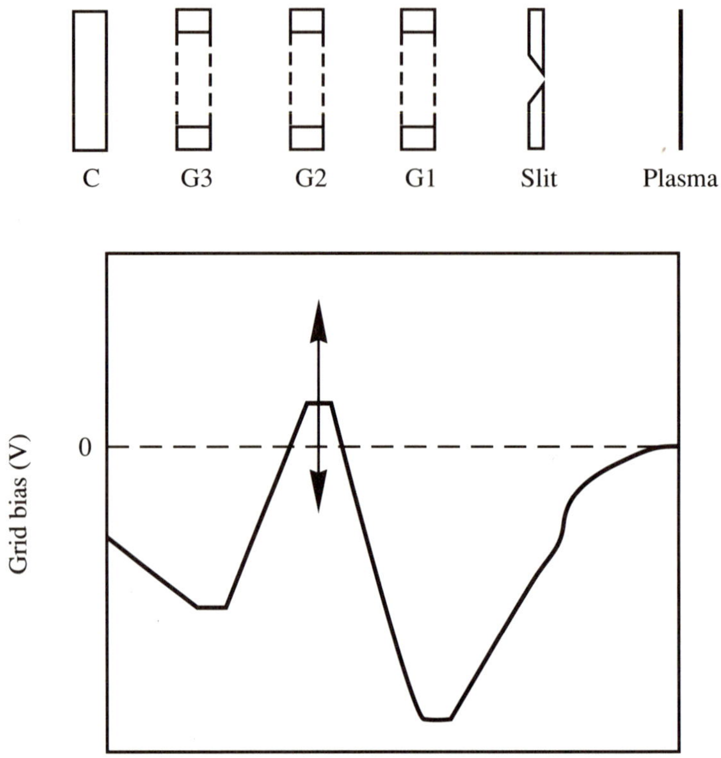

Figure 7.12. Schematic representation of the gridded energy analyzer (top) and the applied grid bias (bottom).

Ion energy analysis

Ion energy distributions cannot be measured by a simple probe technique, since the small ion current to the probe is totally masked by the large electron currents. Some method is therefore required to separate the ion current from the electron current. In discharges in a magnetic field, this can readily be achieved by using the magnetic field to separate the ions and electrons in a small mass spectrometer. This is the operating principle of the ion sensitive probes [350,351]. This same principle can be used to separate electrons and negative ions. In discharges without magnetic fields, probes with small permanent magnets can be used. All of these devices are seldom used.

More common is the use of grids, in the form of a gridded energy analyzer (GEA), shown schematically in figure 7.12. The entrance aperture consists of a grid or slit, dependent on particle flux considerations. The aperture produces a sheath which simultaneously reduces the electron flux to the interior of the analyzer and retains the Maxwellian nature of the ion distribution function [352]. The first grid G_1 is used to repel the unwanted species, in this case electrons. The bias of G_2 is then varied to limit the collection of ions. The last grid is biased to suppress secondary electron emission from the collector. By measuring the current at the collector as a function of the bias at G_2, it is possible to determine T_i in the same way as T_e with single probes. Note that by correct biasing it is also possible to determine electron parameters. The main drawback of this technique is the size of the analyzer, which will create a significant disturbance of the plasma (and the reactor). For more details, the reader is referred to reference [352].

7.7 Signal processing and automation

Up till now many different "sensors" have been described which have some relationship with the progress of etching in the reactor. The sensor is only one part of any "real" endpoint detector. The second part is some processor that determines whether endpoint has been reached. The simplest processor is no doubt a human being. Unfortunately, this processor is quite expensive and we may doubt the reproducibility in time or from one operator to another. Automation is therefore required.

As discussed, the signals obtained from the sensor need to show some variation towards endpoint to be used for detection. Furthermore, for automation it is generally required that the signal be in a specific voltage range – resulting in the need signal amplification. Any practical signal will be noisy. By reducing the noise level, better quality signals and more controlled endpoint detection can be obtained. Three categories of noise can be considered,

- Noise which is fast with respect to the time-scale of the signal.
- Noise on the same time-scale as the signal.
- Noise which is slow with respect to the time-scale of the process.

Fast noise comes from small fluctuations in the plasma and from the electronics in the sensor and amplifier circuitry. This type of noise can generally be filtered out by a low-pass filter. Noise on the same time-scale as the process itself is very difficult to remove. It can for example be caused by rotating electrodes. Noise correlation techniques are then needed. Slow variations are generally caused by drifting electronics and variations in process conditions during processing (such as temperature and pressure). A high-pass filter is generally not practical – process stabilization or calibration for these variations is usually easier to implement.

Assuming that a signal can be obtained which is sufficiently noise-free to give a clear variation towards endpoint, the decision whether endpoint has been reached still has to be made. Three different systems are commonly used for deciding when endpoint has been reached:

Threshold is probably the simplest signal interpreter. Endpoint is detected when the signal comes above or below a predefined value. This is an absolute measurement, hence sensitive to plasma variations, loading effects, contamination, and so on. A more reliable method is obtained by detecting a relative change. This is the oldest and simplest technique.

Slope detection evaluates the derivative of the signal as a function of time. The point of an abrupt change in signal is detected and used for endpoint. This technique is more advanced than threshold detection; a disadvantage is that it is sensitive to sudden changes, which can quite easily be caused by noise rather than the endpoint itself.

Curve recognition is a technique by which the knowledge of how a signal should look is programmed into the interpreter. In most cases this means that slopes and baselines are detected in a certain sequence; endpoint is reached when the whole sequence has been completed. These systems are process specific and very reliable, but difficult to program. A very advanced, general purpose signal recognition interpreter has been reported on [353]. A discussion on the problems associated with interferometric endpoint signals is given in reference [324]. Reference [354] provides an overview of various commercial endpoint detection systems.

REFERENCES

[1] J.R. Hollahan, "Plasma reaction system for reacting a gas with a non-gaseous material", U.S. Patent 3,428,548 (1969).

[2] S.M. Irving, K.E. Lemons, and G.E. Bobos, "Gas plasma vapor etching process", U.S. Patent 3,615,956 (1971).

[3] A. Jacob, "Process and material for manufacturing semiconductor devices", U.S. Patent 3,795,557 (1974).

[4] R.G. Poulsen, "Plasma etching in integrated circuit manufacture – a review", J. Vac. Sci. Technol. **14**, 266 (1977).

[5] N. Hosokawa, R. Matsuzaki, and T. Asamaki, "RF sputter-etching by fluoro-chloro-hydrocarbon gases", in Proc. 6th Int. Vacuum Congr. (Vacuum Society Japan, Kyoto, 1974), Jap. J. Appl. Phys. Suppl. 2, part 1, 435 (1974).

[6] J.A. Bondur, "Dry process technology (reactive ion etching)", J. Vac. Sci. Technol. **13**, 1023 (1976).

[7] J.M. Harvilchuck, J.S. Logan, W.C. Metzger, and P.M. Schaible, "Reactive ion etching of aluminum", U.S. Patent 3,994,793 (1976).

[8] A.R. Reinberg, "Radial flow reactor", U.S. Patent 3,757,733 (1973).

[9] A.R. Reinberg, "Plasma etching in semiconductor manufacture – a review", in Proc. Symp. on Etching for Pattern Definition, The Electrochemical Society, 91 (1976).

[10] D.J. Syverson and G.T. Duranko, "Contamination aspects of anhydrous HF gas processing", Solid State Technol. **31**(10), 101 (1988).

[11] K. Suzuki, S. Hiraoka, and S. Nishimatsu, "Anisotropic etching of polycrystalline silicon with a hot Cl_2 molecular beam", J. Appl. Phys. **64**, 3697 (1988).

[12] M. Rothschild and D.J. Ehrlich, "A review of excimer laser projection lithography", J. Vac. Sci. Technol. B **6**, 1 (1988).

[13] Y. Horiiki, M. Sekine, K. Horioka, T. Arikado, and H. Okano, "Photo-excited dry etching for VLSI's", in Ext. Abstr. 16th Conf. on Solid State Devices and Materials (The Japan Soc. of Appl. Phys., Kobe, 1984), p. 441.

[14] C. Arnone and G.B. Scelsi, "Anisotropic Laser Etching of Oxidized (100) Silicon", Appl. Phys. Lett. **54**, 225 (1989).

[15] I. Haller, Y.H. Lee, J.J. Nocera jr., and M.A. Jaso, "Selective wet and dry etching of hydrogenated amorphous silicon and related materials", J. Electrochem. Soc. **135**, 2042 (1988).

⋆ following the reference number indicates that the abstract of the article is reprinted in the "selected abstracts" section of the book.

[16] D.E. Ibbotson and D.L. Flamm, "Plasma etching for III-IV compound devices: part I", Solid State Technol. **31**(10), 77 (1988).

[17] D.E. Ibbotson and D.L. Flamm, "Plasma etching for III-IV compound devices: part II", Solid State Technol. **31**(11), 105 (1988).

[18] C.B. Cooper III, S. Salimian, and M.E. Day, "Dry etching for the fabrication of integrated circuits in III-IV compound semiconductors", Solid State Technol. **32**(1), 109 (1989).

[19] H. Saeki, A. Shigetomi, and Y. Watanabe, "Mask fabrication by using electron beam resist, chloromethylated polystyrene, and dry etching process", J. Electrochem. Soc. **134**, 3134 (1987).

[20] M.J. Vasile and C.J. Mogab, "Chemically assisted sputter etching of permalloy using CO or Cl_2", J. Vac. Sci. Technol. A **4**, 1841 (1986).

[21] J. Sugiura, W.-J. Lu, K.C. Cadien, and A.J. Steckl, "Reactive ion etching of SiC thin films using fluorinated gases", J. Vac. Sci. Technol. B **4**, 349 (1986).

[22] J.R. Hollahan and A.T. Bell (Eds.), *Techniques and applications of plasma chemistry* (Wiley, New York, 1974).

[23] F.C. Chen, *Introduction to plasma physics* (Plenum, New York, 1977).

[24] W. Elenbaas, *The high pressure mercury vapour discharge* (Interscience, New York, 1951).

[25] G. Francis, *Ionization Phenomena in Gases* (Butterworths, London, 1960).

[26] J.M. Anderson, "Hall effect and electron drift velocities in the plasma of the positive column" Phys. Fluids **7**, 1517 (1964).

[27] D. Bohm, "Minimum ionic kinetic energy for a stable sheath", in *The characteristics of electrical discharges in magnetic fields*, edited by A. Guthrie and R.K. Wakerling (McGraw-Hill, New York, 1949), p. 77.

[28] G.F. Weston, *Cold Cathode Glow Discharge Tubes* (ILIFFE, London, 1968).

[29] S.C. Brown, *Introduction to Electrical Discharges in Gases* (Wiley, New York, 1966).

[30] W.B. Pennebaker, "Influence of scattering and ionization on RF impedance in glow discharge sheaths", IBM J. Res. Develop. **23**, 16 (1979).

[31] P. Bletzinger and M.J. Flemming, "Impedance characteristics of an rf parallel plate discharge and the validity of a simple circuit model", J. Appl. Phys. **62**, 4688 (1987).

[32] B.N. Chapman, *Glow Discharge Processes* (Wiley, New York, 1980).

[33]⋆ W.D. Davis and T.A. Vanderslice, "Ion energies at the cathode of a glow discharge", Phys. Rev. **131**, 219 (1963).

[34] E.W. McDaniel and E.A. Mason, *The mobility and diffusion of ions in gases* (Wiley, New York, 1973).

[35]⋆ D.J. Economou, D.R. Evans, and R.C. Alkire, "A time-average model of the RF plasma sheath", J. Electrochem. Soc. **135**, 756 (1988).

[36] B.E. Thompson, H.H. Sawin, and D.A. Fisher, "Monte Carlo simulation of ion transport through rf glow-discharge sheaths", J. Appl. Phys. **63**, 2241 (1988).

[37]⋆ A.D. Kuypers and H.J. Hopman, "Ion energy measurement at the powered electrode in an rf discharge", J. Appl. Phys. **63**, 1894 (1988).

[38] C.W. Jurgensen and E.S.G. Shaqfeh, "Nonlocal transport models of the self-consistent potential distribution in a plasma sheath with charge transfer collisions", J. Appl. Phys. **64**, 6200 (1988).

[39]⋆ A.J. van Roosmalen, "Electrical characteristics of dry etching reactors", in Comptes Rendus des Travaux du CIPG87, Soc. Française du Vide (Paris, France, 1987), p. 99.

[40]⋆ R.J. Visser, J.A.G. Baggerman, J.P.J. Poppelaars, and E.J.H. Collart, "Dynamics of formation and decay of pulsed nitrogen rf plasmas", in Proc. 9th Int. Symp. on Plasma Chem. Vol. I (Int. Union of Pure and Appl. Chemistry, Pugnochiuso, Italy, Sept 1989), p. 555.

[41] Y. Okuno, S. Yagura, and H. Fujita, "Observation of sheath structure near a rf electrode in a magnetized afterglow plasma", J. Appl. Phys. **65**, 2205 (1989).

[42]⋆ D.L. Flamm, "Frequency effects in plasma etching", J. Vac. Sci. Technol. A **4**, 729 (1986).

[43] A. Metze, D.W. Ernie, and H.J. Oskam, "Application of the physics of plasma sheaths to the modelling of rf plasma reactors", J. Appl. Phys. **60**, 3081 (1986)

[44]⋆ A. Metze, D.W. Ernie, and H.J. Oskam, "The energy distribution of ions bombarding electrode surfaces in rf plasma reactors", J. Appl. Phys. **65**, 993 (1989).

[45] G.A. Hebner and J.T. Verdeyen, "The spatial and temporal evolution of the glow in an RF discharge", IEEE Trans. Plasma Sci. **14**, 132 (1986).

[46]⋆ K.-U. Riemann, "Theoretical analysis of the electrode sheath in rf discharges", J. Appl. Phys. **65**, 999 (1989).

[47] S. Biehler, "Theory of the rf sheath in the regime between the ion and electron plasma frequencies", Appl. Phys. Lett. **54**, 317 (1989).

[48] A.M. Pointu, "Dynamics of a rf sheath in the range between the ionic and the electronic plasma frequencies", Appl. Phys. Lett. **50**, 316 (1987).

[49]⋆ J.W. Coburn and E. Kay, "Positive-ion bombardment of substrates in rf diode glow discharge sputtering", J. Appl. Phys. **43**, 4965 (1972).

[50] Y. Okamoto and H. Tamagawa, "Energy dispersion of positive ions effused from an RF plasma", J. Phys. Soc. Japan **29**, 187 (1970).

[51]⋆ R.A. Gottscho and G.R. Scheller, "Double layers, breathing sheaths, and chemistry in BCl_3/rare gas radio frequency discharges", in Abstr. of the 1987 Spring mtg. of the Am. Phys. Soc., Bull. Am. Phys. Soc. **32**, 1165 (1987).

[52]⋆ A.J. van Roosmalen, W.G.M. van den Hoek, and H. Kalter, "Electrical properties of planar rf discharges for dry etching", J. Appl. Phys. **58**, 653 (1985).

[53]⋆ V.M. Donnelly, D.L. Flamm, and G. Collins, "Laser diagnostics of plasma etching: measurement of Cl_2^+ in a chlorine discharge", J. Vac. Sci. Technol. **21**, 817 (1982).

[54] M.A. Weald and C.B. Wharton, *Plasma Diagnostics with Microwaves* (Wiley, New York, 1965).

[55] A.D. MacDonald, *Microwave Breakdown in Gases* (Wiley, New York, 1966).

[56]⋆ H.R. Koenig and L.I. Maissel, "Application of RF discharges to sputtering", IBM J. Res. Devel. **14**, 168 (1970).

[57] A.M. Pointu, "A model of radio frequency planar discharges", J. Appl. Phys. **60**, 4113 (1986).

[58]⋆ R.J. Visser and C.A.M. de Vries, "Study of the mechanism of the etching of polymers in oxygen and argon plasmas", in Proceedings of the 8th International Symposium on Plasma Chemistry, edited by K. Akashi and A. Kinbara (IUPAC, Tokyo, 1987), p. 1029.

[59]⋆ W.G.M. van den Hoek, C.A.M. de Vries, and M.G.J. Heijman, "Power loss mechanisms in radio frequency dry etching systems", J. Vac. Sci. Technol. B **5**, 647 (1987).

[60]⋆ C.M. Horwitz, "Radio frequency sputtering – the significance of power input", J. Vac. Sci. Technol. A **1**, 1795 (1983).

[61] M.J. Kushner, "Mechanism for power deposition in Ar/SiH_4 capacitively coupled RF discharges", IEEE Trans. Plasma Sci. **14**, 188 (1986).

[62]⋆ D.B. Graves, "Fluid model simulations of a 13.56-MHz rf discharge: time and space dependence of rates of electron impact excitation", J. Appl. Phys. **62**, 88 (1987).

[63]⋆ C.G. Goedde, A.J. Lichtenberg, and M.A. Lieberman, "Self-consistent stochastic electron heating in radio frequency discharges", J. Appl. Phys. **64**, 4375 (1988).

[64]⋆ G.R. Misium, A.J. Lichtenberg, and M.A. Lieberman, "Macroscopic modelling of radio-frequency plasma discharges", J. Vac. Sci. Technol. A **7**, 1007 (1989).

[65]⋆ A.J. Anastassiades and E.G. Sideris, "Energy-loss rate and enhanced diffusion due to a low-frequency plasma oscillation", J. Appl. Phys. **60**, 3422 (1986).

[66]⋆ C.A.M. de Vries, A.J. van Roosmalen, and G.C.C. Puylaert, "Microwave spectroscopic measurement of the electron density in a planar discharge: relation to reactive-ion etching of silicon oxide", J. Appl. Phys. **57**, 4386 (1985).

[67] W.G.M. van den Hoek, "The etch mechanism for Al_2O_3 in fluorine and chlorine based rf dry etch plasmas", in *Plasma processing*, edited by J.W. Coburn, R.A. Gottscho, and D.W. Hess (Mat. Res. Soc. Symp. Proc. Vol. 68, Pittsburgh, 1986), p. 71.

[68]⋆ M.S. Barnes, T.J. Colter, and M.E. Elta, "Large-signal time-domain modelling of low-pressure rf glow discharges", J. Appl. Phys. **61**, 81 (1987).

[69]⋆ R.H. Bruce and B. Gelernt, "Frequency dependence of CCl_4 etching", in Ext. Abstracts of the Electrochem. Soc. Spring mtg., ECS Ext. Abstr. **80-1**, 307 (1980).

[70] N.P. Armstrong and J. Maleham, "A two-stage solution to the anisotropic polysilicon etching problem", Vacuum **33**, 291 (1983).

[71] K.E. Greenberg, G.A. Hebner, and J.T. Verdeyen, "Negative ion densities in NF_3 discharges", Appl. Phys. Lett. **44**, 299 (1984).

[72]⋆ J. Kramer, "The optogalvanic effect in a 13.56-MHz chlorine discharge", J. Appl. Phys. **60**, 3072 (1986).

[73]⋆ R.A. Gottscho and C.E. Gaebe, "Negative ion kinetics in radio-frequency glow discharges", J. Vac. Sci. Technol. A **4**, 1795 (1986).

[74] L.G. Christophorou (ed.), *Electron–molecule interactions and their applications (volume 1)* (Academic Press, Orlando, 1984).

[75] E.W. McDaniel, *Collision phenomena in ionized gases* (Wiley, New York, 1964).

[76] L.G. Christophorou, *Atomic and Molecular Radiation Physics* (Wiley, London, 1971).

[77] J.M. Hollas, *Modern spectroscopy* (Wiley, Chichester, 1987).

[78] G. Herzberg, *Molecular spectra and molecular structure. I. Spectra of diatomic molecules* (van Nostrand Reinhold, New York, 1950).

[79] A. von Engel, *Electric plasmas: their nature and uses* (Taylor & Francis, London, 1983).

[80] G.F. Drukarev, *Collisions of electrons with atoms and molecules* (Plenum, New York, 1987).

[81] S.C. Brown, *Basic data of plasma Physics* (Wiley, New York, 1966).

[82] S. Trajmar, "Electron impact excitation of molecules", in *Electronic and atomic collisions*, edited by D.C. Lorents, W.E. Meyerhof, and J.R. Peterson (North-Holland, Amsterdam, 1986), p. 77.

[83] E.W. McDaniel, M.R. Flannery, H.W. Ellis, F.L. Eisele, W. Pope, and T.G. Roberts, "Compilation of data relevant to rare gas-rare gas and rare gas-monohalide excimer lasers", Tech. Report H-78-1 (1977), vol. 2, pp. 526-587 (U.S. Army Missile Research and Development Command, Redstone Arsenal, Alabama 35809).

[84] G.M. Barrow, *Physical Chemistry* (McGraw-Hill, New York, 1979).

[85] D.G. Thompson, "The vibrational excitation of molecules by electron impact", in *Advances in atomic and molecular physics, volume 19*, edited by D. Bates and B. Bederson (Academic Press, New York, 1983), p. 309.

[86]⋆ M. Capitelli, R. Celiberto, C. Gorse, R. Winkler, and J. Wilhelm, "Electron energy distribution functions in radio-frequency collision dominated nitrogen discharges", J. Phys. D **21**, 691 (1988).

[87]⋆ C.M. Ferreira and J. Loureiro, "Electron excitation rates and transport parameters in high-frequency N_2 discharges", J. Phys. D **22**, 76 (1989).

[88] J.A.G. Baggerman and J.P.J. Poppelaars, "Recombination of Cl in a Cl_2 afterglow", Le Vide, les Couches Minces, Supplement au No. 246, 317 (1989).

[89] M.A.A. Clyne and D.H. Stedman, "Recombination of ground state halogen atoms. Part 2 - kinetics of the overall recombination of chlorine atoms", Trans. Faraday Soc. **64**, 2698 (1968).

[90] F. Kaufmann, "Reactions of oxygen atoms", in *Progress in reaction kinetics, volume 1*, edited by G. Porter (Pergamon, Oxford, 1961), p. 1.

[91] Y. Hatano, "Electron attachment to van der Waals molecules", in *Electronic and atomic collisions*, edited by D.C. Lorents, W.E. Meyerhof, and J.R. Peterson (North-Holland, Amsterdam, 1986), p. 153.

[92] C. Willis, A.W. Boyd, M.J. Young, and D.A. Armstrong, "Radiation chemistry of gaseous oxygen: experimental and calculated yields", Can. J. Chem. **48**, 1505 (1970).

[93] F.A. Houle, "Basic mechanisms in laser etching and deposition", Appl. Phys. A **41**, 315 (1986).

[94] W.M. Holber and R.M. Osgood jr., "Photon assisted plasma etching", Solid State Technol. **30**(4), 139 (1987).

[95]⋆ J.V. Florio and W.D. Robertson, "Chlorine reactions on the Si(111) surface", Surf. Sci. **18**, 398 (1969).

[96] F.R. McFeely, J.F. Morar, N.D. Shinn, G. Landgren, and F.J. Himpsel, "Synchrotron photoemission investigation of the initial stages of fluorine attack on Si surfaces: relative abundance of fluorosilyl species", Phys. Rev. B **30**, 764 (1984).

[97] B. Roop, S. Joyce, J.C. Schultz, N.D. Shinn, and J.I. Steinfeld, "Reactions of XeF_2 chemisorbed on Si(111) 7×7", Appl. Phys. Lett **46**, 1187 (1985).

[98] M. Seel and P.S. Bagus, "Ab initio cluster study of the interaction of fluorine and chlorine with the Si(111) surface", Phys. Rev. B **28**, 2023 (1983).

[99] J. Dieleman, "Angular-resolved time-of-flight mass spectroscopic studies of reactions at surfaces", in *Photon, beam and plasma enhanced processes*, edited by A. Golanski, V.T. Nguyen, and E.F. Krimmel (les éditions de physique, Les Ulis, France, 1987), p. 363.

[100] R. Kelly, "Thermal effects in sputtering", Surf. Sci. **90**, 280 (1979).

[101]⋆ R.J.A.A. Janssen, A.W. Kolfschoten, and G.N.A. van Veen, "Time-of-flight study on the thermal etching of Al with Cl_2", Appl. Phys. Lett. **52**, 98 (1988).

[102] H.F. Winters, "Etch products from the reaction on Cl_2 with Al(100) and Cu(100) and XeF_2 and Nb", J. Vac. Sci. Technol. B **3**, 9 (1985).

[103] H.F. Winters and J.W. Coburn, "Plasma-assisted etching mechanisms: the implications of reaction probability and halogen coverage", J. Vac. Sci. Technol. B **3**, 1376 (1985).

[104]⋆ P.C. Zalm, "Quantitative sputtering", Surf. & Int. Anal. **11**, 1 (1988).

[105] Y. Yamamura and N. Itoh, "Sputtering yield", in *Ion beam assisted film growth*, edited by T. Itoh (Vol. 3 of the Series: "Beam modifications of materials", Elseviers, Amsterdam, 1989), p. 59.

[106]⋆ P. Sigmund, "Sputtering by ion bombardment: theoretical concepts", in *Sputtering by particle bombardment I*, edited by R. Behrisch (Springer, Berlin, 1981), p. 9.

[107]⋆ P. Sigmund, "Theory of sputtering. I. Sputtering yield of amorphous and polycrystalline targets", Phys. Rev. **184**, 383 (1969). [*Errata*: Phys. Rev. **187**, 768 (1969)].

[108]⋆ P.C. Zalm, "Ion-beam assisted etching of semiconductors", Vacuum **36**, 787 (1986).

[109] T.S. Baller and J. Dieleman, "Model experiments for ion and laser beam etching", in Proc. Symp. on Dry Process, The Inst. Electr. Eng. Japan (Tokyo, 1989), p. 1.

[110]⋆ J. Dieleman, "Some recent results of fundamental studies on beam-induced surface processes", in *Laser- and particle-beam chemical processes on surfaces*, edited by A.W. Johnson, G.L. Loper and T.W. Sigmon (Mat. Res. Soc. Symp. Proc. Vol. 129, Pittsburgh, 1989), p. 3.

[111] P.C. Zalm, "Elementary processes in plasma-surface interactions with emphasis on ions", Pure & Appl. Chem. **57**, 1253 (1985).

[112] M.W. Thompson, "II. The energy spectrum of ejected atoms during the high energy sputtering of gold", Phil. Mag. **18**, 377 (1968).

[113] J. Dieleman, F.H.M. Sanders, A.W. Kolfschoten, P.C. Zalm, A.E. de Vries, and A. Haring, "Studies on the mechanism of chemical sputtering of silicon by simultaneous exposure to Cl_2 and low-energy Ar^+ ions", J. Vac. Sci. Technol. B **3**, 1384 (1985).

[114]⋆ Ch. Steinbrüchel, "On the sputtering yield of molecular ions", J. Vac. Sci. Technol. A **3**, 1913 (1985).

[115] D.J. Oostra, R.P. van Ingen, A. Haring, A.E. de Vries, and F.W. Saris, "Time-of-flight measurements of nonlinear effects in sputtering of dimer ions", Phys. Rev. Lett. **61**, 1392 (1988).

[116] S. Hasegawa and P.C. Zalm, "Formation and bonding structure of silicon nitride by 20-keV N^+ ion implantation", J. Appl. Phys. **58**, 2539 (1985).

[117]⋆ S. Tachi and S. Okudaira, "Chemical sputtering of silicon by F^+, Cl^+, and Br^+ ions: reactive spot model for reactive ion etching", J. Vac. Sci. Technol. B **4**, 459 (1986).

[118]⋆ J.W. Coburn and H.F. Winters, "Ion- and electron-assisted gas-surface chemistry – An important effect in plasma etching", J. Appl. Phys. **50**, 3189 (1979).

[119]⋆ G.N.A. van Veen, F.H.M. Sanders, J. Dieleman, P.C. Zalm, D.J. Oostra, and A.E. de Vries, "Extension of the model for Ar^+ ion induced etching of Si by SF_6", Nucl. Instrum. & Methods Phys. Res. B **19/20**, 1022 (1987).

[120]⋆ H.F. Winters, "Phenomena produced by ion bombardment in plasma-assisted etching environments", J. Vac. Sci. Technol. A **6**, 1997 (1988).

[121]⋆ A.W. Kolfschoten, R.A. Haring, A. Haring, and A.E. de Vries, "Argon-ion assisted etching of silicon by molecular chlorine", J. Appl. Phys. **55**, 3813 (1984).

[122] R.A. Haring, A. Haring, F.W. Saris, and A.E. de Vries, "Mass and energy distribution of particles sputter etched from Si in a XeF_2 environment", Appl. Phys. Lett. **41**, 174 (1982).

[123]⋆ D.J Oostra, A. Haring, R.P. van Ingen, and A.E. de Vries, "Mechanisms of sputtering of Si in a Cl_2 environment by ions with energies down to 75 eV", J. Appl. Phys. **64**, 315 (1988).

[124] T.M. Mayer, M.S. Ameen, and D.J. Vitkavage, "Surface chemistry of dry etching processes", in *The chemical physics of solid surfaces and heterogeneous catalysis*, edited by D.A. King and D.P. Woodruff (Vol. 5 of the Series: "Surface properties of electronic materials", Elseviers, Amsterdam, 1988), p. 427.

[125]⋆ J. van Zwol, J. van Laar, A.W. Kolfschoten, and J. Dieleman, "Effects of Ar^+ angle of incidence on the etching of Si with Cl_2 and low-energy Ar^+ ions", J. Vac. Sci. Technol. B **5**, 1410 (1987).

[126] T. Mizutani, C.J. Dale, W.K. Chu, and T.M. Mayer, "Surface modification in plasma-assisted etching of silicon", Nucl. Instrum. & Methods Phys. Res. B **7/8**, 825 (1985).

[127] C.A.M. de Vries and W.G.M. van den Hoek, "Validity of self-bias voltage measurements on insulating electrodes in radio-frequency dry etching systems", J. Appl. Phys. **58**, 2074 (1985).

[128]⋆ J.W. Coburn and E. Kay, "Some chemical aspects of the fluorocarbon plasma etching of silicon and its compounds", IBM J. Res. Develop. **23**, 33 (1979).

[129] W.G.M. van den Hoek, "The influence of the electrode material on the sputter etch characteristics of an argon plasma", in Proc. 4th Symp. Plasma Processing, The Electrochemical Society, ECS Proc. Vol. **83-10**, 40 (1983).

[130] W.C. Johnson, *Transmission lines and networks* (McGraw Hill, New York, 1950).

[131]⋆ H. Norström, "Experimental and design information for calculating impedance matching networks for use in rf sputtering and plasma chemistry", Vacuum **29**, 341 (1979).

[132] N.M. Mazza, "Automatic impedance matching system for RF sputtering", IBM J. Res. Devel. **14**, 192 (1970).

[133] A. Kalnitsky and W.K. Chung, "Characterization and optimization of a single wafer downstream plasma stripper", J. Electrochem. Soc. **135**, 2338 (1988).

[134] N. Hayasaka, H. Okano, and Y. Horiike, "Highly selective etching of Si_3N_4 over SiO_2 employing a downstream type reactor", Solid State Technol. **31**(4), 127 (1988).

[135] C.M. Horwitz, "Rf sputtering-voltage division between two electrodes", J. Vac. Sci. Technol. A **1**, 60 (1983).

[136]⋆ M.A. Lieberman, "Spherical shell model of an asymmetric rf discharge", J. Appl. Phys. **65**, 4186 (1989).

[137] F. Martinet, "DC voltage measurement from an RF plasma during plasma etching and deposition", in Proc. 3rd Symp. on Plasma Processing, The Electrochemical Society, ECS Proc. Vol. **82-6**, 379 (1982).

[138] D.L. Flamm, V.M. Donnely, and D.E. Ibbotson, "Basic chemistry and mechanisms of plasma etching", J. Vac. Sci. Technol. B **1**, 23 (1983).

[139] A.J. van Roosmalen, A.P.M. van Arendonk, H.T. Arends, and F. Schmidt, "Etching of Si in low-frequency Cl_2 discharges", in Ext. Abstracts of the Electrochem. Soc. Fall mtg., ECS Ext. Abstr. **84-2**, 574 (1984).

[140]⋆ J.M.E. Harper, "Ion beam techniques in thin film deposition", Solid State Technol. **30**(4), 129 (1987).

[141] J.M.E. Harper, "Ion beam etching", in *Plasma etching, an introduction*, edited by D.M. Manos and D.L. Flamm (Academic Press, New York, 1989), p. 391.

[142] R.E. Lee "Ion-beam etching (milling)", in *VLSI electronics microstructure science. Volume 8: plasma processing for VLSI*, edited by N.G. Einspruch and D.M. Brown (Academic Press, New York, 1984), p. 341.

[143] B.A. Heath and T.M. Mayer, "Reactive ion-beam etching", in *VLSI electronics microstructure science. Volume 8: plasma processing for VLSI*, edited by N.G. Einspruch and D.M. Brown (Academic Press, New York, 1984), p. 365.

[144] E. Ghanbari, I. Trigor, and T. Nguyen, "A broad-beam electron cyclotron resonance ion source for sputtering etching and deposition of material", J. Vac. Sci. Technol. A **7**, 918 (1989).

[145] P.J. Revell and G.F. Goldspink, "A review of reactive ion beam etching for production", Vacuum **34**, 455 (1984).

[146] S. Takami, "Cassette-to-cassette dry etching raises productivity in VLSI process', Jap. J. Electr. Eng. **20**(9), 82 (1983).

[147] E.D. Wolf, I. Adesida, and J.D. Chinn, "Dry etching for submicron structures", J. Vac. Sci. Technol. A **2**, 464 (1984).

[148] S.J. Fonash, "Advances in dry etching process – a review", Solid State Technol. **28**(1), 150 (1985).

[149] E. Bogle-Rohwer, D. Gates, L. Hayler, H. Kurasaki, and B. Richardson, "Wall profile control in a triode etcher", Solid State Technol. **28**(4), 251 (1985).

[150] L.M. Ephrath and R.S. Bennet, "Directional and selective etching of polysilicon and polycide", in Proc. 1st Symp. VLSI Science and Technology, The Electrochemical Society, ECS Proc. Vol. **82-7**, 108 (1982).

[151] L. Giffen and M.M. Stark, "Plasma trench etch", European Patent 203,560 (1986).

[152] S.P. DeOrnellas and J.D. Ocampo, "The etching of silicon in a multi-frequency discharge reactor", in Proc. 6th Symp. Plasma Processing, The Electrochemical Society, ECS Proc. Vol. **87-6**, 77 (1987).

[153] Y.H. Lee, J.E. Heidenreich III, and G. Fortuno, "Plasma characterization of an electron cyclotron resonance – radio-frequency hybrid plasma reactor", J. Vac. Sci. Technol. A **7**, 903 (1989).

[154] K. Hirobe and H. Azuma, "Reduction of radiation damage on silicon substrates in magnetron reactive ion etching", J. Electrochem. Soc. **132**, 938 (1985).

[155]⋆ A.D. Kuypers and H.J. Hopman, "Measurement of ion energy distributions at the powered rf electrode in a variable magnetic field", J. Appl. Phys. **67**, 1229 (1990).

[156] A.A. Bright, S. Kaushik, and G.S. Oehrlein, "Plasma chemical aspects of magnetron ion etching with CF_4/O_2", J. Appl. Phys. **62**, 2518 (1987).

[157]⋆ Ch. Steinbrüchel, B.J. Curtis, H.W. Lehman, and R. Widmer, "Diagnostics of low-pressure oxygen RF plasmas and the mechanism for polymer etching: a comparison of reactive sputter etching and magnetron sputter etching", IEEE Trans. Plasma Sci. **14**, 137 (1986).

[158]⋆ S.M. Bobbio, "A review of magnetron etch technology", in *Dry processing for sub-micrometer lithography*, edited by J. Bondur and A.R. Reinberg, S.P.I.E. Proc. Vol. **1185**, 262 (1989).

[159] H.S. Okano, T. Yamazaki, Y. Horiiki, and H. Horie, "Dry etching apparatus using reactive ions", Japan Patent 49,142/83 (1983) (*in Japanese*).

[160] W.H. Class, S.D. Hurwitt, M.L. Hill, and M.K. Hutt, "Magnetically enhanced plasma process and apparatus", U.S. Patent 4,422,896 (1983).

[161] L.M. Ephrath, "Review of dry etching technology: recent advances", in Proc. 2nd Symp. ULSI Science and Technology, The Electrochemical Society, ECS Proc. Vol. **89-9**, 394 (1989).

[162]⋆ J. Asmussen, "Electron cyclotron resonance microwave discharges for etching and thin-film deposition", J. Vac. Sci. Technol. A **7**, 883 (1989).

[163] M. Matsuoka and K. Ono, "Low-energy ion extraction with small dispersion from an electron cyclotron resonance microwave plasma stream", Appl. Phys. Lett. **50**, 1864 (1987).

[164]⋆ T. Ono, M. Oda, C. Takahashi, and S. Matsuo, "Reactive ion stream etching utilizing electron cyclotron resonance plasma", J. Vac. Sci. Technol. B **4**, 696 (1986).

[165] K. Suzuki, K. Ninomiya, S. Nishimatsu, and S. Okudaira, "Radio-frequency biased microwave plasma etching technique: a method to increase SiO_2 etch rate", J. Vac. Sci. Technol. B **3**, 1025 (1985).

[166]⋆ R.R. Burke and C. Pomot, "Microwave multipolar plasma for etching and deposition", Solid State Technol. **31**(2), 67 (1988).

[167] I. Lin, D.C. Hinson, W.H. Class, and R.L. Sandstrom, "Low-energy high flux reactive ion etching by rf magnetron plasma", Appl. Phys. Lett. **44**, 185 (1984).

[168]⋆ J.I. Ulacia F., C.J. Petti, and J.P. McVittie, "Crystal-orientation dependent etch rates and a trench model for dry etching", J. Electrochem. Soc. **135**, 1521 (1988).

[169]⋆ D.P. Hamblen and A. Cha-Lin, "Angular etching correlations from RIE. Application to VLSI fabrication and process modeling", J. Electrochem. Soc. **135**, 1816 (1988).

[170]⋆ A.J. van Roosmalen, "Review: dry etching of silicon oxide", Vacuum **34**, 429 (1984).

[171]⋆ G.S. Oehrlein and H.L. Williams, "Silicon etching mechanisms in a CF_4/H_2 glow discharge", J. Appl. Phys. **62**, 662 (1987).

[172]⋆ G.S. Oehrlein, S.W. Robey, J.L. Lindström, K.K. Chan, M.A. Jaso, and G.J. Scilla, "Surface modifications of electronic materials induced by plasma etching", J. Electrochem. Soc. **136**, 2050 (1989).

[173]⋆ D.L. Flamm, V.M. Donnelly, and J.A. Mucha, "The reaction of fluorine atoms with silicon", J. Appl. Phys. **52**, 3633 (1981).

[174]⋆ D.L. Flamm, D.E. Ibbotson, J.A. Mucha, and V.M. Donnelly, "XeF_2 and F-atom reactions with Si: their significance for plasma etching", Solid State Technol. **26**(4), 117 (1983).

[175] M. Kitamura, H. Akiya, and T. Urisu, "Polymer deposition and etching mechanisms in C_2F_6 radio-frequency plasma as studied by laser-induced fluorescence", J. Vac. Sci. Technol. B **7**, 14 (1989).

[176]⋆ L.M. Ephrath and E.J. Petrillo, "Parameter and reactor dependence of selective oxide RIE in CF_4 + H_2", J. Electrochem. Soc. **129**, 2282 (1982).

[177]⋆ Y.W. Hu, "The effect of moisture on the plasma oxide etch", in Proc. 5th Symp. Plasma Processing, The Electrochemical Society, ECS Proc. Vol. **85-1**, 470 (1985).

[178] J.W. Coburn and E. Kay, "Some chemical aspects of the fluorocarbon plasma etching of silicon and its compounds", Solid State Technol. **22**(4), 117 (1979).

[179] P.E. Riley, V.D. Kulkarni, and S.H. Bishop, "Examination of fluorocarbon-based plasmas used in the selective and uniform etching of silicon dioxide by response-surface methodology: effects of helium addition", J. Vac. Sci. Technol. B **7**, 24 (1989).

[180]⋆ G.S. Mathad, "Design considerations for a high pressure, high etch rate single wafer reactor", in Proc. 6th Symp. Plasma Processing, The Electrochemical Society, ECS Proc. Vol. **87-6**, 134 (1987).

[181] P.C. Karulkar and M.A. Wirzbicki, "Characterization of etching of silicon dioxide and photoresist in a fluorocarbon plasma", J. Vac. Sci. Technol. B **6**, 1595 (1988).

[182]⋆ F.H.M. Sanders, J. Dieleman, H.J.B. Peters, and J.A.M. Sanders, "Selective isotropic dry etching of Si_3N_4 over SiO_2", J. Electrochem. Soc. **129**, 2559 (1982).

[183]⋆ Y. Kawamoto, T. Kure, N. Hashimoto, and T. Takaichi, "Highly selective dry etching of Si_3N_4", in Ext. Abstracts of the Electrochem. Soc. Fall mtg., ECS Ext. Abstr. **84-2**, 565 (1984).

[184]⋆ L.M. Loewenstein, "Selective etching of silicon nitride using remote plasmas of CF_4 and SF_6", J. Vac. Sci. Technol. A **7**, 686 (1989).

[185]⋆ G. Fortuño-Wiltshire and G.S. Oehrlein, "Hollow cathode etching of Si and SiO_2 using CF_4 and H_2", J. Electrochem. Soc. **136**, 1447 (1989).

[186]⋆ A.J. van Roosmalen, A.P.M. van Arendonk, H.T. Arends, and F. Schmidt, "Etching of silicon in low-frequency chlorine discharges", in Proc. 5th Symp. Plasma Processing, The Electrochemical Society, ECS Proc. Vol. **85-1**, 527 (1985).

[187]⋆ R.N. Carlile, V.C. Liang, and M.M. Smadi, "High quality trench etches in silicon", Solid State Technol. **32**(4), 119 (1989).

[188] M. Sato and Y. Arita, "Etched shape control of single-crystal silicon in reactive ion etching using chlorine", J. Electrochem. Soc. **134**, 2856 (1987).

[189] R.N. Carlile, V.C. Liang, O.A. Palusinski, and M.M. Smadi, "Trench etches in silicon with controllable sidewall angles", J. Electrochem. Soc. **135**, 2058 (1988).

[190]⋆ G.R. Powell and A.A. Chambers, "Deep Si trench etching with a high rate process through a $n/n^+/p$ type structure, using a triode configured reactor", in Proc. 2nd Symp. ULSI Science and Technology, The Electrochemical Society, ECS Proc. Vol. **89-9**, 498 (1989).

[191]⋆ G.C. Schwartz and P.M. Schaible, "The effects of arsenic doping in reactive ion etching of silicon in chlorinated plasmas", J. Electrochem. Soc. **130**, 1898 (1983).

[192]⋆ Y.H. Lee and M.-M. Chen, "Silicon doping effects in reactive plasma etching", J. Vac. Sci. Technol. B **4**, 468 (1986).

[193]⋆ J.C. Moreno-Marin, J.A. Valles-Abarca, and A. Gras-Marti, "Secondary effects in ion milling", J. Vac. Sci. Technol. B **4**, 322 (1986).

[194] S.W. Pang, J.N. Randall, and M.W. Geis, "Sub-100-nm-wide, deep trenches defined by reactive ion etching", J. Vac. Sci. Technol. B **4**, 341 (1986).

[195]⋆ S. Ohki, M. Oda, H. Akiya, and T. Shibata, "Cavernous undercuts appearing in reactive ion etched submicron-wide deep trenches", J. Vac. Sci. Technol. B **5**, 1611 (1987).

[196] M. Engelhardt and S. Schwarzl, "Deep silicon trench etching with perfect profile control using $CBrF_3/N_2$ gas mixtures", in Proc. 2nd Symp. ULSI Science and Technology, The Electrochemical Society, ECS Proc. Vol. **89-9**, 505 (1989).

[197] A.P.M. van Arendonk and A.J. van Roosmalen, "Method of manufacturing a semiconductor device", European Patent 337,562 (1989).

[198]⋆ K. Ninomiya, K. Suzuki, S. Nishimatsu, and O. Okada, "Role of sulfur atoms in microwave plasma etching of silicon", J. Appl. Phys. **62**, 1459 (1987).

[199] J.H. Thomas III and L.H. Hammer, "An x-ray photoelectron spectroscopy study of sulphur hexafluoride etchant residue on silicon and silicon dioxide", J. Vac. Sci. Technol. B **5**, 1617 (1987).

[200] R. Sellamuthu, J. Barkanic, and R. Jaccodine, "A study of anisotropic trench etching of Si with NF_3-halocarbon", J. Vac. Sci. Technol. B **5**, 342 (1987).

[201]⋆ J.P. McVittie, T.A. Lin, and A.J. Bariya, "Ion suppression for studying etch inhibitor layers", in Proc. 6th Symp. Plasma Processing, The Electrochemical Society, ECS Proc. Vol. **87-6**, 675 (1987).

[202]⋆ S. Tachi, "Low temperature dry etching for ULSI", in Proc. 2nd Symp. ULSI Science and Technology, The Electrochemical Society, ECS Proc. Vol. **89-9**, 381 (1989).

[203]⋆ P. Parrens, "Anisotropic and selective reactive ion etching of polysilicon using SF_6", J. Vac. Sci. Technol. **19**, 1403 (1981).

[204] B. Mahi, Y. Arnal, and C. Pomot, "The etching of silicon in diluted SF_6 plasmas: correlation between the flux of incident species and the etching kinetics", J. Vac. Sci. Technol. B **5**, 657 (1987).

[205]⋆ J.E. Spencer, G.G. Barna, and D.E. Carter, "Rational process optimization and characterization", in Proc. 5th Symp. Plasma Processing, The Electrochemical Society, ECS Proc. Vol. **85-1**, 413 (1985).

[206]⋆ G.C. Schwartz, "Reactive plasma-assisted etching of aluminum and aluminum alloys", in Proc. 5th Symp. Plasma Processing, The Electrochemical Society, ECS Proc. Vol. **85-1**, 26 (1985).

[207]⋆ N.N. Efremow, M.W. Geis, R.W. Mountain, G.A. Lincoln, J.N. Randall, and N.P. Economou, "Anisotropic etching of Al by a directed Cl_2 flux", J. Vac. Sci. Technol. B **4**, 337 (1986).

[208]⋆ D.L. Smith and R.H. Bruce, "Si and Al etching and product detection in a plasma beam under ultrahigh vacuum", J. Electrochem. Soc. **129**, 2045 (1982).

[209]⋆ D.A. Danner, M. Dalvie, and D.W. Hess, "Plasma etching of aluminum. A comparison of chlorinated etchants", J. Electrochem. Soc. **134**, 669 (1987).

[210]⋆ R.A.M. Wolters, "Selectivity aspects in aluminium plasma etching", in Proc. 3rd Symp. Plasma Processing, The Electrochemical Society, ECS Proc. Vol. **82-6**, 293 (1982).

[211] R.W. Light and H.B. Bell, "The relationship between anisotropy and etch time", in Proc. 5th Symp. Plasma Processing, The Electrochemical Society, ECS Proc. Vol. **85-1**, 430 (1985).

[212] J.-H. Lin, W. Ray, and M.D. Welch, "Plasma etch of aluminum alloys", in Proc. 2nd Symp. ULSI Science and Technology, The Electrochemical Society, ECS Proc. Vol. **89-9**, 445 (1989).

[213]⋆ R.H. Bruce and G.P. Malafsky, "High rate anisotropic aluminum etching", J. Electrochem. Soc. **130**, 1369 (1983).

[214] A.L. Keaton and D.W. Hess, "Temperature and flow effects in aluminum etching using bromine-containing plasmas", J. Vac. Sci. Technol. B **6**, 72 (1988).

[215] H.B. Bell, H.M. Anderson, and R.W. Light, "Reactive ion etching of aluminum/silicon in BBr_3/Cl_2 and BCL_3/Cl_2 mixtures", J. Electrochem. Soc. **135**, 1184 (1988).

[216]⋆ J. Maa, "Surface passivation by adding fluorocarbons and chlorofluorocarbons into an Al dry etch reactor", J. Vac. Sci. Technol. B **5**, 652 (1987).

[217] T. Tsukada, K. Yasuda, E. Wani, and K. Ukai, "After corrosion treatment in aluminum alloy reactive ion etching", in Proc. 4th Symp. Plasma Processing, The Electrochemical Society, ECS Proc. Vol. **83-10**, 341 (1983).

[218]⋆ P.C. Chang and K.K. Chao, "Prevention of corrosion during the reactive ion etch of aluminum on TiW", in Proc. 5th Symp. Plasma Processing, The Electrochemical Society, ECS Proc. Vol. **85-1**, 12 (1985).

[219]⋆ J. Maleham, "Investigation of the dependence of reactive ion etching of Al-Si-Cu alloys upon film deposition characteristics", Vacuum **34**, 437 (1984).

[220] J.M. Eldridge, G. Olive, B.J. Luther, and J.O. Moore, "Effects of oxygen on the RIE characteristics of aluminum films", J. Electrochem. Soc. **134**, 1025 (1987).

[221] S.J.H. Brader and S.C. Quinlan, "PECVD oxide step-coverage considerations for use as an interlayer dielectric in CMOS DLM processing", J. Electrochem. Soc. **135**, 2291 (1988).

[222] S. Park, T.N. Rhodin, and L.C. Rathbun, "Halide formation and etching of Cu thin films with Cl_2 and Br_2", J. Vac. Sci. Technol. A **4**, 168 (1986).

[223]⋆ T. Abraham, "Correlation of postetch residues to deposition temperature in plasma etched aluminum alloys", J. Electrochem. Soc. **134**, 2809 (1987).

[224] C.-K. Hu, B. Canney, D.J. Pearson, and M.B. Small, "A process for improved Al(Cu) reactive ion etching", J. Vac. Sci. Technol. A **7**, 682 (1989).

[225]⋆ T.P. Chow and A.J. Steckl, "A review of plasma etching of refractory metal silicides", in Proc. 4th Symp. Plasma Processing, The Electrochemical Society, ECS Proc. Vol. **83-10**, 362 (1983).

[226] S.P. Murarka, *Silicides for VLSI applications* (Academic Press, New York, 1983).

[227]⋆ R.J. Saia, B. Gorowitz, D. Woodruff, and D.M. Brown, "Plasma etching methods for the formation of planarized tungsten plugs used in multilevel VLSI metallizations",

in Proc. 6th Symp. Plasma Processing, The Electrochemical Society, ECS Proc. Vol. **87-6**, 173 (1987).

[228] S. Park, C. Sun, and R.J. Purtell, "A mechanistic study of SF_6/O_2 reactive ion etching of molybdenum", J. Vac. Sci. Technol. B **5**, 1372 (1987).

[229] W.M. Greene, D.W. Hess, and W.G. Oldham, "Ion-bombardment-enhanced plasma etching of tungsten with NF_3/O_2", J. Vac. Sci. Technol. B **6**, 1570 (1988).

[230]⋆ F. Fracassi and J.W. Coburn, "Plasma-assisted etching of tungsten films: a quartz-crystal microbalance study", J. Appl. Phys. **63**, 1758 (1988).

[231] V.M. Donnelly, D.L. Flamm, and D.E. Ibbotson, "Plasma etching of III-V compound semiconductors", J. Vac. Sci. Technol. A **1**, 626 (1983).

[232]⋆ F.Y. Robb, "Plasma etching of oxygen containing titanium silicide films", J. Electrochem. Soc. **131**, 2906 (1984).

[233]⋆ S.P. Sun and S.P Murarka, "Reactive-ion etching (RIE) of $TaSi_2/n^+$ polysilicon bilayers", J. Electrochem. Soc. **135**, 2353 (1988).

[234] M.E. Burba, E. Degenkolb, S. Henck, M. Tabasky, E.D. Jungbluth, and R. Wilson, "Selective dry etching of tungsten for VLSI metallization", J. Electrochem. Soc. **133**, 2113 (1986).

[235]⋆ J. Herrmann, P. Martins, and W. Rust, "A multistep etch process for $TaSi_2$/polysilicon composite films", in Proc. 6th Symp. Plasma Processing, The Electrochemical Society, ECS Proc. Vol. **87-6**, 494 (1987).

[236] W.Y. Hata, "Reactive ion etch characterization of moly/TiW, for first level interconnects", in Proc. 6th Symp. Plasma Processing, The Electrochemical Society, ECS Proc. Vol. **87-6**, 484 (1987).

[237]⋆ B.J. Curtis and H.R. Brunner, "The reactive ion etching of tantalum silicide/polysilicon bilayers in silicon tetrachloride and hydrogen chloride", J. Electrochem. Soc. **136**, 1463 (1989).

[238] D.S. Fischl, G.W. Rodrigues, and D.W. Hess, "Etching of tungsten and tungsten silicide films by chlorine atoms", J. Electrochem. Soc. **135**, 2016 (1988).

[239]⋆ M. Balooch, D.S. Fischl, D.R. Olander, and W.J. Siekhaus, "The kinetics of tungsten etching by atomic and molecular chlorine", J. Electrochem. Soc. **135**, 2090 (1988).

[240]⋆ D.S. Fischl and D.W. Hess, "Molybdenum etching with chlorine atoms and molecular chlorine plasmas", J. Vac. Sci. Technol. B **6**, 1577 (1988).

[241] D.S. Fischl and D.W. Hess, "Plasma-enhanced etching of tungsten and tungsten silicide in chlorine-containing discharges", J. Electrochem. Soc. **134**, 2265 (1987).

[242] T.C. Mele, S.C. Arney, J.P. Krusius, and N.C. MacDonald, "Anisotropic reactive ion etching of $MoSi_2$ and in situ doped n^+ and p^+ polysilicon using Cl_2 and BCl_3", J. Electrochem. Soc. **135**, 2373 (1988).

[243]⋆ F.Y. Robb, "Chlorine-based plasma etching of titanium silicide films", in Proc. 5th Symp. Plasma Processing, The Electrochemical Society, ECS Proc. Vol. **85-1**, 1 (1985).

[244] R.J. Saia and B. Gorowitz, "The reactive ion etching of molybdenum and bilayer metallization schemes containing molybdenum", J. Electrochem. Soc. **135**, 2795 (1988).

[245]⋆ M.A. Hartney, D.W. Hess, and D.S. Soane, "Critical review. Oxygen plasma etching for resist stripping and multilayer lithography", J. Vac. Sci. Technol. B **7**, 1 (1989).

[246]⋆ V. Vukanovic, G.A. Takacs, E.A. Matuszak, F.D. Egitto, F. Emmi, and R.S. Horwath, "Plasma etching of inorganic materials. II. Polyimide etching and passivation downstream of an O_2-CF_4-Ar microwave plasma", J. Vac. Sci. Technol. B **6**, 66 (1988).

[247] J.E. Spencer, R.A. Borel, and A. Hoff, "High rate photoresist stripping in an oxygen afterglow", J. Electrochem. Soc. **133**, 1922 (1986).

[248]⋆ B. Robinson and S.A. Shivashankar, "A study of the mechanism of polyimide etching in a microwave plasma downstream etcher", in Proc. 5th Symp. Plasma Processing, The Electrochemical Society, ECS Proc. Vol. **85-1**, 206 (1985).

[249] N.R. Lerner and T. Wydeven, "Polymer etching in the oxygen afterglow: increased etch rates with increased reactor loading", J. Electrochem. Soc. **136**, 1426 (1989).

[250]⋆ F.Y. Robb, "Hydrogen-plasma etching of organics", in Proc. 4th Symp. Plasma Processing, The Electrochemical Society, ECS Proc. Vol. **83-10**, 382 (1983).

[251]⋆ C.W. Jurgensen, A. Shugard, N. Dudash, E. Reichmanis, and M.J. Vasile, "Experimental tests of the steady-state model for oxygen reactive ion etching of silicon-containing polymers", J. Vac. Sci. Technol. A **6**, 2938 (1988).

[252]⋆ R.J. Visser and J.A.G. Baggerman, "Ion induced etching of polymers in oxygen and argon rf plasmas", in Proc. 9th Int. Symp. on Plasma Chem. Vol. II (Int. Union of Pure and Appl. Chemistry, Pugnochiuso, Italy, Sept 1989), p. 1039.

[253]⋆ H. Hiraoka, "Selective removal of metal atoms in hydrogen reactive ion etching", J. Vac. Sci. Technol. B **4**, 345 (1986).

[254]⋆ S. Crapella and F. Gualandris, "Planarization by two-resist level", J. Electrochem. Soc. **135**, 683 (1988).

[255] A. Schiltz, P. Abraham, and E. Dechenaux, "Improvement of photoresist planarization properties by thermal cure. Role of the PAC", J. Electrochem. Soc. **134**, 190 (1987).

[256]⋆ A. Schiltz and M. Pons, "Two-layer planarization process', J. Electrochem. Soc. **133**, 178 (1986).

[257] K. Machida and H. Oikawa, "SiO_2 planarization technology with biasing and electron cyclotron resonance plasma deposition for submicron interconnects", J. Vac. Sci. Technol. B **4**, 818 (1986).

[258] R.J. Saia and B. Gorowitz, "Dry etching of tapered contact holes using multilayer resist", in Proc. 5th Symp. Plasma Processing, The Electrochemical Society, ECS Proc. Vol. **85-1**, 270 (1985).

[259]⋆ A.G. Nagy, "Sidewall tapering in reactive ion etching", J. Electrochem. Soc. **132**, 689 (1985).

[260]⋆ V. Grewal, H.P. Erb, and P. Mokrisch, "Contact hole and via profiling by high rate isotropic and anisotropic etching of oxides", in Proc. 4th Int. IEEE VLSI Multilevel Interconnect Conf., 298 (1987).

[261]⋆ W.G.M. van den Hoek, T.E. Wicker, B.F. Westlund, and G.B. Powell, "Isotropic plasma etching of doped and undoped silicon dioxide for contact holes and vias", J. Vac. Sci. Technol. A **7**, 670 (1989).

[262]⋆ G.S. Oehrlein, R.M. Tromp, J.C. Tsang, Y.H. Lee, and E.J. Petrillo, "Near-surface damage and contamination after CF_4/H_2 reactive ion etching of Si", in Proc. 5th Symp. Plasma Processing, The Electrochemical Society, ECS Proc. Vol. **85-1**, 587 (1985).

[263] D.J. Vitkavage and T.M. Mayer, "Residue formation on Si surfaces in a CHF_3 discharge environment", J. Vac. Sci. Technol. B **4**, 1283 (1986).

[264] G.S. Oehrlein, J.G. Clabes, and P. Spirito, "Investigation of reactive-ion-etching-related fluorocarbon film deposition onto silicon and a new method for surface residue removal", J. Electrochem. Soc. **133**, 1002 (1986).

[265] T.J. Faith, J.J. O'Neill Jr, R.S. Irven, J.L. Vossen, J.M. Shaw, and J.H. Thomas III, "Comparative investigation of CF_4-plasma, Ar-plasma, and dilute-HF-dip cleaning methods for (Al-Si)/n^+Si contacts", J. Electrochem. Soc. **134**, 665 (1987).

[266]⋆ I.-W. H. Connick, A. Bhattacharyya, K.N. Ritz, and W.L. Smith, "Study of reactive-ion-etch-induced lattice damage in silicon by Ar, CF_4, NF_3, and CHF_3 plasmas", J. Appl. Phys. **64**, 2059 (1988).

[267] D. Misra and E.L. Heasell, "A study of reactive ion etching (CF_4 + O_2 plasma) induced deep levels in silicon", J. Electrochem. Soc. **134**, 956 (1987).

[268] J.M. Heddleson, M.W. Horn, S.J. Fonash, and D.C. Nguyen, "Effects of dry etching on the electrical properties of silicon", J. Vac. Sci. Technol. B **6**, 280 (1988).

[269]⋆ R.G. Frieser, J.A. Bondur, and E.F. Gorey, "Oxygen plasma damage in boron-diffused silicon", J. Electrochem. Soc. **134**, 419 (1987).

[270]⋆ S. Fujimura and H. Yano, "Heavy metal contamination from resists during plasma stripping", J. Electrochem. Soc. **135**, 1195 (1988).

[271]⋆ S. Samukawa, "Damage caused by stored charge during ECR plasma etching", in Proc. 2nd Symp. ULSI Science and Technology, The Electrochemical Society, ECS Proc. Vol. **89-9**, 421 (1989).

[272] K.-H. Rydén, H. Norström, C. Nender, and S. Berg, "Oxide breakdown due to charge accumulation during plasma etching", J. Electrochem. Soc. **134**, 3113 (1987).

[273] R. Pinto, K.V. Ramanathan, and R.S. Babu, "Reactive ion etching in SF_6 gas mixtures", J. Electrochem. Soc. **134**, 165 (1987).

[274]⋆ T.H. Fedynyshyn, G.W. Grynkewich, and T.-P. Ma, "Mask dependent etch rates II. The effect of aluminum vs. photoresist masking on the etch rates of silicon and silicon dioxide in fluorine containing plasmas", J. Electrochem. Soc. **134**, 2580 (1987).

[275]⋆ G.C. Eiden, P.K. Boyer, and J.A. Hughes, "Geometry independent deep trench etching, refill and planarization for isolation of merged Bipolar-CMOS", in Proc. 5th Symp. Plasma Processing, The Electrochemical Society, ECS Proc. Vol. **85-1**, 502 (1985).

[276]⋆ S.E. Bernacki and B.B. Kosicki, "Controlled film formation during CCl_4 plasma etching", in Proc. 4th Symp. Plasma Processing, The Electrochemical Society, ECS Proc. Vol. **83-10**, 505 (1983).

[277] O. Auciello and D.L. Flamm (eds.), *Plasma diagnostics. Volume 1: discharge parameters and chemistry* (Academic Press, New York, 1989).

[278] O. Auciello and D.L. Flamm (eds.), *Plasma diagnostics. Volume 2: surface analysis and interactions* (Academic Press, New York, 1989).

[279] G. Fortunato, "End-point determination by reflected power monitoring", J. Phys. E **20**, 1051 (1987).

[280] D.B. Ilić, "Impedance measurement as a diagnostic for plasma reactors", Rev. Sci. Instrum. **52**, 1542 (1981).

[281] A.J. van Roosmalen, "Plasma parameter estimation from rf impedance measurements in a dry etching system", Appl. Phys. Lett. **42**, 416 (1983).

[282] K.D. Allen, H.H. Sawin, M.T. Mocella, and M.W. Jenkins, "The plasma etching of polysilicon with CF_3Cl/argon discharges", J. Electrochem. Soc. **133**, 2315 (1986).

[283]⋆ J.I. Ulacia F. and J.P. McVittie, "In-situ monitoring of electrical parameters for dry etching", in *Plasma processing and synthesis of materials*, edited by D. Apelian and J. Szekely (Mat. Res. Soc. Symp. Proc. Vol. 98, Pittsburgh, 1987), p. 203.

[284] B. Agdur and B. Enander, "Resonances of a microwave cavity partially filled with a plasma", J. Appl. Phys. **33**, 575 (1962).

[285] T.J. Bisschops, H. Störi, J.A. Engels, F.J. de Hoog, and D.C. Schram, "Electron density measurements in a single wafer etch reactor", in Proc. 7th Int. Symp. on Plasma Chem., edited by C.J. Timmermans (IUPAC, Eindhoven, The Netherlands, July 1985), p. 599.

[286]⋆ R.J. Visser, "Determination of the power and current densities in argon and oxygen plasmas by in situ temperature measurements", J. Vac. Sci. Technol. A **7**, 189 (1989).

[287] G. Fortuño, "Silicon dioxide reactive ion etching dependence on sheath voltage", J. Vac. Sci. Technol A **4**, 744 (1986).

[288] S.A. Shivashankar and B. Robinson, "Calorimetry in thin film processing", J. Vac. Sci. Technol. A **4**, 1826 (1986).

[289]⋆ I. Hussla, K. Enke, H. Grünwald, G. Lorenz, and H. Stoll, "In situ silicon-wafer temperature measurements during RF argon-ion plasma etching via fluoroptic thermometry", J. Phys. D **20**, 889 (1987).

[290] V.M. Donnelly, "Optical diagnostic techniques for low pressure plasmas and plasma processing", in *Plasma diagnostics. Volume 1: discharge parameters and chemistry*, edited by O. Auciello and D.L. Flamm (Academic Press, New York, 1989), p. 1.

[291]⋆ E.O. Degenkolb and J.E. Griffiths, "Simple optical devices for detection of radiofrequency oxygen plasma stripping of photoresists", Appl. Spectroscopy **31**, 40 (1977).

[292]⋆ J.W. Coburn and M. Chen, "Optical emission spectroscopy of reactive plasmas: a method for correlating emission intensities to reactive particle density", J. Appl. Phys. **51**, 3134 (1980).

[293] R. d'Agostino, F. Cramarossa, S. De Benedictis, and G. Ferraro, "Spectroscopic diagnostics of CF_4-O_2 plasmas during Si and SiO_2 etching processes", J. Appl. Phys. **52**, 1259 (1981).

[294]⋆ J.W. Coburn and M. Chen, "Dependence of F atom density on pressure and flow rate in CF_4 glow discharges as determined by emission spectroscopy", J. Vac. Sci. Technol. **18**, 353 (1981).

[295]⋆ R. d'Agostino, F. Cramarossa, S. De Benidictis, and F. Fracassi, "Optical emission spectroscopy and actinometry in CCl_4-Cl_2 radiofrequency discharges", Plasma Chem. Plasma Proc. **4**, 163 (1984).

[296] V.M. Donnelly, D.L. Flamm, W.C. Dautremont-Smith, and D.J. Werder, "Anisotropic etching of SiO_2 in low-frequency CF_4/O_2 and NF_3/Ar plasmas", J. Appl. Phys. **55**, 242 (1984).

[297] R.A. Gottscho and V.M. Donnelly, "Optical emission actinometry and spectral line shapes in rf glow discharges", J. Appl. Phys. **56**, 245 (1984).

[298]⋆ D.L. Flamm and V.M. Donnelly, "Time-dependent excitation in high- and low-frequency chlorine plasmas", J. Appl. Phys. **59**, 1052 (1986).

[299]⋆ D.A.O. Hope, T.I. Cox, and V.G.I. Deshmukh, "Langmuir probe and optical emission spectroscopic studies of Ar and O_2 plasmas", Vacuum **37**, 275 (1987).

[300] B.J. Curtis and H.J. Brunner, "End point determination of aluminum CCl_4 plasma etching by optical emission spectroscopy", J. Electrochem. Soc. **125**, 829 (1978).

[301] E.O. Degenkolb, C.J. Mogab, M.R. Goldrick, and J.R. Griffiths, "Spectroscopic study of radiofrequency oxygen plasma stripping of negative photoresists", Appl. Spectroscopy **30**, 520 (1976).

[302] O. Krogh, H. Slomowitz, Y. Melaku, and H.-O. Blom "Spectroscopic diagnostics of photoresist erosion in an aluminum etch plasma", J. Electrochem. Soc. **134**, 2045 (1987).

[303] D.E. Ibbotson, D.L. Flamm, and V.M. Donnelly, "Crystallographic etching of GaAs with bromine and chlorine plasmas", J. Appl. Phys. **54**, 5974 (1983).

[304] J. Wormhoudt, A.C. Stanton, A.D. Richards, and H.H. Sawin, "Atomic chlorine concentration and gas temperature measurements in a plasma etching reactor", J. Appl. Phys. **61**, 142 (1987).

[305] H.-U. Poll, M. Arzt, and K.H. Wickleder, "Reaction kinetics in the polymerization of thin films on the electrodes of a glow-discharge gap", Eur. Polym. J. **12**, 505 (1976).

[306]⋆ H.-U. Poll, D. Hinze, and H. Schlemm, "In situ analysis of fluorinated gases in plasma etching by infrared spectroscopy", Appl. Spectroscopy **36**, 445 (1982).

[307] T.A. Cleland and D.W. Hess, "In situ FTIR diagnostics of the radio-frequency plasma decomposition of N_2O", Plasma Chem. Plasma Proc. **7**, 379 (1987).

[308]⋆ T.A. Cleland and D.W. Hess, "Detection of dry etching product species with in situ Fourier Transform infrared spectroscopy", J. Vac. Sci. Technol. B **7**, 35 (1989).

[309]⋆ D.K. Doughty, S. Salih, and J.E. Lawler, "Two-step optogalvanic effect using intersecting laser beams: a pinpoint discharge diagnostic", Phys. Lett. A **103**, 41 (1984).

[310]⋆ R. Walkup, R.W. Dreyfus, and P. Avouris, "Laser optogalvanic detection of molecular ions", Phys. Rev. Lett. **50**, 1846 (1983).

[311] R.W. Dreyfus, J.M. Jasinski, R.E. Walkup, and G.S. Selwyn, "Optical diagnostics of low pressure plasmas", Pure Appl. Chem. **57**, 1265 (1985).

[312]⋆ S.W. Downey, A. Mitchell, and R.A. Gottscho, "Photoemission optogalvanic spectroscopy: an in situ method for plasma electrode surface characterization", J. Appl. Phys. **63**, 5280 (1988).

[313] J.J. Cuomo and C.R. Guarnieri, "Photoelectric enhanced plasma glow discharge system and method including radiations means", U.S. Patent 4,664,769 (1987).

[314]⋆ G.S. Selwyn, B.D. Ai, and J. Singh, "Real-time measurements of plasma/surface interaction by plasma-amplified photoelectron detection", Appl. Phys. Lett. **52**, 1953 (1988).

[315] R.W.B. Pearse and A.G. Gaydon, *The identification of molecular spectra* (Chapman & Hall, London, 1984).

[316] C.E. Moore, *Atomic energy levels. As derived from the analysis of optical spectra.* Volume I: ^{1}H–^{23}V (1949). Volume II: ^{24}Cr–^{41}Nb (1952). Volume III: ^{42}Mo–^{57}La and ^{72}Hf–^{89}Ac (1958). National Bureau of Standards, Washington D.C.

[317]⋆ M.J. Goeckner and J. Goree, "Laser-induced fluorescence measurement of plasma ion temperature: corrections for power saturation", J. Vac. Sci. Technol. A **7**, 977 (1989).

[318] R.B. Wright, M.J. Pellin, and D.M. Gruen, "Velocity distribution of sputtered Zr atoms as determined by laser induced fluorescence spectroscopy", Surf. Sci. **110**, 151 (1981).

[319] R.B. Wright, C.E. Young, M.J. Pellin, and D.M. Gruen, "High resolution continuous wave laser induced fluorescence spectroscopy of sputtered Zr atoms", J. Vac. Sci. Technol. **20**, 510 (1982).

[320]⋆ W. Husinsky, "The application of Doppler shift laser fluorescence spectroscopy for the detection and energy analysis of particles evolving from surfaces", J. Vac. Sci. Technol. B **3**, 1546 (1985).

[321] K.E. Greenberg and P.J. Hargis jr., "Detection of sulfur dimers in SF_6 and SF_6/O_2 plasma-etching discharges", Appl. Phys. Lett. **54**, 1374 (1989).

[322] G.S. Selwyn, "Atomic arsenic detection by ArF laser-induced fluorescence", Appl. Phys. Lett. **51**, 167 (1987).

[323]⋆ G.S. Selwyn, "Laser diagnostic techniques for reactive ion etching: plasma understanding to plasma control", J. Vac. Sci. Technol. A **6**, 2041 (1988).

[324]⋆ R. Deaton and A. George, "Quantification of laser interference fringes as applied to plasma etch endpoint detection", in *Laser in Microlithography*, S.P.I.E. Vol. **774**, 162 (1987).

[325]⋆ P.A. Heimann, "Optical etch-rate monitoring using active device areas: lateral interference effects", J. Electrochem. Soc **132**, 2003 (1985).

[326] D.E. Aspnes and R.P.H. Chang, "Spectroscopic ellipsometry in plasma processing", in *Plasma diagnostics. Volume 2: surface analysis and interactions*, edited by O. Auciello and D.L. Flamm (Academic Press, New York, 1989), p. 67.

[327] R.H. Muller, "Present status of automatic ellipsometers", Surf. Sci. **56**, 19 (1976).

[328] B. Drevillon, J. Perrin, R. Marbot, A. Violet, and J.L. Dalby, "Fast polarization modulated ellipsometer using a microprocessor system for digital Fourier analysis", Rev. Sci. Intrum. **53**, 969 (1982).

[329] G.E. Jellison jr. and D.H. Lowndes, "Time-resolved ellipsometry measurements of the optical properties of silicon during pulsed excimer laser irradiation", Appl. Phys. Lett. **47**, 718 (1985).

[330] A.M. Antoine and B. Drevillon, "Influence of ion bombardment on the nucleation and growth of plasma deposited amorphous silicon", in *Photon, beam, and plasma stimulated chemical processes at surfaces*, edited by V.M. Donnelly, I.P. Herman, and M. Hirose (Mat. Res. Soc. Symp. Proc. Vol. 75, Pittsburgh, 1987), p. 333.

[331] B. Drevillon, C. Godet, and A.M. Antoine, "Ion bombardment effect on the growth of microcrystalline germanium", in *Photon, beam, and plasma stimulated chemical processes at surfaces*, edited by V.M. Donnelly, I.P. Herman, and M. Hirose (Mat. Res. Soc. Symp. Proc. Vol. 75, Pittsburgh, 1987), p. 341.

[332] L.J. Hanekamp, S.J.H. Brader, and G.A. Bootsma, "Spectroscopic ellipsometric investigations of the clean and oxygen exposed Ni(110) surface", Surf. Sci. **135**, 383 (1983).

[333] R.P.H. Chang and S. Darack, "Hydrogen plasma etching of GaAs oxide", Appl. Phys. Lett. **38**, 898 (1981).

[334] R.P.H. Chang, S. Darack, E. Lane, C.C. Chang, D. Allara, and E. Ong, "Plasma enhanced beam deposition of thin films at low temperatures", J. Vac. Sci. Technol. B **1**, 935 (1983).

[335] M. Erman, J.B. Theeten, N. Vodjdani, and Y. Demay, "Chemical and structural analysis of the GaAs/AlGaAs heterojunctions by spectroscopic ellipsometry", J. Vac. Sci. Technol. B **1**, 328 (1983).

[336] *Eight peak index of mass spectra* (Mass Spectrometry Data Centre AWRE, the Royal Society of Chemistry, Nottingham, England, 1983).

[337] K. Ensslen and S. Veprek, "Dominant reaction channels and the mechanism of silane decomposition in a H_2-Si(s)-SiH_4 glow discharge", Plasma Chem. Plasma Proc. **7**, 139 (1987).

[338] A. Cantin and R.R.J. Gagné, "Pressure dependence of electron temperature using rf-floated electrostatic probes in rf plasmas", Appl. Phys. Lett. **30**, 316 (1977).

[339] E. Eser, R.E. Ogilvie, and K.A. Taylor, "Measurement of plasma discharge characteristics for sputtering applications", J. Vac. Sci. Technol. **15**, 199 (1978).

[340] M.J. Kushner, "A kinetic study of the plasma-etching process. II. Probe measurements of electron properties in an rf plasma-etching reactor", J. Appl. Phys. **53**, 2939 (1982).

[341] C. Chan, Z.-J. Jin, and C. Whitaker, "Parameter investigation of the sheath potential in a low-frequency rf discharge", J. Appl. Phys. **62**, 1633 (1987).

[342] E.R. Mosburg jr., R.C. Kerns, and J.R. Abelson, "Use of electric probes in silane radio frequency discharges", J. Appl. Phys. **54**, 4916 (1983).

[343]⋆ B.E. Cherrington, "The use of electrostatic probes for plasma diagnostics - a review", Plasma Chem. Plasma Proc. **2**, 113 (1982).

[344]⋆ B. Lipschultz, I. Hutchinson, B. LaBombard, and A. Wan, "Electric probes in plasmas", J. Vac. Sci. Technol. A **4**, 1810 (1986).

[345] N. Hershkowitz, "How Langmuir probes work", in *Plasma diagnostics. Volume 1: discharge parameters and chemistry*, edited by O. Auciello and D.L. Flamm (Academic Press, New York, 1989), p. 113.

[346]⋆ E.O. Johnson and L. Malter, "A floating double probe method for measurement in gas discharges", Phys. Rev. **80**, 58 (1950).

[347]⋆ T. Okuda and K. Yamamoto, "Asymmetrical triple probe method for determining energy distribution of electron in plasma", J. Appl. Phys. **31**, 158 (1960).

[348] S. Aisenberg, "Multiple probe measurements in high-frequency plasma lasers", J. Appl. Phys. **35**, 130 (1964).

[349]⋆ S.-L. Chen and T. Sekiguchi, "Instantaneous direct-display system of plasma parameters by means of triple probe", J. Appl. Phys. **36**, 2363 (1965).

[350] I. Katsumata and M. Okazaki, "Ion sensitive probe – a new diagnostic method for plasma in magnetic fields", Jap. J. Appl. Phys. **6**, 123 (1967).

[351] M. Sato, "Capped probe for ion temperature diagnostic in a magnetized plasma", Phys. Fluids **17**, 1903 (1974).

[352] G.F. Matthews, "A combined retarding field analyser and E×B probe for measurement of ion and electron energy distribution in tokamak edge plasmas", J. Phys. D **17**, 2243 (1984).

[353]⋆ S.B. Dolins, A. Srivastava, and B.E. Flinchbaugh, "Monitoring and diagnosis of plasma etch processes", IEEE Trans. Semicond. Manuf. **1**, 23 (1988).

[354] P.H. Singer, "Diagnosing plasmas and detecting endpoints", Semicond. Int. **11** (9), 66 (1988).

SELECTED ABSTRACTS

33 Ion energies at the cathode of a glow discharge,

by W.D. Davis and T.A. Vanderslice, Phys. Rev. **131**, 219 (1963).

The energy distribution of ions striking the cathode in a glow discharge has been measured for a series of gases including hydrogen, helium, neon, and argon. Standard ultrahigh vacuum techniques were used to maintain a high degree of gas purity. The ions were detected by placing a small pinhole in the cathode of a glow discharge and analyzed for energy by means of a sector-type electrostatic analyzer. The ion species was determined with a conventional magnetic analyzer and the resulting ion beam intensity measured with an electron multiplier. An oscilloscope display gave the ion-energy distribution for a particular ion species. A simple theory involving assumptions that the ions originated in the negative glow and undergo primarily symmetrical charge transfer as they pass through the cathode dark space to the cathode gives results that are in most cases in good agreement with experiment. The charge-transfer cross sections determined are in reasonable agreement with other published data.

35 A time-average model of the RF plasma sheath,

by D.J. Economou, D.R. Evans, and R.C. Alkire, J. Electrochem. Soc. **135**, 756 (1988).

A time-average model of the RF plasma sheath was developed. The ion "fluid" equations were used with a frictional force to account for ion-neutral collisions. Consideration of the collision dynamics showed that the frictional force may be taken as proportional to the square of the ion drift velocity. The sheath model was used to investigate the ion energy and flux on the electrodes of plasma reactors. The dimensionless quantity C_o (collision number) was found to be important in describing the ion motion in the sheath. An analytical expression for the ion bombardment energy was derived, in terms of C_o and the sheath voltage, for the range of parameter values typical of high pressure ($\sim$1 Torr) diode plasma etchers. An application of the model to an oxygen discharge in a parallel plate reactor was considered. Over the parameter range investigated, the ion bombardment energy was found to be only a few tens of eV, much lower than typical sheath voltages ($\sim$200 V). The ion bombardment energy was found to be a function of the sheath electric-field-to-pressure ratio. The model provides a framework that can be incorporated into more general plasma reactor models which consider transport and reaction phenomena along surfaces undergoing etching.

37 Ion energy measurement at the powered electrode in an rf discharge,

by A.D. Kuypers and H.J. Hopman, J. Appl. Phys. **63**, 1894 (1988).

Ion energy measurements have been performed with an electrostatic parallel plate energy analyzer at the powered electrode of a 13.56-MHz rf discharge. Considerable splitting of the ion energy distributions is observed due to rf oscillations. Plasma potential, sheath thickness, and total ion current are derived from the observed energy profiles. Low-pressure operation of the plasma at several mTorr permits collisionless sheath approximation and gives rise to well-defined energy spectra.

39 Electrical characteristics of dry etching reactors,

by A.J. van Roosmalen, in Comptes Rendus des Travaux du CIPG87, Soc. Française du Vide (Paris, France, 1987), p. 99.

Some fundamental aspects of low pressure gas discharges are summarized, such as the characteristic frequencies and the phenomena at the plasma-wall interface. Concepts in terms of dielectric properties, ionic bombardment and the plasma density can be used to analyze the most frequently used dry etch equipments. Parasitic impedances and the interpretation of high frequency impedances are discussed in relation to reactor construction. This discussion is accompanied by a summary of the practical problems encountered with electrical measurements. (*Original abstract in French, article in English*)

40 Dynamics of formation and decay of pulsed nitrogen rf plasmas,

by R.J. Visser, J.A.G. Baggerman, J.P.J. Poppelaars, and E.J.H. Collart, in Proc. 9th Int. Symp. on Plasma Chem. Vol. I (Int. Union of Pure and Appl. Chemistry, Pugnochiuso, Italy, Sept 1989), p. 555.

The kinetics of the formation and decay of pulsed nitrogen plasmas have been followed by time and spatially resolved Optical Emission and Laser Induced Fluorescence spectroscopy. The decay of the N_2^+ ions after the power is switched off is caused by dissociative recombination and ambipolar diffusion to the wall. At high pressures the formation of N_4^+ has to be taken into account. A numerical model of the time evolution of the ion concentration profile is developed, which renders good agreement with experimental data. Kinetical measurements of the rotational (T_r) and vibrational (T_v) temperature of N_2^+ show that T_r and T_v of N_2^+ are in equilibrium with T_r and T_v of the N_2 gas. The much higher value of T_v with respect to T_r is explained by these experiments.

42 Frequency effects in plasma etching,

by D.L. Flamm, J. Vac. Sci. Technol. A **4**, 729 (1986).

The effects of excitation frequency on electrical discharges used for low pressure plasma processing are discussed and illustrated with examples. Frequency alters the spatial distribution of species and electric fields across the discharge, the energies and concentrations of species as a function of time, the characteristic potentials and energy of ions impinging on surfaces, and the electron energy distribution function. Important transitions in the behavior of these quantities occur when the excitation frequency crosses the ion-sheath transit frequency (ITF), the electron energy modulation frequency (ν_u), the electron momentum collision frequency (ν), and the plasma frequency (ω_p).

44 The energy distribution of ions bombarding electrode surfaces in rf plasma reactors,

by A. Metze, D.W. Ernie, and H.J. Oskam, J. Appl. Phys. **65**, 993 (1989).

A model is presented for the energy distribution of ions bombarding electrode surfaces in planar rf plasma reactors at low pressures, where collisions in the sheath may be ignored. Given the waveform for the voltage difference across the sheath adjacent to an electrode, the model predicts the ion bombardment energy distribution in the limits $t_r < 1/f$ and $t_r \gg 1/f$, where t_r is the transit time of the ions across the sheath and f is the frequency of the applied rf voltage. Utilizing sheath voltage waveforms from a previously published equivalent circuit model of a rf reactor [A. Metze, D.W. Ernie, and H.J. Oskam, J. Appl. Phys. **60**, 3081 (1986)], the model shows that the ion bombardment energy distribution is bimodal and skewed toward lower ion bombardment energies. The model also demonstrates one of the mechanisms responsible for narrowing of the ion bombardment energy distribution with increasing rf frequency. These results are compared with experimental measurements.

46 Theoretical analysis of the electrode sheath in rf discharges,

by K.-U. Riemann, J. Appl. Phys. **65**, 999 (1989).

Starting from a previous kinetic theory of the plasma boundary layer (charge exchange model of the presheath), we investigate a collisionless planar electrode sheath of a capacitively coupled rf discharge in the frequency range between the ionic and electronic plasma frequencies ($\omega_{p+} \ll \omega \ll \omega_{p-}$). The resulting model equations are evaluated in an approximation based on thick-sheath assumptions. We calculate the current-voltage characteristics of the sheath and analyze the voltage distribution in a two-electrode system.

49 Positive-ion bombardment of substrates in rf diode glow discharge sputtering,

by J.W. Coburn and E. Kay, J. Appl. Phys. **43**, 4965 (1972).

The plasma potential and the potential of an electrically isolated surface are measured in an rf diode sputtering glow discharge. The influence on these potentials of both the geometry enclosing the discharge volume and of a positively biased auxiliary electrode in contact with the discharge is investigated. It is shown that confining the discharge increases the plasma potential and the energy of positive ions incident on electrically isolated substrates, whereas applying a positive voltage to an auxiliary electrode also increases the plasma potential but does not significantly increase the energy of ions incident on electrically isolated substrates. The effect of rf modulation on the ionic energy distributions is demonstrated. This occurs as the ions pass through the plasma-substrate sheath and results in a large broadening of the energy distributions of low-mass species.

51 Double layers, breathing sheaths, and chemistry in BCl_3/rare gas radio frequency discharges,

by R.A. Gottscho and G.R. Scheller, in Abstr. of the 1987 Spring mtg. of the Am. Phys. Soc., Bull. Am. Phys. Soc. **32**, 1165 (1987).

Using laser-induced fluorescence (LIF) and optical emission spectroscopy, the properties of low frequency (dc to 1 MHz) discharges through mixtures of BCl_3 and He, Ar or Kr have been investigated. LIF is used to monitor the magnitude of the local electric field in the electrode sheaths and the concentrations of BCl and Ar metastables. Feedstocks with more than 5% BCl_3 exhibit double layers at the plasma-sheath boundary, whose position is nearly invariant during a cycle. The formation of double layers is understood in terms of field-dependent attachment and detachment rates. For discharges containing less BCl_3, the sheath fields are monotonic but are observed to "breath" periodically. As a function of Ar an Kr concentration in BCl_3, the BCl density exhibits a maximum at 80% rare gas. Although the BCl density decreases monotonically as He is added to the discharge, the decrease is less than what would be expected from simple dilution. By contrast, the Ar $1s_5$ metastable density decreases dramatically when only trace amounts of BCl_3 are added to the discharge. These findings suggest that energy transfer from rare gas metastables is an efficient mechanism for dissociation of BCl_3.

52 Electrical properties of planar rf discharges for dry etching,

by A.J. van Roosmalen, W.G.M. van den Hoek, and H. Kalter, J. Appl. Phys. **58**, 653 (1985).

Rf impedance, dc self-bias, and ion sheath (dark space) thickness are measured in on O_2 discharge for 7–53 Pa pressure and 50–800 W rf power (13.56 MHz). Special attention is paid to corrections for reactor stray impedances. It is concluded that the discharge can be described as a capacitance (the ion sheath) with both a parallel and a series resistance, the series element being the more important one. Good agreement is found between optical and electrical measurements of the ion sheath thickness. Evidence is presented that the dc potential difference between plasma and ground and rf electrode can be estimated with reasonable accuracy from the dc self-bias and the optical dark space thicknesses. Positive ion acceleration in the ion sheath and the electron-neutral collisions in the bulk of the plasma glow account for only part of the total rf power transfer. It is suggested that significant dissipation takes place near the glow-sheath boundary, although a quantitative description cannot be given yet.

53 Laser diagnostics of plasma etching: measurement of Cl_2^+ in a chlorine discharge,

by V.M. Donnelly, D.L. Flamm, and G. Collins, J. Vac. Sci. Technol. **21**, 817 (1982).

Laser-induced fluorescence has been used to determine the relative density of Cl_2^+ in a chlorine radio frequency discharge as function of pressure, flow, power, frequency, and distance from the electrodes. The conditions studied correspond closely to those used in anisotropic plasma etching of materials such as Si and InP. This technique provides quantitative data on ion densities over a wide range of parameters, as contrasted with qualitative methods, such as optical emission.

56 Application of RF discharges to sputtering,

by H.R. Koenig and L.I. Maissel, IBM J. Res. Devel. **14**, 168 (1970).

The operation of rf discharges is described and the internal distribution of voltages is considered. The significance of this with respect to sputtering, particularly of insulators, is then discussed. An equivalent circuit for the discharge is presented and the influence of such parameters as pressure and magnetic field on the components in this circuit is described. Finally, energy distributions for positive ions, electrons, and negative ions incident at the substrate during deposition are given.

58 Study of the mechanism of the etching of polymers in oxygen and argon plasmas,

by R.J. Visser and C.A.M. de Vries, in Proceedings of the 8th International Symposium on Plasma Chemistry, edited by K. Akashi and A. Kinbara (IUPAC, Tokyo, 1987), p. 1029.

In situ temperature measurements of silicon wafers in an RIE reactor have been used to analyse the processes by which a wafer loses and gains heat during plasma etching. The analysis shows that the energy flux from the positive ions to the powered and grounded electrode can be determined from the equilibrium temperature of a wafer in a plasma. The kinetic energy and the flux of the ions have been obtained from the energy flux, the D.C. bias potential and the thickness of the sheath. These data have been used to analyse the etching of novolak, polystyrene and PMMA in oxygen and argon plasmas at pressures between 1.3 and 13 Pa. It is shown that in an oxygen plasma the ratio of the etch rate and the energy flux is constant at all pressures. The yield per ion has been determined.

59 Power loss mechanisms in radio frequency dry etching systems,

by W.G.M. van den Hoek, C.A.M. de Vries, and M.G.J. Heijman, J. Vac. Sci. Technol. B **5**, 647 (1987).

The relation between the total radio frequency (rf) generator output power and the power that is actually dissipated in the plasma in rf dry etching systems was investigated. Power loss processes are described, their relation with the construction of the etch system is demonstrated, and their influence on the discharge is discussed. A method to estimate the power losses occurring in the etch system is described. For three different dry etching systems it is demonstrated that the stray capacitance of the system plays an important role in the power loss processes. In asymmetrical systems the dc self-bias voltage is shown to be a reliable parameter in determining the actual power dissipated in the plasma.

60 Radio frequency sputtering – the significance of power input,

by C.M. Horwitz, J. Vac. Sci. Technol. A **1**, 1795 (1983).

Radio frequency (rf) sputtering is used for the deposition and etching of thin layers. In both cases the target etch rate must be controlled, and often power input has been used as one of the controlling parameters. An earlier paper has shown that the applied rf target voltage (V_{pp}) was a more useful parameter, and here new data is presented which indicates that the peak-to-peak voltage remains the preferable parameter for both etch rate control and in transferring between machines. However, a new method is described here which obtains the discharge power by subtracting losses from the total power reading. This discharge power is shown to be related to the etch rates, and can thus be used as an aid in the scaling of system sizes. In addition, this discharge power exhibits interesting behavior as the pressure is varied, at constant applied rf voltage. Three main regions are evident when etching Si with CF_4: at low pressures directional ion-induced etching is obtained; in an intermediate region the input power rises rapidly with pressure and the etching is less directional resulting in overhang profiles; and at high pressures an isotropic etching component results in undercut profiles. The two extreme regions correlate with the commonly identified regimes of: (a) low power input, low pressure reactive ion etching, and (b) high power input, high pressure plasma etching.

62 Fluid model simulations of a 13.56-MHz rf discharge: time and space dependence of rates of electron impact excitation,

by D.B. Graves, J. Appl. Phys. **62**, 88 (1987).

Results from fluid model simulations of a 13.56-MHz rf discharge qualitatively reproduce many of the experimentally observed features of time and space resolved electron impact excitation in several gases (oxygen, nitrogen, and silane). The shape of the excitation rate waveform, its direction of propagation, time of occurrence in the rf period, and the initial increase in intensity followed by attenuation are all observed experimentally as well as in the simulation. Secondary electrons play no direct role in the excitation waveform in the simulation, although secondary electron creation is included in the model. The excitation waveform is the result of the combination of electron motion, electric field profiles, and the electron energy balance in the discharge. Electron heating by the rf field peaks at the plasma-sheath boundary, resulting in a local rise in the electron mean energy there. It is suggested that this electron heating mechanism is common to all high-frequency rf discharges in electropositive or weakly electronegative gases, and that this mechanism is responsible for increased rates of electron impact molecular dissociation around the plasma-sheath boundary. The qualitative agreement between simulation results and experimental measurements implies that the fluid model captures essential elements of rf discharge physics, and that the fluid models are useful for the interpretation of discharge diagnostics.

63 Self-consistent stochastic electron heating in radio frequency discharges,

by C.G. Goedde, A.J. Lichtenberg, and M.A. Lieberman, J. Appl. Phys. **64**, 4375 (1988).

Fermi acceleration is considered as an underlying mechanism for electron heating in rf discharges, in which the heating arises from the reflection of electrons from moving sheaths. By examining the dynamics of the electron collisions with the sheath, the map that describes the electron motion is derived. For high-frequency discharges ($\omega/2\pi > 50$ MHz), the electron motion is shown to be stochastic. By combining these dynamics with collisional effects in the bulk plasma and incorporating self-consistent physical constraints, a self-consistent model of the discharge is developed. The model is used to calculate physically interesting quantities, such as the electron temperature and average lifetime, and to predict the minimum pressure necessary to sustain the plasma. The distribution of electron energies is shown to be non-Maxwellian. These results can be applied to experimentally interesting parallel-plate rf plasma discharges to predict the operating conditions necessary for stochastic heating to occur.

64 Macroscopic modelling of radio-frequency plasma discharges,

by G.R. Misium, A.J. Lichtenberg, and M.A. Lieberman J. Vac. Sci. Technol. A **7**, 1007 (1989).

We describe a self-consistent model of a symmetric plane parallel rf discharge. The model is built upon basic laws such as conservation of particles and energy, and is capable of predicting rapidly the important discharge parameters from a processing point of view, such as the ion energy and flux to the electrodes. The following physics is incorporated into the model: energy-dependent electron–neutral ionization, excitation and elastic scattering; nonuniform, self-consistent collisionless and collisional rf sheaths; electron Ohmic heating by elastic scattering in the sheaths and bulk plasma stochastic heating by the oscillating fields in the sheaths; electron energy losses to neutrals through collisions and to the electrodes; ambipolar ion diffusion; and total rf power balance. A set of equations describing this dynamics has been obtained and used in a code to simulate different discharges. The model has proven to be useful in comparing the effect of varying parameters on the discharge. Comparisons with experimental results show a good agreement between predicted and measured parameters.

65 Energy-loss rate and enhanced diffusion due to a low-frequency plasma oscillation,

by A.J. Anastassiades and E.G. Sideris, J. Appl. Phys. **60**, 3422 (1986).

We have studied the relation of the energy-loss rate and the enhanced diffusion with the amplitude of spontaneously excited low-frequency plasma oscillation in a current-carrying argon plasma. The experimental results (radial distributions of the electron density, electric floating potential, and electric field) and the theoretical model that is developed showed the important role of the low-frequency plasma oscillations on the additional power losses of the plasma. Also, the energy-loss rate has been studied as a function of the plasma radius.

66 Microwave spectroscopic measurement of the electron density in a planar discharge: relation to reactive-ion etching of silicon oxide,

by C.A.M. de Vries, A.J. van Roosmalen, and G.C.C. Puylaert, J. Appl. Phys. **57**, 4386 (1985).

It is demonstrated that microwave cutoff can be used to measure absolute electron densities for CHF_3, Ar, and O_2 plasmas in a batch-mode planar reactor excited at 13.56 MHz. The densities obtained are between 10^{15} and 10^{16} m^{-3} for all gases and a large set of conditions (pressure, rf power, electrode spacing). A comparison with selective etching of SiO_2 over Si in the same reactor shows a qualitative agreement between the substrate ion flux calculated from the electron density and the etch rate of silicon oxide; the apparent sputter yield ranges from 0.4 to 1.4 molecules per ion.

68 Large-signal time-domain modelling of low-pressure rf glow discharges,

by M.S. Barnes, T.J. Colter, and M.E. Elta, J. Appl. Phys. **61**, 81 (1987).

Large-signal time-domain modeling (simulation) of rf glow discharges is a very useful and potentially accurate tool for the study of low-pressure (50-500 mT) gaseous electronics at high frequencies. Unfortunately, the computational limitations imposed for stability, accuracy, and efficiency can often hinder the production of useful, cost-effective results. This paper describes a self-consistent argon rf glow-discharge simulation at 13.56 MHz for equal- and unequal-area parallel-plate electrode geometries. Some of the numerical problems associated with this type of simulation are identified and the numerical methods used to overcome them are described. To illustrate the usefulness of this modelling scheme, the plasma potential and the cathode dc bias are examined as functions of the electrode area ratio and rf power.

69 Frequency dependence of CCl_4 etching,

by R.H. Bruce and B. Gelernt, in Ext. Abstracts of the Electrochem. Soc. Spring mtg., ECS Ext. Abstr. **80-1**, 307 (1980).

Carbon tetrachloride is commonly employed for plasma etching of aluminum, and less commonly, as an anisotropic etchant for silicon. We have measured etch rates and monitored mass spectra of the CCl_4 system at several excitation frequencies from 20 kHz to 27 MHz is a parallel plate reactor. The etch rate of silicon exhibits a large frequency dependence with fast rates at low frequencies but with negligible rates at high frequencies where polymer is observed to form on the substrates. Quite differently, the etch rates of aluminum at high frequencies are fast while at low frequencies the rates are generally much slower.

72 The optogalvanic effect in a 13.56-MHz chlorine discharge,

by J. Kramer, J. Appl. Phys. **60**, 3072 (1986).

Irradiation of the bulk plasma region in a 13.56-MHz Cl_2 discharge with an XeCl laser increased the current through the discharge. The perturbed current waveform had the shape of an amplitude modulated waveform with a period long with respect to the rf cycle. This optogalvanic effect has been attributed to the photodetachment of electrons from Cl^-, which perturbs the electron density. The optogalvanic signal depended on the square root of the pressure and was almost independent of the discharge current. The optogalvanic current has been correlated with the dependence of the Cl^- density on pressure and discharge current as well as the changes in the electric field induced by the perturbation. Saturation of the signal at high laser intensity has been accounted for by the impedance mismatch induced by the increased electron density. From the relationship between the saturation and the impedance mismatch, the ratio of Cl^- to the electron density was calculated. Probing various parts of the discharge with a laser beam having a smaller cross section than the discharge did not reveal any significant spatial variations in the discharge.

73 Negative ion kinetics in radio-frequency glow discharges,

by R.A. Gottscho and C.E. Gaebe, J. Vac. Sci. Technol. A **4**, 1795 (1986).

Although the use of electronegative gases in rf plasma processing of electronic materials is widespread, the role which negative ions play in such processes has not been considered in detail. While negative ions must exist in these discharges, several questions have remained unanswered. Do anions impact surfaces? If so, how might the effects of negative ion bombardment differ from positive ion bombardment? How does the application of an rf field affect anion kinetics? Do answers to the above questions depend upon operating conditions? Using temporally and spatially resolved laser spectroscopic techniques, we have been able to answer some of these questions. The presence, approximate concentration, and kinetics of negative ions in rf discharges are probed using the photodetachment technique. The sheath electric field is measured as a function of position and time using the Stark-mixing technique.

86 Electron energy distribution functions in radio-frequency collision dominated nitrogen discharges,

by M. Capitelli, R. Celiberto, C. Gorse, R. Winkler, and J. Wilhelm, J. Phys. D **21**, 691 (1988).

Electron energy distribution functions (EEDFS) in RF collision dominated N_2 discharges have been calculated by solving the time-dependent Boltzmann equation including elastic, inelastic and superelastic collisions. The results have been obtained at different electric field amplitudes E_0/p_0 ($10 \leq E_0/p_0 \leq 20$ V cm^{-1} Torr^{-1}), at different frequencies ω/p_0 ($\pi \times 10^6 \leq \omega/p_0 \leq \pi \times 10^8$ s^{-1} Torr^{-1}) and different concentrations of excited (vibrational and electronic) states. It has been found that the presence of excited states strongly modifies the modulation tail of EEDFS with strong consequences on the collision frequencies of processes having high energy thresholds. The role of field frequency is studied for the understanding of both the temporal behaviour of EEDFS and the averaged (over the period) values of EEDFS and of macroscopic quantities. We obtain that at fixed E_0/p_0 the increase of frequency decreases the tail of EEDFS, strongly reducing the collision frequency of the ionisation process. Finally the importance of the time-dependent solution of the Boltzmann equation is tested by comparing present results with corresponding ones obtained by using effective field approximation. Differences up to several orders of magnitude are obtained in this comparison, the differences being reduced at high E_0/p_0, high ω/p_0 and large concentrations of excited states.

87 Electron excitation rates and transport parameters in high-frequency N_2 discharges,

by C.M. Ferreira and J. Loureiro, J. Phys. D **22**, 76 (1989).

The electron energy distribution function, transport parameters and rate coefficients for excitation and ionisation in N_2 under the action of a uniform HF field were calculated by solving the Boltzmann equation together with a set of rate balance equations for the populations in the $N_2(V,v)$ vibrational levels. The treatment is restricted to situations in which the electrons do not lose appreciable energy during a cycle of field oscillation but applies to a broad range of applied frequencies, ω, bridging the gap between the limiting cases of few and many collisions per oscillation. The case of DC applied field previously investigated is obtained from the present formulation for $\omega = 0$. The effects of changes in ω and in vibrational excitation on the electron kinetics are investigated in detail, thus opening the way to firmly based theoretical interpretations of the properties of N_2 discharges sustained by HF fields.

95 Chlorine reactions on the Si(111) surface,

by J.V. Florio and W.D. Robertson, Surf. Sci. **18**, 398 (1969).

Auger electron emission spectroscopy has been used to identify and, with suitable calibration, to quantitatively specify chlorine surface concentrations on Si(111) surfaces. Adsorption rates of Cl_2 on Si(111) 7 and Si(111) $\sqrt{19}$ surfaces have been measured as a function of the substrate temperature, and sticking probabilities have been subsequently calculated. The initial sticking probability is $\sim$0.1. Adsorbed chlorine is observed to exist in two states, one, weakly bound and the other, strongly bound to the surface. Electron beam desorption of chlorine from Si(111) surfaces can be significant with a focussed electron beam. Starting with an ordered (1×1) chlorine structure, the result of beam desorption is a disordered surface. The electron beam desorption cross-section for chlorine was calculated to be $\sim 2 \times 10^{-19}$ cm^2. Various LEED structures were observed for chlorine on Si(111) surfaces. These include two with chlorine monolayer concentrations, Si(111) 7*-Cl and Si(111) 1*-Cl, and a third with only a trace (4.5% of a monolayer) of chlorine present, Si(111) 1. The 1×1 structures at high and low coverage are different, as shown by the voltage dependence of the reflected beam intensities. The reaction between chlorine and the Si(111) surface follows second order kinetics in its dependence on chlorine surface concentration. The desorption reaction product is $SiCl_4$. From the temperature dependence of the rate constant, an activation energy of 35 kcal/mol for the desorption process was measured. A reaction mechanism, involving the surface species [$>$ $SiCl_2$] which is formed on the surface ledge at the junction of (001) and (111) planes, is suggested.

101 Time-of-flight study on the thermal etching of Al with Cl_2,

by R.J.A.A. Janssen, A.W. Kolfschoten, and G.N.A. van Veen, Appl. Phys. Lett. **52**, 98 (1988).

A pseudorandom cross-correlation technique has been used to study the thermal etching of aluminum with molecular chlorine. Time-of-flight (TOF) distributions of particles desorbing from the substrate are measured by modulating the ejected product beam. Modulation is achieved by a rotating chopping disc with slots in a pseudorandom sequence on its periphery. TOF distributions and temperature-dependent reaction product yields are measured for the products desorbing from the chlorinated aluminum substrate in the range from 300 to 900 K. A reaction mechanism will be discussed in detail.

104 Quantitative sputtering,

by P.C. Zalm, Surf. & Int. Anal. **11**, 1 (1988).

This review discusses experimental techniques used for determining the variety of observables in ion-bombardment induced erosion of solid surfaces. Boundary conditions for performing reproducible and accurate sputtering yield measurements are formulated. To this end, an inventory is made of the observed systematics in several phenomena accompanying sputtering and of the more regular exceptions to the general trends, also, a brief treatment of some theoretical predictions is given, based on the available literature. Following this extensive survey, the merits and limitations of a number of measurement methods, tailored to study one or more different observables, are discussed. Techniques that seem widely applicable or highly promising are emphasized.

106 Sputtering by ion bombardment: theoretical concepts,

by P. Sigmund, in *Sputtering by particle bombardment I*, edited by R. Behrisch (Springer, Berlin, 1981), p. 9.

This chapter is an introduction to sputtering theory rather than a review. The first section contains a brief historical survey as well as an attempt to both define and classify sputtering processes. In the second section, some pertinent results from atomic scattering, penetration, and collision cascade theory are summarized. The central section is an updated summary of the linear cascade theory of sputtering, based in part on still unpublished work by the author more than a decade ago. The limitations of linear cascade theory in the isotropic limit are discussed, and comments are made on feasible ways to go beyond this limit. The discussion of spike phenomena in the fourth section was no longer up to date at the time of the page proofs, but a substantial revision was not made because the field is still moving rapidly. On the other hand, an effort was made to clarify and specify the predictions and implications of linear cascade theory with regard to sputtering of multicomponent targets in the last section. (*Introduction to the chapter*).

107 Theory of sputtering. I. Sputtering yield of amorphous and polycrystalline targets,

by P.Sigmund, Phys. Rev. **184**, 383 (1969). [*Errata*: Phys. Rev. **187**, 768 (1969)].

Sputtering of a target by energetic ions or recoil atoms is assumed to result from cascades of atomic collisions. The sputtering yield is calculated under the assumption of random slowing down in an infinite medium. An integrodifferential equation for the yield is developed from the general Boltzmann transport equation. Input quantities are the cross sections for ion-target and target-target collisions, and atomic binding energies. Solutions of the integral equation are given that are asymptotically exact in the limit of high ion energy as compared to atomic binding energies. Two main stages of the collision cascade have to be distinguished: first, the slowing down of the primary ion and all recoiling atoms that have comparable energies – these particles determine the spatial extent of the cascade; second, the creation and slowing down of low-energy recoils that constitute the major part of all atoms set in motion. The separation between the two stages is essentially complete in the limit of high ion energy, as far as the calculation of sputtering yield is concerned. High-energy collisions are characterized by Thomas-Fermi-type cross sections, while a Born-Mayer-type cross section is applied in the low-energy region. Electronic stopping is included when necessary. The separation of the cascade into two distinct stages has the consequence that two characteristic depths are important for the qualitative understanding of the sputtering process. First, the scattering events that eventually lead to sputtering take place within a certain layer near the surface, the thickness of which depends on ion mass and energy and on ion-target geometry. In the elastic collision region, this thickness is a sizable fraction of the ion range. Second, the majority of sputtered particles originate from a very thin surface layer ($\approx$5 Å), because small energies dominate. The general sputtering-yield formula is applied to specific situations that are of interest for comparison with experiment. These include backsputtering of thick targets by ion beams at perpendicular and oblique incidence and ion energies above $\approx$100 eV, transmission sputtering of thin foils, sputtering by recoil atoms from α-active atoms distributed homogeneously or inhomogeneously in a thick target, sputtering of fissionable specimens by fission fragments, and sputtering of specimens that are irradiated in the core of a reactor or bombarded with a neutron beam. There is good agreement with experimental results on polycrystalline targets within the estimated accuracy of the data and the input parameters entering the theory. There is no need for adjustable parameters in the usual sense, but specific experimental setups are discussed that allow independent checks or accurate determination of some input quantities.

108 Ion-beam assisted etching of semiconductors,

by P.C. Zalm, Vacuum **36**, 787 (1986).

Many groups have investigated the interaction of surfaces with ion beams, with and without simultaneous exposure to a reactive gas, as a model simulation for plasma etching. On the basis of their experimental evidence current understanding of ion beam assisted etching of semiconductors is reviewed. The discussion is restricted to technologically important ions with atomic numbers $\mathcal{Z} \geq 6$ in the energy range 0.25–25 keV. This allows for a meaningful comparison with predictions from linear cascade theory. First, physical sputtering by noble gas ions will be discussed. Next, reactive and molecular ions are treated. It is shown that their sputtering yield behaviour is not dramatically different from that of inert gas ions and that chemical contributions can reasonably be accounted for. Finally, chemical sputtering of materials by simultaneous exposure to ion bombardment and a reactive gas flow is examined. Observed systematics in absolute etch rates and energy distributions of emitted reaction products point to a number of parallel acting mechanisms. Adsorption and subsequent ion beam mixing of the reactive gas into the surface, ion induced desorption, electronic excitation of the target, surface temperature and the chemical activity of the gas can each dominate the detailed etch behaviour. In spite of the fact that the broad trends in chemical sputtering apparently agree with a collision-cascade-like ejection mechanism, this multitude of interactions presently prohibits development of quantitative models.

110 Some recent results of fundamental studies on beam-induced surface processes,

by J. Dieleman, in *Laser- and particle-beam chemical processes on surfaces*, edited by A.W. Johnson, G.L. Loper and T.W. Sigmon (Mat. Res. Soc. Symp. Proc. Vol. 129, Pittsburgh, 1989), p. 3.

Recent results of studies on the mechanisms of processes induced by the interaction of beams of (reactive) particles with surfaces in UHV, often combined with concurrent bombardment of the surface with low-energy ion or excimer laser beams, are reviewed. Angular-resolved mass spectrometry combined with time-of-flight studies on the desorbing products is used as a key diagnostic. A more complete picture is obtained using several other diagnostic tools to characterize the surface before, during or after the interaction. Interactions at Si and Cu surfaces will be emphasized. The review will deal successively with data for interaction with a single beam of Cl_2, low-energy noble gas ions or excimer laser pulses, followed by data on the concurrent interaction of crossed beams of Cl_2 and low-energy noble gas ions or excimer laser pulses with these surfaces. Some conclusions will be drawn.

114 On the sputtering yield of molecular ions,

by Ch. Steinbrüchel, J. Vac. Sci. Technol. A **3**, 1913 (1985).

It is shown that the sputtering yield of a molecular ion can be expressed in terms of the sputtering yields of the constituent atoms, irrespective of the details of the process by which the molecular ion dissociates upon impact on the target surface. It then follows that, in general, sputtering yields on a clean surface should be about the same for molecular and atomic ions of the same mass and energy, exceptions being very light targets or very heavy projectiles, in which case sputtering yields for molecular ions should be somewhat higher. For ions which are chemically reactive with respect to a target and thus change its surface composition, these conclusions need to be modified. Sputtering yields at very low ion energy, i.e., near the sputtering threshold, are also discussed and compared for atomic and molecular ions.

117 Chemical sputtering of silicon by F^+, Cl^+, and Br^+ ions: reactive spot model for reactive ion etching,

by S. Tachi and S. Okudaira, J. Vac. Sci. Technol. B **4**, 459 (1986).

Chemical sputtering yields of silicon resulting from the incidence of isotopically pure $^{19}F^+$, $^{35}Cl^+$, and $^{81}Br^+$ ion beams were measured in the 100–3000 eV energy range. It is found for these cases that the chemical sputtering yields saturate in the high energy range of more than 600 eV. The saturated yields (atoms/ion) are 0.18 for F^+/Si, 0.45 for Cl^+/Si, and 0.23 for Br^+/Si. The yields increased with increasing ion energy in the 100–500 eV region with different slopes. The steepest slope is observed for the case of Br^+/Si, and the smallest is the F^+/Si case. This difference in the slope indicates that the chemical reaction between Br^+ and Si cannot be activated without high acceleration of Br^+, and the same occurs for Cl^+/Si. On the other hand the F^+ ion is assumed to react with Si with a relatively high probability even when the energy is low. On the basis of these results, a reactive spot model is proposed for the ion-assisted chemical etching process. A generalized method to get anisotropic profiles is also presented for reactive ion etching of Si with fluorine-, chlorine-, and bromine-containing gases.

118 Ion- and electron-assisted gas-surface chemistry – An important effect in plasma etching,

by J.W. Coburn and H.F. Winters, J. Appl. Phys. **50**, 3189 (1979).

The extent to which gas-surface chemical reactions can be enhanced by energetic radiation (primarily ions and electrons) incident on the surface is described. Emphasis is placed on chemical systems which lead to volatile reaction products. In particular, the reactions of Si, SiO_2, and Si_3N_4 with XeF_2, F_2, and Cl_2 are examined experimentally. Possible mechanisms for the radiation-induced enhancement are discussed and some technological implications of this process in plasma etching technology and lithography are considered.

119 Extension of the model for Ar^+ ion induced etching of Si by SF_6,

by G.N.A. van Veen, F.H.M. Sanders, J. Dieleman, P.C. Zalm, D.J. Oostra, and A.E. de Vries, Nucl. Instrum. Methods Phys. Res. B **19/20**, 1022 (1987).

Recently energy spectra of reaction products emitted during Ar^+ ion bombardment induced etching of Si by SF_6 condensed on top of it have been reported. The Si(SF_6;Ar^+) system differs from all others studied to date in that multilayer adsorption of reactive gas occurred during the experiments. This is shown here to induce an additional ejection mechanism. The matrix from which reaction products are finally emitted changes from a "SF_6-mixed-into-Si" into a "Si-mixed-into-SF_6" type and finally into pure SF_6 when the SF_6 flux is increased or the substrate temperature is lowered. The profound changes observed in the kinetic energy distributions of the specific emitted products upon variation of experimental conditions are reinterpreted and attributed largely to an alteration of the thickness of the condensed SF_6 "buffer" layer.

120 Phenomena produced by ion bombardment in plasma-assisted etching environments,

by H.F. Winters, J. Vac. Sci. Technol. A **6**, 1997 (1988).

The emphasis of this review paper will be to summarize recent experimental results, describe and evaluate contemporary concepts in the etching field, and to provide references to recent work.

121 Argon-ion assisted etching of silicon by molecular chlorine,

by A.W. Kolfschoten, R.A. Haring, A. Haring, and A.E. de Vries, J. Appl. Phys. **55**, 3813 (1984).

Argon-ion assisted etching of silicon by molecular chlorine has been investigated. The masses and some kinetic energy distributions of the neutral particles emitted from the surface have been determined by mass spectrometry and time-of-flight spectra. It is found that an important part of the silicon-containing particles consists of SiCl and $SiCl_2$ molecules which have kinetic energies in the eV region. The observations exclude a simple evaporation of $SiCl_4$ at the target temperature. A tentative model – consisting of a collision cascade like process parallel to thermal evaporation at the substrate temperature induced by a reduction of the effective surface binding energy during a sputtering event – is given to explain the observed kinetic energy distributions qualitatively.

123 Mechanisms of sputtering of Si in a Cl_2 environment by ions with energies down to 75 eV,

by D.J Oostra, A. Haring, R.P. van Ingen, and A.E. de Vries, J. Appl. Phys. **64**, 315 (1988).

Sputtering of Si in a Cl_2 environment by Ar^+ and Xe^+ ions with energies down to 75 eV has been investigated. Mass spectra and time-of-flight distributions of the sputtered species have been measured. Under 75-eV Ar^+-ion bombardment of the Si target, SiCl, $SiCl_2$, $SiCl_3$, and/or $SiCl_4$ are sputtered. When increasing the ion energy the $SiCl_4$ contribution decreases in comparison with SiCl. This is caused by the fact that the newly formed Si-Cl compounds are sputtered at a high rate compared to the rate of the $SiCl_4$ formation. Time-of-flight distributions indicate that under 100-eV Ar^+-ion bombardment the species are not sputtered by a collision-cascade mechanism. The spectra can be fitted by a Maxwell-Boltzmann distribution at a high (> 2000 K) temperature. Increasing the Ar^+-ion energy to approximately 250 eV the time-of-flight spectra of the sputtered species change from Maxwell-Boltzmann-like into spectra as expected for a collision-cascade mechanism. For low-energy Xe^+ ion bombardment the sputtered species also show Maxwell-Boltzmann time-of-flight distributions. The change from Maxwell-Boltzmann to collision-cascade distributions occurs at higher ion energies than for Ar^+-ion bombardment. The results obtained for low ion energies are discussed in terms of evaporation from an ion-induced hot spot.

125 Effects of Ar^+ angle of incidence on the etching of Si with Cl_2 and low-energy Ar^+ ions,

by J. van Zwol, J. van Laar, A.W. Kolfschoten, and J. Dieleman, J. Vac. Sci. Technol. B **5**, 1410 (1987).

Experiments have been performed in a beam apparatus where an Ar^+ beam and a Cl_2 beam are directed simultaneously at a Si target. The apparatus allows independent variation of the ion angle-of-incidence and the detection angle. Time-of-flight (TOF) distributions of the reaction product SiCl are measured for various Cl_2 to Ar^+ flux ratios and various angles-of-incidence. These distributions can for the larger part well be fitted with collision cascade distributions, showing that at the parameter values chosen, the etch process in $Si(Cl_2,Ar^+)$ is a chemically enhanced physical sputtering process. The observed variations of the binding energy U_0 obtained from the TOF distributions supports this mechanism. A discussion in relation to similar studies on the etch rate and the composition of the surface and the altered layer leads to a consistent picture in support of the above model.

128 Some chemical aspects of the fluorocarbon plasma etching of silicon and its compounds,

by J.W. Coburn and E. Kay, IBM J. Res. Develop. **23**, 33 (1979).

Mass spectrometric sampling of fluorocarbon glow discharges and in situ measurements of the etch rate of Si and SiO_2 with quartz crystal microbalances have been used to provide additional insight concerning the chemistry involved when additive gases such as O_2, H_2, N_2, H_2O, and C_2F_4 are injected into a CF_4 glow discharge. The results obtained in our low-pressure, long-residence-time system indicate that the etching behavior of the discharge is not significantly influenced by the molecular structure of the injected gas molecules but is determined primarily by the elemental composition of the glow discharge. This phenomenologically observed result can be used to predict qualitatively the relative etching behavior of a large class of gaseous etchant mixtures as well as the role of various electrode or wall materials in the plasma etching process. Although this oversimplified interpretation should not be extended to short-residence-time plasma systems, it is believed to be valid for some of the configurations used in plasma etching and reactive ion etching.

131 Experimental and design information for calculating impedance matching networks for use in rf sputtering and plasma chemistry,

by H. Norström, Vacuum **29**, 341 (1979).

To calculate impedance matching networks for capacitively coupled rf excited plasmas (as used for sputtering and some type of chemical reactors) it is necessary to know the values of the plasma parameters, i.e. the capacitance and the conductance of the discharge and the variation of these with the operating conditions of the system, i.e. the pressure, the frequency and the input power. The paper presented here gives information on the foregoing and shows how the results can be used to design suitable matching networks. To provide basic information on the glow discharge conditions experiments have been carried out using an rf power supply with one side of the matching network connected via a blocking capacitor to an electrode and the remaining side grounded. This system has been operated under various conditions, i.e. the frequency has been varied between 2 and 14 MHz, the air pressure from 1×10^{-1} down to 5×10^{-3} torr and the input power density from 0.5 to 2 W cm^{-2}. The results obtained for 14 MHz are in close agreement with those reported for 13.56 MHz by Logan et al. From the values obtained for the plasma parameters different types of matching network (L, Π, T and tuned transformer) have been calculated. The design equations for these networks are presented with calculated and tested examples to demonstrate their application. The procedure used, although common electrical engineering practice, should with the measured data be of value to those undertaking rf system design without experience of matching conditions.

136 Spherical shell model of an asymmetric rf discharge,

by M.A. Lieberman, J. Appl. Phys. **65**, 4186 (1989).

A spherical model is used to study ion transport and bias formation in asymmetric, capacitive rf discharges, which have unequal areas A and glow-to-electrode voltage V

at the powered (a) and grounded (b) electrodes. Ions are generated by thermal electron ionization and are lost by ambipolar diffusion in the glow. Resonant charge transfer with a constant cross section is assumed to dominate the ion transport. We obtain the density ratio scaling $n_a/n_b \propto (A_b/A_a)^{7/24}$, where n is the density at the glow-sheath edge. Three electrode sheath models are considered: collisionless ion, collisional (constant mobility) ions, and a constant-ion cross-section collisional law. Using these and the continuity of the rf current flow, we obtain the scaling of the electrode voltage ratio with the electrode area ratio: $V_a/V_b \propto (A_b/A_a)^q$. For typical rf materials processing discharges, the constant cross section law yields $q \approx 2.21$. The effect of secondary electron ionization and local ionization near the sheaths due to stochastic heating are shown to further reduce the value of q.

140 Ion beam techniques in thin film deposition,

by J.M.E. Harper, Solid State Technol. **30**(4), 129 (1987).

The importance of ion bombardment in thin film deposition has long been recognized. Plasma-based processes such as rf bias sputtering, magnetron sputtering, and plasma enhanced CVD show pronounced changes in film properties as the degree of ion bombardment of the growing film is varied. Recently, ion beam techniques have been developed which provide quantitative information on the importance of ion energy, ion flux, and the ratio of arrival rates of ions to film atoms. Several ion beam techniques are reviewed and examples are given of property modification which can be achieved by controlled ion bombardment.

155 Measurement of ion energy distributions at the powered rf electrode in a variable magnetic field,

by A.D. Kuypers and H.J. Hopman, J. Appl. Phys. **67**, 1229 (1990).

High-resolution energy distributions of ions, accelerated by the sheath at the powered electrode of a low-pressure 13.56-MHz gas discharge, have been measured. The observed spectra are compared to existing models. Excellent agreement between measured and calculated spectra is obtained. Detailed information on rf sheath behavior is derived from the observed energy profiles and from the measured total ion current densities towards the electrode surface. Analogous to the case of dc discharges, a decrease of sheath thickness is observed when a homogeneous variable magnetic field ($0 \leq B \leq 315$ G) is applied. However, the product of magnetic-field strength B and sheath thickness d is found to be independent of sheath voltage. This leads to the conclusion that in rf discharges, sheath contraction under influence of a magnetic field proceeds by a different mechanism than in dc discharges. It is suggested that the value of the product Bd is determined by the (virtually constant) temperature of the plasma electrons, rather than by the energy of secondary electrons that have been liberated from the electrode surface by ion bombardment. The decrease of sheath thickness d with magnetic-field strength B leads to a changing capacitive-voltage division of the applied generator voltage over the discharge. When the magnetic-field strength is sufficiently high, this may result in a sign reversal of the electrode self-bias voltage.

157 Diagnostics of low-pressure oxygen RF plasmas and the mechanism for polymer etching: a comparison of reactive sputter etching and magnetron sputter etching,

by Ch. Steinbrüchel, B.J. Curtis, H.W. Lehman, and R. Widmer, IEEE Trans. Plasma Sci. **14**, 137 (1986).

We have compared low-pressure oxygen RF plasmas and the etching of photoresist in a reactive sputter etch reactor and in a magnetron etch reactor using Langmuir probe, optical emission actinometry, and mass spectrometry measurements. The Langmuir probe data allow the determination of the plasma ion density and electron temperature, and thus the ion flux onto the substrate. The optical data yield information on the presence of O atoms and O_2^+ ions. Stable reactant and product species are monitored with a mass spectrometer. The main difference between the two reactors is that in magnetron sputter etching (MSE), the ion flux to the substrate is about an order of magnitude higher, under comparable plasma conditions, than in reactive sputter etching (RSE). This accounts for the higher etch rate in MSE. However, the etch yield per ion is higher in RSE because of the higher ion energy. Etch rates correlate neither with the ion flux to the substrate nor with the density of O atoms in the plasma, but change in parallel with the consumption of reactant gas. We conclude that in etching a polymer in a low-pressure oxygen plasma, the main neutral reactant species are O_2 molecules, and an important role of the ions is to remove reaction products from the substrate surface.

158 A review of magnetron etch technology,

by S.M. Bobbio, in *Dry processing for submicrometer lithography*, edited by J. Bondur and A.R. Reinberg, S.P.I.E. Proc. Vol. **1185**, 262 (1989).

Magnetic confinement of a low pressure rf discharge is useful in generating high density plasmas with low sheath voltages at the bounding surfaces. Thus, substrates exposed to such plasmas receive large fluxes of low energy ions. Additionally, since source gas fragmentation is substantially higher than in conventional systems, the etching chemistry is also very different. The discussion will focus on designs in which the magnetic field is generally parallel to the substrate and highlight the physical properties of these configurations which lead to low impedance and high plasma density. Examples of etching characteristics in polymer and silicon etch chemistries will be given. In many cases, excellent uniformity, high etch rate, and very flexible profile control have been demonstrated; and magnetron systems have proven to be entirely appropriate for in situ wafer at a time processing applications. The future of these systems is very bright, although there are still a few unresolved issues, such as oxide etching which is selective to silicon. The nature of the problems, as well as potential solutions, will also be discussed in this presentation.

162 Electron cyclotron resonance microwave discharges for etching and thin-film deposition,

by J. Asmussen, J. Vac. Sci. Technol. A **7**, 883 (1989).

A recent, important development in low-pressure and low-temperature plasma processing is the microwave electron cyclotron resonance (ECR) discharge. Its lack of electrodes and its ability to create high densities of charged and excited species at low pressures ($\leq 10^{-4}$ Torr) make it an attractive processing discharge in etching and thin-film deposition applications. This article reviews the basic physics of ECR discharges and reviews the associated microwave system and applicator technologies. Waveguide and cavity ECR applicators are compared and are described in detail. Several ECR plasma processing reactors are also described. Methods of processing large surfaces are outlined, and typical experimentally measured ECR discharge characteristics are presented.

164 Reactive ion stream etching utilizing electron cyclotron resonance plasma,

by T. Ono, M. Oda, C. Takahashi, and S. Matsuo, J. Vac. Sci. Technol. B **4**, 696 (1986).

Reactive ion stream etching has been developed for highly accurate etching with little bombardment induced damage by utilizing an electron cyclotron resonance (ECR) plasma. The ion energy during etching is controlled by utilizing the interaction between the ECR plasma and a divergent magnetic field, in a low energy range from 20 to 50 eV at low gas pressures of 10^{-2} Pa. Highly accurate submicron patterns of polysilicon and molybdenum were obtained with high selectivities to SiO_2, larger than 30, by using Cl_2 gas as the main etching gas.

166 Microwave multipolar plasma for etching and deposition,

by R.R. Burke and C. Pomot, Solid State Technol. **31**(2), 67 (1988).

Microwave Multipolar Plasmas (MMP) are plasmas which are confined by multipolar magnetic fields and excited by 2.45 GHz microwaves. Two main features distinguish MMPs from conventional plasmas: the decoupling of plasma excitation from plasma-surface interactions and the absence of self-bias, allowing independent control of substrate bias and ion impact energy. Microwave excitation is applied in three ways: localized (i.e. conventional) electron cyclotron resonance (ECR), surface wave, and distributed ECR (DECR). All three methods have been used to investigate silicon etching by fluorine, but only DECR seems technologically promising. Silicon homoepitaxy has been done on 100 mm wafers at temperatures down to 400 °C in silane MMPs excited by DECR.

168 Crystal-orientation dependent etch rates and a trench model for dry etching,

by J.I. Ulacia F., C.J. Petti, and J.P. McVittie, J. Electrochem. Soc. **135**, 1521 (1988).

This paper presents two separate but related topics. The first is the <111> to <100> crystal-orientation effects in plasma trench etching where the ratio of etch rates between the orientations is approximately 1:0.63. The second is a trench-etching model in which the etch rate is proportional only to the ion flux incident on the wafer. This requirement is met at high pressures and in inhibitor-layer-assisted processes. In deep trenches, the ion flux on the surface is reduced as the aspect ratio increases. The only fitting parameter is the ion angular standard deviation which is a function of the number of collisions encountered by an ion while crossing the plasma sheath, as found with Monte Carlo simulations that represent this phenomenon. The simulated results predict both the experimental etch rates at the center of the trench and the general profile; however, differences are apparent at the corners and sidewalls of the trench.

169 Angular etching correlations from RIE. Application to VLSI fabrication and process modeling,

by D.P. Hamblen and A. Cha-Lin, J. Electrochem. Soc. **135**, 1816 (1988).

Etch-rate dependency as a function of ion angle of incidence on topological features common to VLSI devices is measured under RIE conditions. Investigated are oxide, nitride, Si, poly-Si, and Kodak micro positive resist 820 (820 resist) using fluoroform, oxygen, and other gaseous ions. Construction of a special cathode electrode allowed samples to be mounted at different angles with respect to the positive ion flux at gas pressures of 10–20 mtorr. Depending on the substrate and ionic species, angular etching correlations resolved three distinct curve shapes: (*i*) that showing chemisorption followed by desorption of chemical-induced by-product, which resembles a cosine function at low ion energy; (*ii*) a combination of chemical plus high-angle sputtering produced at increased ion energies; and (*iii*) that characteristic due to knock-on bombardment. Incorporation of angular etching data into a modelling program is shown to be useful for two-dimensional simulation of etched features important to fabrication of microelectronic structures.

170 Review: dry etching of silicon oxide,

by A.J. van Roosmalen, Vacuum **34**, 429 (1984).

Recent developments in anisotropic, selective dry etching of silicon oxide over silicon are reviewed. Various approaches in reactor and process design are discussed, together with concise treatment on current mechanisms for etching in freon plasmas. Special attention is paid to plasma-induced surface damage. Finally, some examples are given of structures that can be obtained with presently available techniques.

171 Silicon etching mechanisms in a CF_4/H_2 glow discharge,

by G.S. Oehrlein and H.L. Williams, J. Appl. Phys. **62**, 662 (1987).

Fluorocarbon film deposition onto Si and its influence on the measured Si etch rate in CF_4/H_2 reactive ion etching in a symmetric two electrode reactor has been studied as a function of CF_4/H_2 feed gas composition, total gas flow, and applied rf power. For reactive ion etching, the fluorocarbon film thickness on Si increases as the percentage x of H_2 in $CF_4/x\%$ H_2 is increased. The fluorocarbon film thickness depends on the total gas flow and is greater for greater gas flows. The observed Si etch rate is controlled by the fluorocarbon film. The Si etch rate is directly proportional to the inverse of the F,C-film thickness for fluorocarbon films thicker than $\approx$10 Å, which may indicate a diffusion-limited mechanism. Both in-diffusion of fluorine and out-diffusion of SiF_4 etch product through the fluorocarbon film are consistent with the decrease of the Si etch rate. The relative importance of the lowering of the atomic F concentration in the gas phase on the Si etch rate, e.g., by the H scavenging mechanism, has also been studied. This reaction can be important for conditions where either the inner walls of the etching apparatus can be maintained free of C,F film throughout the etching experiment, and/or a low hydrogen concentration ($\leq$20%) in the CF_4/H_2 feed gas is used. For a fixed gas composition of CF_4/40% H_2, the rf-power dependence of the C,F-film thickness and of the Si etch rate was studied. In reactive ion etching, i.e., if rf power is supplied to the bottom (substrate) electrode, at first a monotonic rise in deposited fluorocarbon film thickness with increasing rf power is observed; at high rf-power levels a dramatic decrease in the C,F-layer thickness occurs, which is concomitant with a greater intensity of near surface lattice disorder (from ion channeling studies) and Si etching. Silicon etching is not observed for lower rf powers. In cases where rf power was supplied to the top electrode only, C,F-film deposition has been observed and no Si etching. These data are consistent with a recombinant model of etch anisotropy.

172 Surface modifications of electronic materials induced by plasma etching,

by G.S. Oehrlein, S.W. Robey, J.L. Lindström, K.K. Chan, M.A. Jaso, and G.J. Scilla, J. Electrochem. Soc. **136**, 2050 (1989).

A review of in-situ x-ray photoelectron spectroscopy studies of fluorine-based plasma etching induced surface modifications of several electronic materials, but primarily single-crystal silicon, is presented. Post-plasma surface analysis of vacuum-transferred etched specimens has been conducted in an ultra-high vacuum surface analysis chamber connected via a load-lock to the etching chamber. The achievement of SiO_2/Si etch selectivity because of different formation rates of a passivating fluorocarbon film which controls the etching behavior and in general the substrate dependence of the fluorocarbon film formation rate is discussed. The post-plasma in situ surface analysis methodology has been extended to the study of realistic semiconductor microstructures after reactive ion etching. Spatially resolved in situ x-ray photoemission spectroscopy of the surface chemistry of a photoresist-patterned silicon oxide on silicon contact hole etching process has been undertaken by utilizing the topography and conduction characteristics of the realistic microstructure. Often, dry etching processes are followed by a cleaning treatment in order to prepare device quality surface conditions. The dependence of the efficiency of conventional oxygen plasma/wet oxide removal cleaning treatments of reactive ion etched (RIE) silicon surfaces on RIE operating conditions has been studied. The composition and extent of reacted surface layers, e.g., the competitive formation of fluorides or oxides in CF_4/O_2 discharges, have also been reviewed in the present article.

173 The reaction of fluorine atoms with silicon,

by D.L. Flamm, V.M. Donnelly, and J.A. Mucha, J. Appl. Phys. **52**, 3633 (1981).

Fluorine atoms etch silicon with a rate,
$R_{F(Si)} = (2.91 \pm 0.20) \times 10^{-12} n_F T^{1/2} e^{-0.108eV/kT}$ Å/min,
where n_F is the atom concentration. This etching is accompanied by a chemiluminescent continuum in the gas phase which exhibits the same activation energy. These phenomena are described by the kinetics: (1) $F_{(g)} + Si_{surf} \rightarrow SiF_{2(g)}$, (2) $SiF_{2(g)} + F_{(g)} \rightarrow SiF^{\star}_{3(g)}$, (3) $SiF_{2(g)} + F_{2(g)} \rightarrow SiF^{\star}_{3(g)} + F_{(g)}$, (4) $SiF^{\star}_{3(g)} \rightarrow SiF_{3(g)} + h\nu_{continuum}$ where formation of SiF_2 is the rate-limiting step. A detailed model of silicon gasification is presented which accounts for the low atomic fluorine reaction probability (0.00168 at room temperature) and formation of SiF_2 as a direct product. Previously reported etch rates of SiO_2 by atomic fluorine are high by a constant factor. The etch rate of SiO_2 is

$R_{F(SiO_2)} = (6.14 \pm 0.49) \times 10^{-13} n_F T^{1/2} e^{-0.163eV/kT}$ Å/min
and the ratio of Si to SiO_2 etching by F atoms is $(4.74 \pm 0.49)e^{-0.055/kT}$

174 XeF_2 and F-atom reactions with Si: their significance for plasma etching,

by D.L. Flamm, D.E. Ibbotson, J.A. Mucha, and V.M. Donnelly, Solid State Technol. **26**(4), 117 (1983).

The reaction of fluorine atoms with Si and SiO_2 outside of a plasma mimics the rates and selectivity encountered in isotropic etching by fluorine atom etchant plasmas. Silicon etching is almost always isotropic in these plasmas, hence model experiments using this etchant chemistry are of limited value for characterizing anisotropic process mechanisms. New data show that analogous isotropic etching reactions with XeF_2 as a fluorine source exhibit a drastically different behavior; elemental steps in the ion-induced etching of Si by XeF_2 will not apply generally to anisotropy in plasma processing. Model experiments on ion-induced etching in Cl_2 under similar conditions are not in accord with the large anisotropic etch rates observed in chlorine plasmas. Additional research is necessary to clarify the relationship between these data and ion-induced etch rates with a large flux of neutral chlorine atoms and molecules.

176 Parameter and reactor dependence of selective oxide RIE in $CF_4 + H_2$,

by L.M. Ephrath and E.J. Petrillo, J. Electrochem. Soc. **129**, 2282 (1982).

Highly selective etching of SiO_2 with respect to silicon and resist is obtained by reactive ion etching in $CF_4 + H_2$. The maximum etch rate ratio depends in a sensitive way on gas residence time with etch rate ratio increasing with decreasing residence time. The reproducibility of high etch rate ratios was found to decrease with increasing pressure. The cleanliness of etching is determined by the material of the electrode and the electrode to wall area ratio. A process was developed by optimizing etching conditions and reactor design and then was tested electrically. Contact holes were etched to 0.25 μm junctions using full RIE in $CF_4 + H_2$ and buffered HF. The same low junction leakage and contact resistivity, 10^{-9} amps/cm^2 and $10^{-6}\Omega$-cm^2, were measured for both. Implications for scaling $CF_4 + H_2$ RIE into larger etching reactors as well as its use to pattern features at 1 μm ground rules are discussed.

177 The effect of moisture on the plasma oxide etch,

by Y.W. Hu, in Proc. 5th Symp. Plasma Processing, The Electrochemical Society, ECS Proc. Vol. **85-1**, 470 (1985).

A linear relation exists between the plasma oxide etching rate and the moisture content measured by the intensity of the hydrogen emission line. Data also indicated that the moisture effect is probably caused by a combined effect from both the hydrogen and the oxygen species. This result also explains many phenomena observed in the plasma oxide etching process.

180 Design considerations for a high pressure, high etch rate single wafer reactor,

by G.S. Mathad, in Proc. 6th Symp. Plasma Processing, The Electrochemical Society, ECS Proc. Vol. **87-6**, 134 (1987).

In recent years, there has been an increasing interest in the use of single wafer plasma reactors which etch one wafer at a time as compared to several wafers in a batch reactor. This migration has been prompted by several factors. First, single wafer etching (SWE) provides individual wafer process control. Improved control is becoming a major consideration as the circuit density of devices continue to increase with consequent decrease in lithographic dimensions. With near-zero etch bias requirements to achieve the dimensional control, over-etching needs to be kept to a minimum. This is easily accomplished in SWE since the variations in film thickness considered pertain to only one wafer rather than a batch. Second motivation for SWE is the reduced risk of losing wafers if a process mishap or a system failure has to occur during etching. The loss is minimized to at most one in the case of SWE. With the ever increasing cost of fabrication, this can be a significant factor. A third advantage of single wafer processing is the elimination of process scale-up and the accompanying savings in process development time and cost. Other advantages like simpler automation and the throughput being independent of wafer size (this assumes that the etch rates can be scaled up for larger wafers) are obvious. For single wafer reactors to be economically competitive with batch reactors, the etch rate in SWR's have to equal the batch size multiple of the rates obtainable in a batch system. Most of the designs available today are aimed at this particular objective. The object of this paper is to discuss one approach and the improvements to conventional designs for achieving even higher etch rates. (*Introduction to the article*)

182 Selective isotropic dry etching of Si_3N_4 over SiO_2,

by F.H.M. Sanders, J. Dieleman, H.J.B. Peters, and J.A.M. Sanders, J. Electrochem. Soc. **129**, 2559 (1982).

The addition of a few percent of bromine or a component containing bromine like CF_3Br to a plasma in a mixture of CF_4 and O_2 in a barrel reactor reduces the dry etch rates of Si, SiO_2, Si_3N_4, and the resist. The etch rate of SiO_2 is reduced much more rapidly with an increasing percentage of CF_3Br than that of Si_3N_4. The selectivity of dry etching of Si_3N_4 over SiO_2 can be enhanced from about 2 for a pure CF_4–O_2 plasma to values far above 10 when a few percent of CF_3Br are added. At a selectivity of 10 the dry etch rate of Si_3N_4 is reduced by only 30%. For optimally chosen conditions the etch uniformity is better than $\pm 5\%$ over a batch of 25 3 in. wafers.

183 Highly selective dry etching of Si_3N_4,

by Y. Kawamoto, T. Kure, N. Hashimoto, and T. Takaichi, in Ext. Abstracts of the Electrochem. Soc. Fall mtg., ECS Ext. Abstr. **84-2**, 565 (1984).

Silicon nitride film is used in various VLSI production steps, beginning with the thermal oxidation mask. The film is usually etched by CF_4+O_2 in a barrel reactor, or by CF_4+H_2 in a parallel plate reactor. When silicon nitride film is etched isotropically in a barrel reactor, its selectivity over SiO_2 is only 3–4. This selectivity can be improved to 10–20 by adding CF_3Br to the $CF_4 + O_2$, but the etched profile is still isotropic. On the other hand, the etched profile is anisotropic in a parallel plate reactor, but the selectivity of Si_3N_4 over SiO_2 is only about unity using CF_4+H_2. So far, high selectivity and anisotropy have not been simultaneously realized in Si_3N_4 dry etching. However, it has been found that such etching characteristics for Si_3N_4 can be achieved using different gas, either CH_2F_2 or CH_3F, in a parallel plate reactor. In this report, etch rates for Si_3N_4, SiO_2, and polysilicon in CH_2F_2 and CH_3F gas plasma are presented under various discharge conditions. To characterize the high selectivity of Si_3N_4, x-ray photoelectron spectra (XPS) of the etched surfaces in various gas plasmas are also examined. (*Introduction to the article*)

184 Selective etching of silicon nitride using remote plasmas of CF_4 and SF_6,

by L.M. Loewenstein, J. Vac. Sci. Technol. A **7**, 686 (1989).

Silicon nitride is etched using a remote plasma discharge of CF_4 or SF_6, diluted in a He or N_2 carrier gas. By itself, this process chemistry does not etch silicon nitride selectively, a desirable property of a useful process; however, by including a source of abstractable hydrogen, e.g., H_2 or CH_4, in the process mix, the Si_3N_4:Si and Si_3N_4:SiO_2 etch rate ratios can be considerably increased. The hydrogen source serves to consume atomic fluorine, enhancing the preferential deposition of a fluorocarbon layer on Si and SiO_2 (in the case of CF_4). This fluorocarbon film impedes Si and SiO_2 etching by atomic F. For SF_6, some form of etch-impeding surface layer also may be formed.

185 Hollow cathode etching of Si and SiO_2 using CF_4 and H_2,

by G. Fortuño-Wiltshire and G.S. Oehrlein, J. Electrochem. Soc. **136**, 1447 (1989).

Si and SiO_2 films have been etched using CF_4 and H_2 mixtures in a hollow cathode etching (HCE) reactor. We have measured etch rates and characterized the Si surface after HCE by x-ray photoemission spectroscopy (XPS). Addition of H_2 to CF_4 did not significantly enhance the etch rate ratio of SiO_2 to Si. The fluorocarbon film, which is responsible for the achievement of SiO_2/Si etch selectivity in reactive ion etching (RIE) reactors, has not been observed by XPS for Si surfaces etched in our HCE reactor at high RF power levels. This result is similar to that found in magnetron ion etching (MIE) at high power. C—C and C—H groups were the predominant species detected on Si surfaces etched in the HCE reactor; only small amounts of CF_x were present. Low fluorination of the carbonaceous film is believed to be due to high dissociation rates of the parent molecules in both the HCE and MIE reactors. The low fluorination of the carbonaceous deposits on SiO_2 surfaces results in low etch rates for SiO_2. When both carbon and fluorine arrive at the surface as in RIE, the oxygen in the SiO_2 can form volatiles by reaction with carbon, while the Si can be removed by F atoms. When mostly carbon is deposited on the surface, as in HCE and MIE, the Si in the SiO_2 is removed at a relatively slower rate. The thicknesses of the deposits on the Si surface were found to be smaller for the HCE reactor than for RIE reactor at high H_2 percentages. Slow deposition of carbonaceous species on Si can account for the relatively high Si etch rates. The high Si etch rates and the effect of low fluorination of the carbonaceous deposits on SiO_2 for CF_4/H_2 can account for the low etch rate ratios of SiO_2 to Si in the HCE reactor.

186 Etching of silicon in low-frequency chlorine discharges,

by A.J. van Roosmalen, A.P.M. van Arendonk, H.T. Arends, and F. Schmidt, in Proc. 5th Symp. Plasma Processing, The Electrochemical Society, ECS Proc. Vol. **85-1**, 527 (1985).

Dry etching of mono and poly-Si is studied in a high-pressure, low-frequency planar reactor. Etch anisotropy and selectivity to SiO_2 are shown to be strongly influenced by the presence of small amounts of chlorocarbon in the discharge, originally from additives (CCl_4 or CH_4) or photoresist erosion. The formation of ear-like residues along photoresist-masked poly-Si lines, and overcutting of deep trenches etched in mono-Si are explained from simultaneous angle-dependent etching and angle-independent deposition.

187 High quality trench etches in silicon,

by R.N. Carlile, V.C. Liang, and M.M. Smadi, Solid State Technol. **32**(4), 119 (1989).

A method is described for producing quality trench etches in Si using a $CHCl_3$ chemistry. The sidewall angle is determined by the temperature of the wafer and etch rate is primarily determined by pressure. The etch sidewalls will be planar and smooth providing wafer temperature is held constant and the mask has anisotropic sidewalls and is devoid of residual material. The addition of N_2 is required to control the formation of polymer.

190 Deep Si trench etching with a high rate process through a n/n$^+$/p type structure, using a triode configured reactor,

by G.R. Powell and A.A. Chambers, in Proc. 2nd Symp. ULSI Science and Technology, The Electrochemical Society, ECS Proc. Vol. **89-9**, 498 (1989).

A process consisting of $Cl_2/SiCl_4/O_2$ has been developed in a triode configured reactor to etch deep trenches in the order of 10 μm into single crystal silicon. The etching procedure comprises a one or two step process, has a typical etch rate of 1 μm/min and results in positively sloped profiles with a readily adjustable taper angle. However, on applying the process to a n/n$^+$/p structure, massive undercutting along the highly doped n$^+$ buried layer was observed unless appropriate precautions were taken. For typical trench etch processes the selectivity to the SiO_2 mask is >40:1, with an etch uniformity on 125 mm wafers of $< \pm 4\%$. In order to achieve a successful trench etch, free from "grass" residue, an in-situ wafer pre-clean process is essential. It was also found that a modification of the latter process prevents the accumulation of unwanted deposits in the reaction chamber.

191 The effects of arsenic doping in reactive ion etching of silicon in chlorinated plasmas,

by G.C. Schwartz and P.M. Schaible, J. Electrochem. Soc. **130**, 1898 (1983).

The results of a study of the influence of arsenic doping of silicon and Cl_2 and CCl_4 reactant concentrations on etch rate and etch profile are described. From approximately 10^{16} to 10^{20} As/cm^3, the etch is anisotropic and the etch rate does not increase in either Cl_2 or CCl_4 plasmas. At concentrations above 10^{20} As/cm^3, mask undercutting occurs and the etch rate rapidly increases (approximately 300% for 1.5×10^{21} As/cm^3), although the functional form of of the relationship between etch rate and As concentration depends on the reactant, reactant supply, and silicon load. Temperature effects are also described.

192 Silicon doping effects in reactive plasma etching,

by Y.H. Lee and M.-M. Chen, J. Vac. Sci. Technol. B **4**, 468 (1986).

Silicon etch rates in CF_4/O_2 plasma have been studied in a diode reactor which allows both reactive ion etching (RIE) and plasma etching. A fluorine population was also measured from the intensity of optical emission in conjunction with argon actinometry, in order to separate etch rates contributed by the ion enhanced etching and the spontaneous chemical etching. A heavily doped n-type silicon etches faster than undoped silicon. Although this doping effect is present in both ion enhanced etching and chemical etching, it is a more severe influence in the isotropic chemical etching. The thermal activation energy for the chemical etching was measured to be 0.10 eV, independent of the chemical and electrical properties of dopants. The Coulomb attraction between uncompensated donors (As^+) and chemisorbed halogens (F^-) enhances etch rates in a heavily doped n-type silicon, whereas the Coulomb repulsion between uncompensated acceptors (B^-) and chemisorbed halogens (F^-) inhibits etch rates in a heavily doped p-type silicon.

193 Secondary effects in ion milling,

by J.C. Moreno-Marin, J.A. Valles-Abarca, and A. Gras-Marti, J. Vac. Sci. Technol. B **4**, 322 (1986).

The contribution of secondary processes such as redeposition, resputtering, and ion reflection at the sidewalls, to the transfer by sputter etching of a pattern to a substrate is analyzed. An analytical theory developed by R. Smith and J.M. Walls [Philos. Mag. A **42**, 235 (1980)] is used, and the evolution profiles are parametrized.

195 Cavernous undercuts appearing in reactive ion etched submicron-wide deep trenches,

by S. Ohki, M. Oda, H. Akiya, and T. Shibata, J. Vac. Sci. Technol. B **5**, 1611 (1987).

The mechanism for the cavernous undercuts appearing on submicron-wide Si trench sidewalls etched in Cl_2 plasma is investigated. Etching with the inclined Si substrate implies that the ions scattered from the opposing mask edge plane are responsible for the undercut. Angular and energy distributions of ions scattered from the mask edges are simulated with a Monte Carlo method. The number and energies of the scattered ions are calculated as being high enough to etch sidewalls. Etching profile simulation incorporating the ion scattering effect shows good agreement with experimental results concerning the amount and the location of the undercut. These results apparently show that ion scattering at a sloped mask edge is the dominant origin of the cavernous undercut.

198 Role of sulfur atoms in microwave plasma etching of silicon,

by K. Ninomiya, K. Suzuki, S. Nishimatsu, and O. Okada, J. Appl. Phys. **62**, 1459 (1987).

The Si etch rate in an ($F_2 + O_2$) microwave plasma has been measured as a function of O_2 mixing ratio at a fixed total pressure of 5.3×10^{-2} Pa. The etch rate significantly decreases with the mixing ratio. This etch rate decrease is due primarily to surface oxidation. When sulfur is added to the ($F_2 + O_2$) plasma, the Si surface is much less oxidized and the etch rate increases by about a factor of 4. Such sulfur-containing species as S atoms react with O atoms or ions in the plasma and form O atom-containing species, such as SO_2, SO^+, SOF^+, and SOF_2^+, thereby reducing the O atom and O^+ ion concentration in the plasma. As a result, the Si surface is scarcely oxidized, so that the etching reaction can easily proceed. Sulfur atoms inhibit surface oxidation and promote Si etching. Sulfur atoms contained in SF_6, which is usually used in microwave plasma etching of Si, are expected to have the same role.

201 Ion suppression for studying etch inhibitor layers,

by J.P. McVittie, T.A. Lin, and A.J. Bariya, in Proc. 6th Symp. Plasma Processing, The Electrochemical Society, ECS Proc. Vol. **87-6**, 675 (1987).

Etch inhibitor layers are key to anisotropy for a number of dry etch processes, yet little is known about these layers because of the difficulty in analyzing them on the side walls where they form. In this paper we will show that a Al grid suspended above an etching surface can be used to suppress ion bombardment on the etch surface and allow the inhibitor to form on large horizontal surfaces which can easily be analyzed. This method was used to look at the polymeric inhibitor formed during Si etching in SF_6/C_2ClF_5 with and without the presence of photoresist.

202 Low temperature dry etching for ULSI,

by S. Tachi, in Proc. 2nd Symp. ULSI Science and Technology, The Electrochemical Society, ECS Proc. Vol. **89-9**, 381 (1989).

Low-temperature dry etching including low-temperature microwave plasma etching and reactive ion etching is described for ULSI materials. Highly selective anisotropic etching at a high rate, which implies dry etching without tradeoff, is found to be possible without changing the discharge parameters. This etching is only achieved at reduced wafer temperatures with highly reactive gas plasmas. The etching model, etching system and properties obtained below 0 °C are reviewed.

203 Anisotropic and selective reactive ion etching of polysilicon using SF_6,

by P. Parrens, J. Vac. Sci. Technol. **19**, 1403 (1981).

Experiments performed in a reactive ion etch system are discussed with the purpose of displaying the influence of reactive gas nature, ion energy, and the nature of the cathode on etching anisotropy and selectivity. Using SF_6 etching, isotropy or anisotropy is believed to be related to the enthalpy of product formation reaction. Cathodes consisting of elements that form volatile fluorides lead to perfect anisotropic etching and low selective etching of polysilicon over SiO_2; whereas cathode material which does not react leads to higher selectivity but undercutting appears with an overetch. The possibility of achieving anisotropic and selective etching of polysilicon over thermal SiO_2 (etch rate ratio = 10:1) is demonstrated for NMOS technology. Examples of polysilicon gates of submicron (0.6 μm) MOST are given.

205 Rational process optimization and characterization,

by J.E. Spencer, G.G. Barna, and D.E. Carter, in Proc. 5th Symp. Plasma Processing, The Electrochemical Society, ECS Proc. Vol. **85-1**, 413 (1985).

High yield and efficient production of semiconductor devices require that each step in the process flow operate at peak performance. Two main phases of process implementation are optimization and characterization. These can be achieved efficiently with the aid of two complementary experimental techniques: the sequential simplex algorithm and the Plackett-Burman experimental design. The simplex algorithm provides empirical feedback to arrive at a process optimum. The Plackett-Burman design is a statistically generated set of experiments that will correlate process parameters with process observables. This paper describes the use of these techniques for development of plasma etch processes. Photoresist ash and polysilicon etch processes are optimized with the simplex algorithm. The optimized polysilicon etch process is then characterized by the Plackett-Burman design.

206 Reactive plasma-assisted etching of aluminum and aluminum alloys,

by G.C. Schwartz, in Proc. 5th Symp. Plasma Processing, The Electrochemical Society, ECS Proc. Vol. **85-1**, 26 (1985).

Reactive plasma-assisted etching processes for aluminum and aluminum alloys have become increasingly important in the manufacture of advanced semiconductor devices. The technical literature on this subject has been reviewed. The subjects that have been covered in this review are: (A) reactors used for (a) plasma etching in symmetrical parallel plate systems, (b) reactive ion etching (reactive sputter etching) including the use of magnetron discharge, single wafer high rate adaptations, and a hollow cathode configuration, and (c) reactive ion beam etching, (B) initiation time, (C) etchants, (D) corrosion, (E) mechanisms of anisotropic etching, (F) end-point detection, (G) masking, (H) loading, frequency, temperature, and film substrate effects, (I) uniformity, (J) selectivity to the substrate, and (K) etching of aluminum alloys containing silicon and/or copper.

207 Anisotropic etching of Al by a directed Cl_2 flux,

by N.N. Efremow, M.W. Geis, R.W. Mountain, G.A. Lincoln, J.N. Randall, and N.P. Economou, J. Vac. Sci. Technol. B **4**, 337 (1986).

A new Al etching technique is described that uses an ion beam from a Kaufman ion source and a directed Cl_2 flux. The ion beam is used primarily to remove the native oxide to allow the Cl_2 to spontaneously react with the Al film forming volatile Al_2Cl_6. By controlling both the flux equivalent pressure of Cl_2 and the ion beam current, this etching technique makes possible the anisotropic etching of Al with etch rates from 100 nm min^{-1} to nearly 10 μm min^{-1} with a high degree of selectivity.

208 Si and Al etching and product detection in a plasma beam under ultrahigh vacuum,

by D.L. Smith and R.H. Bruce, J. Electrochem. Soc. **129**, 2045 (1982).

The study of plasma etching mechanisms is made difficult by the inaccessibility of the wafer surface to analysis during etching. In the present work, a "plasma beam" is extracted from a miniaturized plasma etching chamber through an array of molecular-flow orifices in an Mg-foil electrode. The beam effuses into ultrahigh vacuum and onto either a mass spectrometer for ionic and neutral plasma species analysis or onto an Si or Al film on a quartz crystal thickness monitor for etching studies. Desorbing etch product can be mass-analyzed before encountering gas or wall collisions. The plasma beam source and its rf power supply can de d-c biased relative to the mass spectrometer and the etching film to accelerate or retard the ionic component without affecting neutral plasma species, thus allowing energy analysis of effusing ions and measurement of ion energy effect on etching. For 50 Pa Cl_2 plasma maximum ion energy was $\cong$ 100 eV at 27 MHz and $\cong$ 400 V at 100 kHz. Beam etching of Si using $CF_4 + O_2$ and Cl_2 plasma (27 MHz) and of Al using Cl_2 plasma has been carried out, and SiF_4, $SiCl_4$, $SiCl_2$, and $AlCl_3$ desorption has been observed. Ion bombardment is required to initiate etching in all cases. Subsequently, the effect of ion energy variation from zero (ions reflected) to 500 eV on etching rate and product desorption rate is large ($\cong \times 18$ enhancement) for Si in Cl_2, much less ($\times 4$) for Si in $CF_4 + O_2$ and zero for Al in Cl_2. These results correlate with the degree of anisotropy obtained for these etching systems in conventional plasma etching, assuming that the anisotropy is produced by vertical etching rate enhancement by the normally incident ions. They also explain the much higher etching rate observed for Si and Cl_2 at the lower frequency, where ion energy is higher.

209 Plasma etching of aluminum. A comparison of chlorinated etchants,

by D.A. Danner, M. Dalvie, and D.W. Hess, J. Electrochem. Soc. **134**, 669 (1987).

The plasma-assisted etching of aluminum in chlorine containing RF glow discharges has been studied. Use of a single parallel plate reactor permitted a direct comparison of etch results using BCl_3, BCl_3/Cl_2, CCl_4, and $SiCl_4$. Separation of aluminum etching into native oxide reduction and water vapor/oxygen scavenging, and metal film etching allowed the likely rate-limiting processes in the etch cycle to be ascertained for the different etch gases. The longer initiation period observed with CCl_4 and $SiCl_4$ compared to BCl_3 appeared to be due to etch gas dissociation effects. Metal etching was believed to be limited by the removal of CCl_x and $SiCl_x$ residues with CCl_4 and $SiCl_4$ and by etchant generation with BCl_3.

210 Selectivity aspects in aluminium plasma etching,

by R.A.M. Wolters, in Proc. 3rd Symp. Plasma Processing, The Electrochemical Society, ECS Proc. Vol. **82-6**, 293 (1982).

The effect of resist erosion on profile development and the selectivity against SiO_2 and Si (contact windows) in an aluminium etching process using $CCl_4 + Cl_2$ were investigated. Using HPR 204 as a masking material the process is in general anisotropic. When, however, SiO_2 is used as a mask the etching is isotropic. Evidently, the erosion products of the photoresist participate in the chemical process of aluminium etching. The etching rate of the underlying oxide was found to be strongly dependent on the composition of the gas mixture used. The selectivity against silicon is poor. To derive full benefit of the plasma process a good etch barrier in the contact windows is necessary. For this purpose silicides of Mo, Co and Pt were investigated. Only PtSi was found to be an excellent barrier against Al etching.

213 High rate anisotropic aluminum etching,

by R.H. Bruce and G.P. Malafsky, J. Electrochem. Soc. **130**, 1369 (1983).

A high rate anisotropic etching process, suitable for plasma etching one wafer at a time, is discussed. Results indicate that the etch rate is overridingly dependent on the Cl_2 concentration and is independent of the rf power used to drive the discharge. Several additives are used to control the etching process. BCl_3 is added to initiate etching, and $CHCl_3$ is included to control anisotropy. Large amounts of He assist photoresist preservation. Parametric studies supporting the roles of the additives have been made.

216 Surface passivation by adding fluorocarbons and chlorofluorocarbons into an Al dry etch reactor,

by J. Maa, J. Vac. Sci. Technol. B **5**, 652 (1987).

Adding fluorocarbon or chlorofluorocarbon compounds into an Al etch reactor can alter the induction time, change the Al etch rate, and inhibit Al etching from the grounded electrode and chamber wall. These phenomena are explained by a surface passivation mechanism. During an etching process, the aluminum can react with chlorine-containing radicals and fluorine-containing radicals simultaneously to form either a volatile aluminum chloride or a stable passivation layer. The important parameters which control the etch and passivation rates, for a given system geometry, are the flow rate, the fluorine atomic concentration, and the bias voltage of the bottom electrode.

218 Prevention of corrosion during the reactive ion etch of aluminum on TiW,

by P.C. Chang and K.K. Chao, in Proc. 5th Symp. Plasma Processing, The Electrochemical Society, ECS Proc. Vol. **85-1**, 12 (1985).

Reactive ion etching of aluminum interconnections is becoming increasingly popular as integrated circuit geometries shrink. Chlorine residues left on the device after the etching process can corrode the aluminum, however. The use of TiW as a barrier layer accentuates the problem because of the strong adhesion of chlorides to titanium and tungsten. A postetch process is being successfully used to remove all traces of chlorine from the device.

219 Investigation of the dependence of reactive ion etching of Al-Si-Cu alloys upon film deposition characteristics,

by J. Maleham, Vacuum **34**, 437 (1984).

Aluminium-silicon-copper alloy films deposited in loadlocked and non-loadlocked sputtering systems with and without substrate bias are etched in a number of commercially-available reactive ion etching machines. Marked differences in etchability are observed depending upon the deposition conditions. The structures of the various films are studied by scanning electron microscopy, transmission electron microscopy, electron probe microanalysis and Auger electron spectroscopy, and the results compared with electrical resistivity and step resistance measurements. It is shown that satisfactory etching can be achieved when films are deposited at high rates in clean vacuum systems, or by using substrate bias to reduce impurity gas levels and minimize the formation of microcracks at substrate steps. If these deposition conditions are not fulfilled it is likely that residual fillets of unetched metal will be left along the edges of steps unless excessive overetching is used to remove them.

223 Correlation of postetch residues to deposition temperature in plasma etched aluminum alloys,

by T. Abraham, J. Electrochem. Soc. **134**, 2809 (1987).

A study of the dependence of postetch residues on deposition temperature was carried out on Al(1%Si), Al(2%Cu), Al(4%Cu) and Al(1%Si + 0.5%Ti). The postetch residues were analyzed using Auger electron spectroscopy, and correlated to the film structure which was characterized using Auger depth profiling, and cross-sectional and planar transmission electron microscopy. The residues seen for the Al(1%Si) and Al(Cu) alloys could be explained on the basis of the size and distribution of a secondary phase involving the alloying element. The results for the Al(1%Si + 0.5%Ti) also indicated a dependence on film structure. In general, residue free etching could be obtained for alloys deposited at temperatures below 200 °C.

225 A review of plasma etching of refractory metal silicides,

by T.P. Chow and A.J. Steckl, in Proc. 4th Symp. Plasma Processing, The Electrochemical Society, ECS Proc. Vol. **83-10**, 362 (1983).

The present status of plasma etching of refractory metal silicides is reviewed in the context of high-resolution patterning. Etching of various silicides (cf. $MoSi_2$, WSi_2, $TaSi_2$) is discussed in terms of etch rate, anisotropy, etch selectivity over SiO_2, resist and other thin film materials. Key issues addressed include choice of etchant (cf. CF_4, NF_3, CCl_4), additive gas (O_2, Ar, He) and reactor configuration (planar plasma, RIE, flexible diode, triode). Factors influencing the control of the edge profiles in silicide and polycide gate structures are described. Basic mechanisms in these plasma processes, such as desorption of reaction products, are discussed with respect to their effect on the silicide etching characteristics.

227 Plasma etching methods for the formation of planarized tungsten plugs used in multilevel VLSI metallizations,

by R.J. Saia, B. Gorowitz, D. Woodruff, and D.M. Brown, in Proc. 6th Symp. Plasma Processing, The Electrochemical Society, ECS Proc. Vol. **87-6**, 173 (1987).

The vias between two levels of metal interconnect in a multilevel metal system are filled using either selective or non-selective chemical vapor deposited (CVD) tungsten. Excess tungsten is then removed from the surface of the interlayer dielectric using planarization and etchback which leaves the vias filled with tungsten plugs. The uniformity of the etchback process can be affected by "micro-loading" which causes local accelerated etching of the tungsten in the vias. Other nonuniformities in etch rate, resulting in trenching around the periphery of the tungsten via, can be caused by contaminants in the CVD reactor chamber which are incorporated in the deposited film. Methods for eliminating such problems and optimizing coverage of the plugs by the overlying metal are described.

230 Plasma-assisted etching of tungsten films: a quartz-crystal microbalance study,

by F. Fracassi and J.W. Coburn, J. Appl. Phys. **63**, 1758 (1988).

The plasma-assisted etching of magnetron-sputtered polycrystalline tungsten films in CF_4–H_2 and CF_4–O_2 glow discharges has been studied as a function of ion energy using quartz-crystal microbalance methods supplemented by vacuum-transfer Auger electron spectroscopy and actinometric emission spectroscopy.

232 Plasma etching of oxygen containing titanium silicide films,

by F.Y. Robb, J. Electrochem. Soc. **131**, 2906 (1984).

Anisotropic etching of titanium silicide is essential if its low resistivity gate properties are to be utilized on scaled MOS devices. CCl_4 and CF_4 plasmas, as well as ion milling, were used to etch composite titanium silicide/polysilicon films, with each displaying definite drawbacks. CCl_4 plasmas yielded rough etching and etch residue, presumably due to the high oxygen content of the silicide film and the relatively high selectivity (low oxide etch rate) of the etch. CF_4 reactive ion etching and ion milling produced smooth etching but lacked the selectivity to stop on thin underlying gate oxides. A combination of these processes, however, provided successful anisotropic patterning of such structures. A nonselective, smooth etching CF_4 reactive ion etch was used to etch through the silicide layer, followed by a higher selectivity Cl_2 plasma etch through the underlying polysilicon down to the thin gate oxide.

233 Reactive-ion etching (RIE) of $TaSi_2/n^+$ polysilicon bilayers,

by S.P. Sun and S.P Murarka, J. Electrochem. Soc. **135**, 2353 (1988).

Dry etching of a binary layer of cosputtered 2900 Åthick $TaSi_2$ on 3900 Åthick phosphorous doped polycrystalline silicon films in $CF_4 + O_2$ has been studied. The influence of oxygen concentration, RF power, and sample temperature, on etch rate was investigated. The surface of the etched silicide was examined using field emission scanning electron microscope (FESEM) and scanning Auger electron spectroscopy (AES). It is found that the etch rates of $TaSi_2$ (*i*) increased with increasing oxygen percentage initially, reached a maximum value near 4% O_2, then started to decrease with increasing oxygen in $CF_4 + O_2$ gas at applied power at 275 W, sample temperature 60 °C, total gas flow rate of 20 sccm, and chamber pressure of 40 mtorr, and (*ii*) increased nonlinearly with the RF power in $CF_4 + 4\%O_2$ gas at chamber pressure of 40 mtorr and total gas flow rate of 20 sccm. Etch rates followed an Arrhenius-type temperature dependence with activation energy of 0.08 eV. The $TaSi_2$ etching in $CF_4 + O_2$ has been related to the large differences in the metal fluoride and silicon fluoride vapor pressures. Also, based on Auger electron spectroscopy data from the etched surface it is suggested that the tantalum enrichment on the surface during silicide etching is responsible for the reduced steady-state etch rate.

235 A multistep etch process for $TaSi_2$/polysilicon composite films,

by J. Herrmann, P. Martins, and W. Rust, in Proc. 6th Symp. Plasma Processing, The Electrochemical Society, ECS Proc. Vol. **87-6**, 494 (1987).

Tantalum/polysilicon structures (polycides) have been etched in a planar reactor using a three-step process. In the first step, $TaSi_2$ is etched with SF_6 and CCl_2F_2. The second step is a fluorine residue removal step. Finally, polysilicon is etched with Cl_2 and CO_2. Anisotropic structures are obtained which are free of undercutting or "notching" at the silicide/polysilicon interface. Refractory silicide films are generally under tensile stress in the as-deposited state. A notch between the film and underlying polysilicon can cause the film to lift and peel when it is subsequently annealed. The second step for fluorine removal and sidewall passivation assisted by CO_2 for the polysilicon etch prevents lateral etching and ensures vertical profiles for the two-layer structure.

237 The reactive ion etching of tantalum silicide/polysilicon bilayers in silicon tetrachloride and hydrogen chloride,

by B.J. Curtis and H.R. Brunner, J. Electrochem. Soc. **136**, 1463 (1989).

Bilayer structures of $TaSi_2$ and poly-Si have been patterned with good resolution using either an $SiCl_4$ or an HCl plasma. End-point detection and diagnostics were carried out by optical emission spectroscopy, laser reflectometry, and mass spectrometry. Either an oxygen plasma or Caro's acid were used for resist stripping. Structures that had apparently been cleanly stripped were etched in SF_6 to remove the remaining $TaSi_2$ and Si. Residues remaining after this treatment provide strong evidence for a sidewall protection mechanism of anisotropic etching.

239 The kinetics of tungsten etching by atomic and molecular chlorine,

by M. Balooch, D.S. Fischl, D.R. Olander, and W.J. Siekhaus, J. Electrochem. Soc. **135**, 2090 (1988).

The reactions of atomic and molecular chlorine with tungsten were studied by modulated beam-mass spectrometric methods over the temperature range 300–1350 K. The atomic beam was generated by an RF plasma discharge. With both atomic and molecular beams, the main reaction product up to about 1000 K was WCl_4. The reaction probability with atomic chlorine was a factor of approximately ten higher than that obtained with molecular chlorine. The reaction was nonlinear with respect to Cl_2 intensity at low beam fluxes but approached linearity at high beam intensities. Above 1000 K the main reaction product was atomic chlorine. Its reaction probability increased rapidly with temperature; at 1300 K nearly complete dissociation of Cl_2 was observed. A kinetic model based on the Eley-Rideal mechanism was proposed and compared with the data.

240 Molybdenum etching with chlorine atoms and molecular chlorine plasmas,

by D.S. Fischl and D.W. Hess, J. Vac. Sci. Technol. B **6**, 1577 (1988).

Thin films of molybdenum were etched both within and downstream from a Cl_2 plasma at 200 mTorr pressure and temperatures below 180 °C. When samples were positioned downstream from the discharge, etching proceeded solely by chemical reaction of the film with chlorine atoms. Without a discharge, molecular chlorine did not etch molybdenum. Downstream and in-discharge etch rates were <20 nm/min and etching by atoms was not observed below 100 °C. The chemical reaction between chlorine atoms and molybdenum was proportional to the gas phase Cl atom mole fraction.

243 Chlorine-based plasma etching of titanium silicide films,

by F.Y. Robb, in Proc. 5th Symp. Plasma Processing, The Electrochemical Society, ECS Proc. Vol. **85-1**, 1 (1985).

Anisotropic etching of both cosputtered and plasma-enhanced CVD (PECVD) titanium silicide films was achieved in chlorine-based plasmas. The etch rate, residue and anisotropy is reported for both CCl_4 and pure Cl_2 plasma etching. Because rough etching of oxygen-containing silicides yielded etch residue, a two-step etch was developed. A very low selectivity initial etch step was used to smoothly "mill" through the oxygen-contaminated silicide layer, eliminating the residue. CCl_4 plasmas, which reportedly produce sidewall polymers, yielded excellent anisotropy for all films. In contrast, Cl_2 plasmas produced dramatic undercutting on PECVD films, while lateral etching of cosputtered films was blocked by both the deposition of an inorganic material on feature sidewalls and the presence of slow etching oxides in the film.

245 Critical review. Oxygen plasma etching for resist stripping and multilayer lithography,

by M.A. Hartney, D.W. Hess, and D.S. Soane, J. Vac. Sci. Technol. B **7**, 1 (1989).

Oxygen-based plasmas commonly used in resist stripping and multilayer resist patterning are contrasted to highlight the differences involved in these applications. Mechanisms for polymer etching are reviewed, with particular emphasis on silicon-containing resists proposed for bilayer lithography. While silicon-containing materials offer a simpler process than trilayer schemes for improving lithographic resolution, considerable differences in etch behavior among these materials have been observed. Further characterization and fundamental understanding are required before widespread acceptance of silicon-containing resists is achieved.

246 Plasma etching of inorganic materials. II. Polyimide etching and passivation downstream of an O_2-CF_4-Ar microwave plasma,

by V. Vukanovic, G.A. Takacs, E.A. Matuszak, F.D. Egitto, F. Emmi, and R.S. Horwath, J. Vac. Sci. Technol. B **6**, 66 (1988).

Parameters which influence the etching of polyimide (PI) downstream of an O_2–CF_4–Ar microwave (MW) plasma were studied. Of particular importance is the influence of the ratio of atomic number densities O/F in the plasma and the gas flow velocity, which caused an almost linear increase in the etch rate within the investigated limits of the experiment. No measurable etching of PI was found downstream of an O_2–Ar plasma, although the presence of oxygen atoms in the vicinity of the substrate was verified by chemiluminescent reaction with NO. Similar to previously reported radio frequency plasma studies, a fluorinated layer at the surface of (PI) was formed by exposure of the substrate downstream of a fluorine-rich O_2–CF_4–Ar MW plasma. Changing the plasma composition to the optimum etching conditions (an oxygen-rich plasma of about 20% CF_4) removed the fluorinated layer after a period of time. Etching then proceeded with an oxygenated layer at the surface. The reported experiments confirmed the formation of a fluorinated layer, the removal of this layer, and the formation of an oxygenated one was caused only by the interaction with neutrals, without the influence of ion bombardment.

248 A study of the mechanism of polyimide etching in a microwave plasma downstream etcher,

by B. Robinson and S.A. Shivashankar, in Proc. 5th Symp. Plasma Processing, The Electrochemical Society, ECS Proc. Vol. **85-1**, 206 (1985).

A microwave plasma downstream etcher has been built to determine the etching characteristics of polyimide in O_2/CF_4 discharges. The etch rate of polyimide at room temperature increases by more than an order of magnitude with the addition of a few percent CF_4 to an oxygen discharge. The addition of only 0.5% CF_4 to an oxygen discharge reduces the activation energy for etching polyimide to 0.10 eV from 0.58 eV for etching in pure oxygen discharge. It is proposed that atomic fluorine produced in the discharge abstracts hydrogen atoms from the polymeric structure, greatly facilitating oxidation. At higher concentrations of CF_4 in the discharge, etching is retarded by polymeric film formation on the surface of the polyimide.

250 Hydrogen-plasma etching of organics,

by F.Y. Robb, in Proc. 4th Symp. Plasma Processing, The Electrochemical Society, ECS Proc. Vol. **83-10**, 382 (1983).

The use of hydrogen plasmas for reactive ion etching of organic films has been characterized. It was found that anisotropic etching could be easily achieved, but with relatively slow etch rates (< 250 Å/min). The effect of reactor variables (such as pressure, temperature and power) and chemistry changes (such as the addition of O_2, and N_2) were studied and are discussed. In addition, the application of contact hole etching and trilevel resist processing are considered.

251 Experimental tests of the steady-state model for oxygen reactive ion etching of silicon-containing polymers,

by C.W. Jurgensen, A. Shugard, N. Dudash, E. Reichmanis, and M.J. Vasile, J. Vac. Sci. Technol. A **6**, 2938 (1988).

A steady-state model based on silicon material balance has been proposed to predict the oxygen reactive-ion-etching resistance of organosilicon polymers. This model assumes that the rate determining step is sputtering of SiO_2 film that forms on the surface of the organosilicon polymer. It predicts the etching rate of organosilicon polymers relative to the sputtering rate of SiO_2, based on the mass density of silicon in the polymer. The steady-state etching rate of a silyl novolac polymer is accurately predicted by the model over a wide range of etching conditions. Silyl methacrylates etch at the predicted rate under low-bombardment-energy conditions. Surface analysis shows that the SiO_2 film thickness continues to increase with time under these conditions, invalidating the steady-state approximation. Low silicon content (4.1 wt.%) polymers do not etch according to the model but form highly porous oxides that continuously accumulate on the surface of the polymer.

252 Ion induced etching of polymers in oxygen and argon rf plasmas,

by R.J. Visser and J.A.G. Baggerman, in Proc. 9th Int. Symp. on Plasma Chem. Vol. II (Int. Union of Pure and Appl. Chemistry, Pugnochiuso, Italy, Sept 1989), p. 1039.

The etching of polymers has been studied in oxygen and argon rf plasmas in an RIE reactor. The etch rate is determined by the ion bombardment and does not depend on gas pressure and temperature. Yields have been obtained as a function of the ion energy between 150 and 700 eV. The chemical enhancement factor of oxygen is about six. The identity and the consumption of the oxidizing species on the surface and in the gas phase are discussed.

253 Selective removal of metal atoms in hydrogen reactive ion etching,

by H. Hiraoka, J. Vac. Sci. Technol. B **4**, 345 (1986).

A short exposure of polymer films containing Si, Al, Bi, and other metals to hydrogen reactive ion etching (RIE) reduces their oxygen RIE resistance dramatically. Preferential removal of Si, Al, or other metal atoms in hydrogen RIE is responsible for this reduction of the etch resistance. Selective removal of these atoms by hydrogen RIE may be important in dry stripping of organometallic resist films. Positive tone polymer images can be fabricated by selective removal of metal atoms. Conversely, patternwise deposition of an organometallic compound in H^+ beam-induced polymerization or after H^+ beam-induced decarboxylation gave rise to negative or positive tone polymer images, respectively, by dry image development.

254 Planarization by two-resist level,

by S. Crapella and F. Gualandris, J. Electrochem. Soc. **135**, 683 (1988).

A new two-layer resist planarization method is explained in detail. This very simple procedure gives a good planarization and solves interfacial problems that usually represent the weak point of the multilevel methods. The key factor in such a process is recognized to be the gradient of viscosity between the two layers. Experimental results and SEM pictures of the most significant situations are shown. A physical model based on the density and viscosity properties of the layers is also reported. According to this model, it is possible to justify all the experimental results.

256 Two-layer planarization process,

by A. Schiltz and M. Pons, J. Electrochem. Soc. **133**, 178 (1986).

A two-layer planarization (TLP) process has been developed to solve the profile variation problems encountered in conventional planarization. All the topographic features are smoothed down regardless of their size and spacing. The TLP process involves first, a positive photoresist layer spun onto the substrate and exposed through a negative of the mask of the pattern. After development and cure, a second layer is spun on in order to flatten the residual undulations. The final result is a completely flat surface.

259 Sidewall tapering in reactive ion etching,

by A.G. Nagy, J. Electrochem. Soc. **132**, 689 (1985).

This paper reports for the first time the existence of a sidewall tapering phenomenon in reactive ion etching (an oxide process in a commercially available hex-type system) which is akin to that seen in sputter etching or ion milling. A facet forms in the photoresist mask because of an angular dependence of the resist loss rate. The facet angle is about 16° off vertical, and its downward propagation rate is more than twice the vertical etch rate of the resist. If this facet is allowed to reach the etching material during the course of the etch, it will replicate significantly into this material. When an initially vertical unmasked silicon step is exposed to the same oxide etch process, an identical facet is formed. However, when a bare silicon dioxide step is exposed to the process, no faceting is seen.

260 Contact hole and via profiling by high rate isotropic and anisotropic etching of oxides,

by V. Grewal, H.P. Erb, and P. Mokrisch, in Proc. 4th Int. IEEE VLSI Multilevel Interconnect Conf., 298 (1987).

In integrated circuit production, it is important to have suitable contact hole and via profiles for good step coverage of subsequent metallization. This paper describes a two step process consisting of a high rate isotropic etch followed by an anisotropic etch in a single wafer etcher to obtain almost ideal profiles. With the isotropic etch step the lateral to vertical etch ratio can be varied between 0.5–1. Uniformity for both etch steps is within 5%. Good selectivities to resist and silicon as well as low ($< 0.1\mu$m) conversion differences from the resist to etched pattern have been achieved.

261 Isotropic plasma etching of doped and undoped silicon dioxide for contact holes and vias,

by W.G.M. van den Hoek, T.E. Wicker, B.F. Westlund, and G.B. Powell, J. Vac. Sci. Technol. A **7**, 670 (1989).

Isotropic plasma etching of doped and undoped oxides has been studied as an alternative to the "wet" isotropic step in the "wet-dry" contact etch approach. Czochralski grown p-type <100> 100-mm-diam silicon substrates with either 0.1-μm low-temperature chemical vapor deposition oxide (LTO) and 0.5-μm borophosphosilicate glass (BPSG, 2.5 wt.% B and 7 wt.% P), or 0.6-μm thermally oxidized silicon were used. These substrates were masked with a resist pattern with contact holes whose diameters were between 10.0 and 0.75 μm. Subsequently, they were isotropically etched in a Matrix System One etcher. In this step 0.3 μm of the oxide was removed. The remaining oxide was anisotropically etched in an AME 8110 hexode etcher in a $CHF_3/CO_2/He$ mixture at 72 mTorr and 1250 W. The isotropic etch process was characterized using a multilevel statistical experimental design with pressure, 13.56-MHz rf power, and total gas flow (50% NF_3 in He) as the variables. Optimized etch conditions resulted in an etch rate uniformity of less than ± 3.0% (one sigma divided by the average for 9 points), a BPSG etch rate in excess of 0.3 μm/min, a thermally oxidized silicon etch rate of 0.12 μm/min, and a resist etch rate of <14 nm/min. The scanning electron microscope measurements showed a critical dimension variation of <0.1 μm total in the contact hole diameter at the LTO-silicon interface as compared to the contact hole diameter at the resist-BPSG interface before etch.

262 Near-surface damage and contamination after CF_4/H_2 reactive ion etching of Si,

by G.S. Oehrlein, R.M. Tromp, J.C. Tsang, Y.H. Lee, and E.J. Petrillo, in Proc. 5th Symp. Plasma Processing, The Electrochemical Society, ECS Proc. Vol. **85-1**, 587 (1985).

Silicon surfaces which had been exposed to $CF_4/40\%H_2$ reactive ion etching have been characterized by X-ray photoelectron emission spectroscopy, He ion channeling, H profiling and Raman scattering techniques. Plasma exposure of a clean Si surface leads to the deposition of a thin (~30-50 Å thick) C,F-containing film. The near-surface region (~30-50 Å) of the Si substrate is heavily disordered ("amorphized") as found by ion channeling and Raman scattering. A modified, less damaged Si region extends from about 30-50 Å from the surface to a depth extent of 250 Å. This layer contains a high concentration (~5 at. %) of H as shown by hydrogen profiling techniques. From the observation of Si–H and Si-H_2 vibrational modes by Raman scattering it has been shown that the H is bonded to the Si lattice. In ion scattering the extended modified Si layer appears to cause a strongly enhanced background in the energy spectra. Results of Monte Carlo range calculations are reported and compared with the damage depth found experimentally.

266 Study of reactive-ion-etch-induced lattice damage in silicon by Ar, CF_4, NF_3, and CHF_3 plasmas,

by I.-W. H. Connick, A. Bhattacharyya, K.N. Ritz, and W.L. Smith, J. Appl. Phys. **64**, 2059 (1988).

Reactive-ion-etched-induced damage is silicon has been investigated using transmission electron microscopy (TEM), Rutherford backscattering (RBS) ion channeling, and laser-induced thermal waves (TW). A correlation has been found between lattice damage in silicon due to reactive ion etching and leakage current properties of thermal oxide films subsequently grown on the damaged silicon. The silicon wafers were plasma etched using Ar, CF_4, NF_3, and CHF_3 etch gases at dc bias voltages ranging from 150 V to 450 V. Lattice damage at the silicon surface, as determined by TEM and RBS, was found to depend on both the dc bias voltage and the etch chemistry. Subsequent leakage current measurements of the silicon oxides show that the samples with more silicon substrate damage prior to oxidation also have correspondingly higher leakage. The thermal wave technique also indicates a damage dependence on dc bias and on etch chemistry; however, the thermal wave measurements indicate a damage dependence on etch chemistry different from TEM and RBS measurements. The source of this difference is not yet understood.

269 Oxygen plasma damage in boron-diffused silicon,

by R.G. Frieser, J.A. Bondur, and E.F. Gorey, J. Electrochem. Soc. **134**, 419 (1987).

Transistor p-n junctions are observed to have leaky electrical characteristics and increased resistivity after exposure to oxygen plasma. Heating the devices above 180 °C for at least 10 min in vacuum, nitrogen, forming gas, or air eliminates the problem and returns the desired electrical characteristics. Experiments indicate that the increase in resistivity is associated with a boron displacement from silicon lattice sites. Additional experiments indicate that this phenomenon can be caused by reactively ion etching or plasma ashing a bare silicon surface which is doped with boron.

270 Heavy metal contamination from resists during plasma stripping,

by S. Fujimura and H. Yano, J. Electrochem. Soc. **135**, 1195 (1988).

A new mode of heavy metal contamination has been investigated. Generally used commercial positive photoresists containing heavy metal less than 0.5 ppm decreased minority carrier lifetime of the silicon substrate under the resist layer during resist stripping with a plasma asher. Since this contamination is induced by direct plasma exposure, it can be avoided by using plasmaless resist removal methods, such as wet stripping or down flow ashing. However, even in such resist stripping methods, a wafer temperature higher than 300 °C introduces this problem. This contamination is reduced by both covering the silicon surface with a silicon oxide layer and by purifying the resists. Neither of these methods are perfect.

271 Damage caused by stored charge during ECR plasma etching,

by S. Samukawa, in Proc. 2nd Symp. ULSI Science and Technology, The Electrochemical Society, ECS Proc. Vol. **89-9**, 421 (1989).

The influence of stored charge during ECR plasma exposure is estimated by evaluating leakage current increase in CMOS inverter circuits. Stored charge causes gate insulator degradation and leakage current increase in CMOS inverter circuits. Leakage current increase is caused by ion density increase, rf power turning off method and electrostatic chuck cooling during ECR plasma exposure. When ion density is over 10 mA/cm^2 and rf power is turned off at the same time of microwave power off, stored charge concentrates on the gate insulator and causes insulator degradation. Also, ECC cooling brings about leakage current increase. The leakage current increase caused by electrostatic chuck cooling method during ECR plasma exposure was due to stored charge in the field oxide. Therefore, a gas cooling system is suitable for wafer cooling method during ECR plasma etching.

274 Mask dependent etch rates II. The effect of aluminum vs. photoresist masking on the etch rates of silicon and silicon dioxide in fluorine containing plasmas,

by T.H. Fedynyshyn, G.W. Grynkewich, and T.-P. Ma, J. Electrochem. Soc. **134**, 2580 (1987).

Plasma etch rates of silicon and silicon dioxide when masked with photoresist and aluminum are compared. It was observed that the aluminum masked silicon had an enhanced etch rate relative to photoresist masking with a variety of fluorocarbon (CF_4, CHF_3, and C_2F_6) and nonfluorocarbon (NF_3 and SF_6) containing etch gases. It is also shown that the aluminum masked silicon dioxide etch rate is depressed relative to that of photoresist masked but that this effect is limited to the fluorocarbon etch gases. It is shown that the aluminum mask is responsible for affecting etch rates and not the photoresist mask. Evidence is presented that favors a mechanism by which excess fluorine radicals are catalytically produced at the aluminum mask surface. The relative amount of silicon etch rate enhancement is strongly linked to the oxygen concentration present in the etch gas mixture. This is consistent with a competition between fluorine and oxygen for reaction with and attachment to the aluminum surface. It is postulated that AlF_3 is a catalyst for fluorine production and Al_2O_3 is not.

275 Geometry independent deep trench etching, refill and planarization for isolation of merged Bipolar-CMOS,

by G.C. Eiden, P.K. Boyer, and J.A. Hughes, in Proc. 5th Symp. Plasma Processing, The Electrochemical Society, ECS Proc. Vol. **85-1**, 502 (1985).

A geometry independent method for silicon etching, refill of the etched grooves and planarization of the refill material will be reported. The method may be applied to a CMOS process, a bipolar process or a merged bipolar-CMOS process. Both CCl_4 and C_2F_5Cl were evaluated as plasma etchants of single crystal silicon. Conformality and stress effects will be reported for the refill materials evaluated (LPCVD polysilicon, atmospheric pressure deposited oxide, Low Temperature Oxide, and Plasma Enhanced CVD oxynitride). Planarization quality of spun-on organic films on refilled trenches will also be discussed. Trenches 2 μm deep have been 85% planarized, that is, the initial step height (trench depth) of 2 μm has been reduced to $\sim$0.30 μm.

276 Controlled film formation during CCl_4 plasma etching,

by S.E. Bernacki and B.B. Kosicki, in Proc. 4th Symp. Plasma Processing, The Electrochemical Society, ECS Proc. Vol. **83-10**, 505 (1983).

Carbon tetrachloride plasma etching can be performed in a regime in which a film-like product is simultaneously formed on vertical surfaces but inhibited from forming on horizontal surfaces. If the photoresist profile is properly tailored this process is capable of etching polysilicon and layered molysilicide films with virtually no change in linewidth from the developed photoresist image. We discuss the influence of the masking profile on the etched film profile and present conditions necessary for controlled film formation.

283 In-situ monitoring of electrical parameters for dry etching,

by J.I. Ulacia F. and J.P. McVittie, in *Plasma processing and synthesis of materials*, edited by D. Apelian and J. Szekely (Mat. Res. Soc. Symp. Proc. Vol. 98, Pittsburgh, 1987), p. 203.

In this paper, external electrical measurements and a circuit model are used to obtain indirectly the plasma potential, electron density, ion current density, and sheath thickness. In a revised and extended circuit model, each element describes a current-transport mechanism. Each mechanism is described by an analytical expression in terms of previously specified plasma parameters. The model is applicable in capacitive discharges with equal-area electrodes, and it can be expanded to consider other systems. To verify its applicability, electrical measurements of voltage, current and phase angle in an SF_6:O_2 discharge are used to calculate the plasma impedance. Electrical plasma measurements yield time-averaged values for the impedance; the values obtained for each device reflect this dependency. After the calculation of each electrical device, SPICE 3.7a simulations separated the individual contribution of current transport by each mechanism and validated the assumptions. The plasma parameters obtained by this technique agree well with a Langmuir-probe measurement, solutions of the Boltzmann transport equation, and data published in the literature.

286 Determination of the power and current densities in argon and oxygen plasmas by in situ temperature measurements,

by R.J. Visser, J. Vac. Sci. Technol. A **7**, 189 (1989).

In situ temperature measurements of silicon wafers in argon and oxygen plasmas are presented. The processes which determine the heat flow from and to the wafer have been established. It is found that conduction of heat to the electrode does not contribute to the heat loss of the wafer. Radiation of heat and conduction of heat by the gas are the main loss processes. The heat flux q_p from the plasma to the wafer is independent of time and temperature. This is attributed to the flux of high-energy ions to the wafer. By measuring the plasma potential, the energy of the ions is estimated. The ion current density is calculated from q_p and the energy of the ions. The value which is obtained is compared to those obtained with other methods. It is concluded that the power density on the electrode can differ by almost an order of magnitude from an estimation using the dissipated rf power and the electrode area.

289 In situ silicon-wafer temperature measurements during RF argon-ion plasma etching via fluoroptic thermometry,

by I. Hussla, K. Enke, H. Grünwald, G. Lorenz, and H. Stoll, J. Phys. D **20**, 889 (1987).

In situ silicon-wafer temperature measurements during RF argon-ion plasma etching are reported for four different wafer cooling environments: (*i*) lying on the electrode; (*ii*) clamped to the electrode; with additional helium gas between wafer and electrode in both (*i*) and (*ii*), in order to enhance heat transfer. Fluoroptic thermometry is applied to measure the local surface temperature of the wafer during plasma processing. The measured heating and cooling curves can be fitted using a model which considers conductive heat losses only. Deviations are observed when higher RF power is applied and surface temperatures higher than 120 °C are reached. We report on effective cooling via helium gas of the rear side of the wafer which is lying on an electrode without any clamping.

291 Simple optical devices for detection of radiofrequency oxygen plasma stripping of photoresists,

by E.O. Degenkolb and J.E. Griffiths, Appl. Spectroscopy **31**, 40 (1977).

In an earlier paper, a spectroscopic technique was presented for studying the removal of negative photoresists from silicon wafers using radiofrequency (RF) oxygen plasma. The potential application of the technique to a wide variety of manufacturing situations became quite evident especially for monitoring and controlling various stripping and etching operations induced by highly energetic plasma. It was also evident that if the technique was to enjoy wide usage in the manufacture of various integrated circuits, for example, a much more compact detection system would be necessary. Two rather versatile devices satisfying this need are described briefly here. (*Introduction to the article*)

292 Optical emission spectroscopy of reactive plasmas: a method for correlating emission intensities to reactive particle density,

by J.W. Coburn and M. Chen, J. Appl. Phys. **51**, 3134 (1980).

The addition of a small concentration of suitably chosen noble gas to a reactive plasma is shown to permit the determination of the functional dependence of reactive particle density on plasma parameters. Examples illustrating the simplicity of this method are presented using F atomic emission from plasma-etching discharges and a comparison is made to available data in the literature.

294 Dependence of F atom density on pressure and flow rate in CF_4 glow discharges as determined by emission spectroscopy,

by J.W. Coburn and M. Chen, J. Vac. Sci. Technol. **18**, 353 (1981).

The dependence of the density of atomic fluorine as a function of pressure and flow rate in CF_4 glow discharges has been measured using emission spectroscopy. The emission intensity of the 7037 Å line of atomic fluorine and of the 7504 Å line of argon, which is added to the CF_4 discharge in small quantities (1%–2%), is used to deduce the F atom density. It is found that the pressure dependence of the F atom emission intensity alone is a poor indicator of the relative F atom concentration because the excitation efficiency of the CF_4 discharge decreases rapidly with increasing pressure. The F atom density decreases with increasing flow rate as is to be expected if the lifetime of the F atom is determined largely by the vacuum pump. However, at low flow rates the F atom density is determined primarily by contaminant gases, in this case CO and CO_2, through several mechanisms which are discussed.

295 Optical emission spectroscopy and actinometry in CCl_4-Cl_2 radiofrequency discharges,

by R. d'Agostino, F. Cramarossa, S. De Benidictis, and F. Fracassi, Plasma Chem. Plasma Proc. **4**, 163 (1984).

Radiofrequency discharges fed with CCl_4–Cl_2 mixtures have been studied in the pressure range 0.3 to 0.6 torr by means of emission spectroscopic actinometry with Ar, He, and

N_2 as actinometers. Two different reactors, a parallel plate and a capacitively coupled tubular one, have been utilized for this study to obtain information for a large range of electron energy distributions. Analysis of the experimental results demonstrates the following: the utilization of actinometry and its range of validity, the importance of electron attachment to CCl_x species during the plasma decomposition process, and the effects of the presence of chlorine and "glow-polymer" in the discharge medium.

298 Time-dependent excitation in high- and low-frequency chlorine plasmas,

by D.L. Flamm and V.M. Donnelly, J. Appl. Phys. **59**, 1052 (1986).

Time-resolved voltage, current and optical emission from Cl, Cl^+, Cl_2^+, and small amounts of added Ar were studied in 0.3 Torr chlorine discharges at 13.2 MHz and 220 kHz, above and below the ion transit frequency (ITF). Emissions, measured as a function of position, were deconvoluted to correct for the finite fluorescence lifetimes and extract the electron impact excitation rates. At 220 kHz, emission falls to zero at the voltage zero point crossings due to electron energy relaxation and attachment. Emission and excitation in a sheath peak sharply about 70 ns before electrode voltage reaches its positive maximum and are stimulated by electron current. Emission then falls to zero and reaches a second maximum ~300 ns after the electrode voltage reaches its negative peak. This emission is excited by a secondary electron avalanche from ions crossing the sheath. Excitation maxima in the center of the discharge lag the voltage peaks by $\sim 25°$. Analysis of time and spatially resolved ratios of emission from Cl and Ar show that actinometric techniques can give an erroneous measure of atom concentrations unless the viewing area and phase are taken into account. At 13.2 MHz, ions cannot respond to the time-varying fields, and excitation in the sheath peaks in phase with the anodic part of the cycle with minimal excitation during the cathodic phase. Emission from Cl^+ was not observable. The lack of high-energy ion bombardment drastically reduces secondary emission and excitation in the cathodic sheath. In the center of the discharge, excitation peaks twice per cycle, but unlike the low-frequency case, it does not fall to zero. The electron energy relaxation frequency (ν_u) is estimated to be 64 MHz.

299 Langmuir probe and optical emission spectroscopic studies of Ar and O_2 plasmas,

by D.A.O. Hope, T.I. Cox, and V.G.I. Deshmukh, Vacuum **37**, 275 (1987).

The variation of electron temperature with pressure in argon and oxygen plasmas generated in a parallel plate dry etcher has been determined by Langmuir probes and optical emission spectroscopy. The good agreement between the two techniques indicates that emission spectroscopy can be reliably employed to monitor the state of the plasma. Consequently, it has been used to confirm that the plasma composition remains constant while the dc bias between the substrate and the plasma is varied at constant power and pressure. This is achieved by varying the power ratio between substrate and counter electrodes. In this way, the effect on etch anisotropy of only one process parameter, the dc bias, has been studied for the etching of polyimide in oxygen. A simple model having no adjustable parameters is proposed and good agreement with experiment is found.

306 In situ analysis of fluorinated gases in plasma etching by infrared spectroscopy,

by H.-U. Poll, D. Hinze, and H. Schlemm, Appl. Spectroscopy **36**, 445 (1982).

Infrared absorption spectroscopy proves to be a useful tool in the evaluation of plasma chemical conversions of etch gas mixtures during plasma etching for microelectronics. The partial pressures of the various fluorinated gas components are obtained from infrared spectra with sufficient accuracy and may yield information about the actual state of the system. Methods of spectra recording and partial pressure computation are discussed. To demonstrate the applicability of the method, the conversion of $CF_4 + O_2$, and C_2F_4 in a glow discharge are investigated.

308 Detection of dry etching product species with in situ Fourier transform infrared spectroscopy,

by T.A. Cleland and D.W. Hess, J. Vac. Sci. Technol. B **7**, 35 (1989).

In situ Fourier transform infrared (FTIR) spectroscopy has been utilized to detect gas-phase species resulting from the dry etching of aluminum and heavily doped n-type polycrystalline silicon (n^+ poly-Si) in Cl_2. The products of aluminum etching in Cl_2 were investigated in the presence and absence of a plasma. With the plasma off, Al_2Cl_6 was the only infrared-active etch product detected at both low (50 °C) and high (>120 °C) sample temperatures. With the discharge on, the spectrum was more complex, indicating partial fragmentation of the dimer etch product in the plasma forming $AlCl_3$ and, perhaps, AlCl. Silicon tetrachloride was the only infrared-absorbing product detected during the Cl_2 plasma etching of n^+ poly-Si. Unsaturated silicon chlorides ($SiCl_x$, $x = 1 - -3$) were not present at detectable levels. These results are compared with those of ultra-high vacuum (UHV) beam studies; in comparing UHV results with those obtained in a processing plasma, allowance must be made for the considerable difference in pressure, and therefore, in species' fluxes in these two experimental environments. Unlike UHV studies, in situ FTIR measurements permit detection of etch products under typical processing conditions. When interpreting the results, however, the possibility that the products leaving the sample surface react further upon entering the plasma must be considered.

309 Two-step optogalvanic effect using intersecting laser beams: a pinpoint discharge diagnostic,

by D.K. Doughty, S. Salih, and J.E. Lawler, Phys. Lett. A **103**, 41 (1984).

Electric fields in the cathode fall region of a glow discharge are accurately measured in a small volume defined by the intersection of two laser beams. The fields are determined from linear Stark effects in Rydberg levels. The Rydberg atoms are produced via a two-step laser excitation and detected with optogalvanic effects.

310 Laser optogalvanic detection of molecular ions,

by R. Walkup, R.W. Dreyfus, and P. Avouris, Phys. Rev. Lett. **50**, 1846 (1983).

This letter reports the first optogalvanic detection of molecular ions (N_2^+, CO^+). The present results indicate a new optogalvanic mechanism involving direct alteration of ion mobility by laser excitation, due to difference in charge-exchange collision rates for excited- vs. ground-state ions. The technique provides a sensitive probe of ions in the cathode dark space. Ion kinetic energies are obtained from Doppler shifts and internal energy distributions from rotational line intensities.

312 Photoemission optogalvanic spectroscopy: an in situ method for plasma electrode surface characterization,

by S.W. Downey, A. Mitchell, and R.A. Gottscho, J. Appl. Phys. **63**, 5280 (1988).

Photoemission optogalvanic spectroscopy (POGS) is shown to be useful for plasma electrode surface characterization. A pulsed ultraviolet laser is used to induce photoemission from the electrode surface in a radio frequency plasma reactor and the increase in plasma current is detected. The photoemission process is first characterized in vacuum and then compared to that in several plasma gases using Al and Si electrodes. In vacuum, the laser-induced photoemission signal is generally consistent with space-charge-limited current. When below the space-charge limit, the magnitude of the photoemission signal depends upon laser wavelength and power, surface composition, and film thickness. The removal of SiO_2 from Si and the contamination of Al in fluorine-containing plasmas is monitored using this technique. A large increase in the POGS signal is observed as the oxide is removed from Si or when a fluorinated Al surface is exposed to an O_2 plasma. The POGS signal decreases with fluorine exposure with both Al and Si. We are continuing to explore the utility of this technique as an endpoint detector and in situ contamination monitor.

314 Real-time measurements of plasma/surface interaction by plasma-amplified photoelectron detection,

by G.S. Selwyn, B.D. Ai, and J. Singh, Appl. Phys. Lett. **52**, 1953 (1988).

A new method, based on the photoelectric effect, is described for real-time, in situ monitoring of metal or semiconductor surfaces during plasma exposure. As an example of the application of this technique, the effect of both sputter and reactive gas plasma exposure is studied for graphite, silicon, and aluminum surfaces. Results are consistent with the formation of a surface-passivating layer of fluoride on aluminum and penetration of fluorine into the silicon bulk during exposure to the CF_4 + Ar etching plasma. An application of this technique for endpoint detection monitoring is described.

317 Laser-induced fluorescence measurement of plasma ion temperature: corrections for power saturation,

by M.J. Goeckner and J. Goree, J. Vac. Sci. Technol. A **7**, 977 (1989).

When using high-power pulsed tunable dye lasers to measure plasma ion temperatures, it is important to attenuate the laser intensity. The temperature is found from the

Doppler broadening of a spectral line. This may be obscured by saturation broadening, an instrumental effect encountered when too much laser intensity is used. Three useful experimental methods for determining the optimum pulsed laser intensity are found from a semiclassical atomic physics model. As an example, an Ar II transition pumped by a 1-GHz bandwidth laser is examined. The fluorescence linewidth of room-temperature ions broadens from 1.50 to 2.87 GHz when the homogeneous laser intensity is increased from 50 kW/m^2 to 5 MW/m^2.

Reprinted with permission of the American Institute of Physics, New York.

320 The application of Doppler shift laser fluorescence spectroscopy for the detection and energy analysis of particles evolving from surfaces,

by W. Husinsky, J. Vac. Sci. Technol. B **3**, 1546 (1985).

Doppler shift laser fluorescence spectroscopy (DSLFS) has been developed in the last few years as a powerful tool to measure densities and energy distributions of neutral particles. Its main application described in this paper is the study of particles ejected from surfaces under impact of atoms, ions or electrons. Complications and controversies in the interpretation of DSLFS spectra will be discussed. The major achievements and applications of the method over the last few years will be presented. One of the most important results has been a deeper insight into the composition of the particle fluxes ejected from surfaces under ion bombardment in respect to the degree of surface oxidation. This has been possible due to the fact that DSLFS enables us to measure the densities and energy distributions of neutral particles in specific electronic states. As a direct consequence, a large amount of information on sputtered excited atoms has recently been obtained by DSLFS.

Reprinted with permission of the American Institute of Physics, New York.

323 Laser diagnostic techniques for reactive ion etching: plasma understanding to plasma control,

by G.S. Selwyn, J. Vac. Sci. Technol. A **6**, 2041 (1988).

Reactive ion etching is the current process of choice for production of high-speed integrated circuits; yet the development and use of this technology has evolved largely through empirical efforts. Beam/surface studies have proven instrumental in examining fundamental issues in etching; however, extension of these results to process plasmas requires considerable extrapolation. Laser diagnostic measurements may be used as a bridge between UHV measurements and the process plasma, a role which is possible due to the high specificity and sensitivity of these techniques. Spatially resolved laser studies provide three-dimensional in situ measurement of concentration gradients for reactive etchants and ions in the plasma. Key plasma etchants such as O and Cl may be studied using two-photon excitation processes. Finally, the emerging use of lasers for process control applications is discussed and compared with conventional approaches, such as optical emission end-point detection.

Reprinted with permission of the American Institute of Physics, New York.

324 Quantification of laser interference fringes as applied to plasma etch endpoint detection,

by R. Deaton and A. George, in *Laser in Microlithography*, S.P.I.E. Vol. **774**, 162 (1987).

As integrated circuit feature sizes have continued to decrease, plasma etching has become the method of choice for achieving the resolution and process control required for device fabrication. To maintain tight control, the precise detection of etch endpoint is very important. A variety of endpoint detection schemes are available. These include optical emission spectroscopy, mass spectroscopy, monitoring of chamber pressure and dc bias, and laser interferometry. This paper deals with the application of laser interferometry for endpoint detection in plasma or dry etching, which is the removal of a film in a plasma or low pressure gaseous discharge. Etching consists of the following steps. The exposed film, i.e. not covered by resist, is removed by chemical and/or physical processes which are determined by the type of etch. In plasma etching, the etching is done by ions and highly reactive chemical species called free radicals. There are many different types of dry etch methods. In this case, the process was RIE, reactive ion etching, where chemical and physical effects interact to produce the etch. (*Introduction to the article*)

325 Optical etch-rate monitoring using active device areas: lateral interference effects,

by P.A. Heimann, J. Electrochem. Soc **132**, 2003 (1985).

The technique of optical (laser) etch-rate monitoring is proving useful for integrated circuit manufacture and process development. This technique detects the change in film thickness during etching by monitoring the reflectance of the film on the integrated circuit wafer. When the reflectance is monitored over a region of the wafer containing both etched and unetched areas, the results are affected by lateral interference, which is interference between the light reflected from the different regions. This lateral interference has only a minor effect when less than 30% of the region being monitored is covered by an etch-resistant layer (typical for metal-type lithography levels), but it can cause severe distortions when a larger fraction of the region is covered by the resist. The effects of lateral interference have been examined using numerical simulations. The results are in good agreement with experimental etch-rate monitor signals from the etching of patterns with resist coverage ranging from 10 to 80% of the total area.

343 The use of electrostatic probes for plasma diagnostics - a review,

by B.E. Cherrington, Plasma Chem. Plasma Proc. **2**, 113 (1982).

The use of electrostatic, or Langmuir, probes for plasma diagnostics is reviewed. The emphasis is on experimental implementation and current techniques, and particular attention is paid to sources of error in theoretical interpretation as well as to experimental problems that can occur in complex, reactive plasmas.

344 Electric probes in plasmas,

by B. Lipschultz, I. Hutchinson, B. LaBombard, and A. Wan, J. Vac. Sci. Technol. A **4**, 1810 (1986).

This paper provides a background for the use of Langmuir and gridded energy analyzer probes in diagnosing plasmas with varied characteristics. Theory is illustrated which governs the analysis of data from, and the design of these probes. Several probe analysis techniques and some of their typical problems are presented.

346 A floating double probe method for measurement in gas discharges,

by E.O. Johnson and L. Malter, Phys. Rev. **80**, 58 (1950).

Measurements with a Langmuir probe in varying or decaying plasmas are defeated by the fact that the plasma potential in these cases follows that of the most positive electrode it can contact. The difficulties can be obviated by the use of a pair of probes joined by a variable potential source. The double probe system "floats" with respect to the discharge system. From the measurements of probe current vs. differential probe voltage, electron temperature and plasma densities can be determined. The method is also applicable to "going" discharges where it has the advantage over the single probe of exerting a negligible influence on the discharge.

347 Asymmetrical triple probe method for determining energy distribution of electron in plasma,

by T. Okuda and K. Yamamoto, J. Appl. Phys. **31**, 158 (1960).

A new probe method is proposed. This method is an improvement of the floating triple probe method, which was proposed by us in a previous paper. In this method, the dimension of one operating probe is different from the other, so that the smaller probe behaves as the single probe. Thus, measurable range of electron energy can be extended as much as that in the single probe method. The asymmetrical triple probe method has a merit that a disturbance, which is inevitable in the single probe method, can be eliminated in this method. The electron energy distribution obtained by the new probe method is not always in accord with that obtained by the usual single probe method. The hump often observed at high energy range in the single probe method does not appear in the new method, illustrating the latter is free from erroneous estimation due to the disturbance.

349 Instantaneous direct-display system of plasma parameters by means of triple probe,

by S.-L. Chen and T. Sekiguchi, J. Appl. Phys. **36**, 2363 (1965).

A new electrostatic probe method utilizing the triple probe is proposed in which no voltage or frequency sweep (or switching) is required. This feature enables us to determine the instantaneous values of the electron temperature (T_e), as well as the electron density (n_e), within a short time which is of the order of the intrinsic response time of the probe itself ($\leq 1\mu$sec). Moreover, the system allows the direct display of the T_e values as well as the semidirect display of the n_e values on appropriate display units, thus permitting us to eliminate almost all procedures usually required for data processing. In view of the features mentioned, this system may be a useful diagnostic tool not only for stationary plasmas but also for rapidly varying time-dependent plasmas of various types. This paper presents theoretical considerations for the instantaneous direct-display system using a symmetrical triple probe. Discussions are also presented for the estimation of errors caused by the variation of ion saturation current, for the application of magneto-plasmas, and for the time and spatial resolutions. Experimental confirmation was made through the measurements of stationary magneto- as well as non-magnetoplasmas. As an example of the application to time-dependent plasmas, the electron temperature waves accompanied by the moving striations in hydrogen discharge were observed.

353 Monitoring and diagnosis of plasma etch processes,

by S.B. Dolins, A. Srivastava, and B.E. Flinchbaugh, IEEE Trans. Semicond. Manuf. **1**, 23 (1988).

Plasma etching is an important process in semiconductor manufacturing. This process removes material from a silicon wafer by applying power and gases in a chamber. As material is removed from a wafer, the amount of particular chemicals given off can be measured; the technique is called emission spectroscopy and the measurements are called endpoint traces. An expert system that automatically interprets the traces has been designed and built. The system combines signal-to-symbol transformations for data abstraction and rule-based reasoning for diagnosis. The system detects problems as soon as they occur and also determines their causes.

INDEX